# Lecture Notes in Control and Information Sciences 305

Editors: M. Thoma · M. Morari

Springer

*Berlin*
*Heidelberg*
*New York*
*Hong Kong*
*London*
*Milan*
*Paris*
*Tokyo*

Alexander Nebylov

# Ensuring Control Accuracy

With 80 Figures

Springer

**Author**

Prof. Alexander Nebylov
International Institute for Advanced Aerospace Technologies (IIAAT)
St. Petersburg State University of Aerospace Instrumentation
67, Bolshaya Morskaya
190000 St. Petersburg
Russian Federation

ISSN 0170-8643

ISBN 3-540-21876-9 Springer-Verlag Berlin Heidelberg New York

Library of Congress Control Number: 2004104955



Springer-Verlag is a part of Springer Science+Business Media

springeronline.com


Printed in Germany



Typesetting: Data conversion by the author.
Final processing by PTP-Berlin Protago-TeX-Production GmbH, Berlin
Cover-Design: design & production GmbH, Heidelberg
Printed on acid-free paper 62/3020Yu - 5 4 3 2 1 0

## Ensuring control accuracy

The methods of analysis and synthesis of nonadaptive automatic control systems and other dynamic systems, capable of providing an acceptable or the highest guaranteed level of operation accuracy under the allowable uncertainty of excitation properties and system characteristics are stated. Control accuracy is estimated by the upper bound of the total r.-m.-s. or maximal error. When giving the class of input excitations, the restriction of possible values of derivatives, generalized moments of spectral densities, width of a spectrum and other numerical characteristics, which are easily controllable in practice are used. The research on frequency and time domains are based on mathematical results of the general problem of moments, the theory of disturbance accumulation and numerical methods.

This book is aimed towards specialists in designing and researching control systems, engineers, postgraduate, and students.

# Preface

Dynamic systems, described by definite differential or difference equations, are the universal model for investigation of laws in behavior of different natural processes related to the motion of material objects, information transfer, and the development of social, biological, economic and other structures. Their application in the investigation of control processes in technical and other systems forms the fundamental idea of automatic control theory.

Estimations of dynamic system accuracy greatly depend on the choice of initial models of actions. Those models give input for actuating the right member of appropriate differential equations. Variability and partial knowledge of real action properties trouble the formation of spectral-correlation and other full models of action, described by the analytical functions of nontrivial mode. This is due to the loss of practical efficiency of investigation (due to low validity of such models) which exceeds the possible gain of more accurate system adjustment on the definite operating regime. Therefore it is rational to use for description of action properties the nonparametric classes of functions. Their satisfactory width provides the required validity of description. This approach requires the investigation and development of special methods for analysis and synthesis of dynamic systems with the ensured accuracy characteristics.

Ensuring good protection of results from errors in the initial data is important not only in researching control accuracy. A similar requirement applicable to mathematical statistics problems was clearly formulated by P. Huber [61]. It was indicated as a "robustness" term in the sense of its insensitivity to low deviation from initial suppositions. The robustness in the modern theory of automatic control is often coupled to the ensuring of system stability at the definite scatter of its parameters (V. L. Kharitonov theorem, discovered by J. Z. Tsypkin and his school). Therefore the more widespread explanation of its concept became convenient. If the system or algorithm posses the high efficiency at the nominal operation conditions and good efficiency at the deviation from the nominal conditions in the preset accessible limits [3, 8, 27, 40, 54] then it is considered robust.

These limits can be determined by the accepted classes of external and parametric disturbances. Ensuring control accuracy in this case can be treated as providing the robust accuracy, and then appropriate control systems are called robust.

The robustness concept is actually not new. It is followed by the tendency to give nonadaptive systems the property of holding the preset characteristics in the admissible limits at possible variation of their operating conditions, without demanding the best quality for some fixed conditions. Highly experienced designers have always been working in that way. Their works were based not only upon any mathematical theories but also upon the prudence and good intuition in major cases. The theoretical methods of dynamic systems investigation in their development could not comprise all the features of practical design problems. Some "residual" in theoretical and practical approaches to design process always existed. It stimulated the improvement of theories and created witty stories about the loss of mutual understanding between theorists and empirics. The limited possibilities of theoretical investigations are coupled to excessive formalization and idealization of problem statements. It can also be the fee for the possibility of finding the strict solution. The robust approach is the attempt to smooth the sharpness of exposed problems in the account of the more rough, approximate description of initial information about the conditions of system operation, assuming the possibility of normality in such conditions.

This book does not comprise all known methods for ensuring control accuracy, which are various enough and can hardly be joined with anything to form a single theory, except with the robust concept used in all of them.

The chosen stated methods suppose mostly the investigation in the frequency domain. The main arguments in advantage of such choice are their relative simplicity and high validity of initial conditions being appreciated by all specialists and practitioners. The direct interaction between those methods and classical frequency domain methods for investigation of automatic control systems quality is also very important. It simplifies the understanding of material for a large group of readers, including students. Therefore, a list of questions is given at the end of every chapter. The statement style can satisfy system designers and designers of automatic control devices. The author hopes that specialists in the sphere of control theory will read this book, although many of them suppose the frequency domain to be some kind of trampled out field of knowledge, where it is hard to find something new. But the author was not seeking the new but required in practice, investigation methods, which are improved by the given formulae and examples.

The main list of references contains minimum basic publications that are close to the subject of the book. Additional literature is indicated by particular references in the text.

Most of the ideas shown in this book were formulated by the authors' cooperation with his scientific teacher V. A. Besekerski. He became the co-author when publishing the first book devoted to robust control systems [8] based on Besekerski fundamental treatises [6, 7]. The author gratefully appreciates Prof. Besekerski for an excellent education and support.

Professor A. A. Zinger was the first to confirm the authors' suppositions about the possibility of investigating the accuracy of linear filtration of signals with the limited variances of derivatives on the basis of using the current problem apparatus by Chebyshev-Markov. He also recommended the perfect treatise by M.G. Krein and A.A. Noodelman. By virtue of those treatise the author has discovered the new vision on familiar verity.

Professor I.B. Chelpanov with his opinions on earlier authors' publications, and with his speeches and books during many years has helped the author to keep certainty in the efficiency of chosen investigation methods.

Professor E.N. Rosenwasser gave much useful advice on the material of this book and approved it conceptually. It has improved the authors' resolution to accomplish this book till the end.

Academician J. Z. Tsypkin has always been the best authority and kindest tutor to the author. J.Z. Tsypkin has greatly influenced the statement of given problems in this book. As an active supporter of the robust approach to dynamic systems investigation he played a great role in spreading these methods in Russia and in the world on IFAC line.

Academician F.L. Chernousko has greatly supported the author at the most difficult period of the book's completion, which predetermined the possibility of publishing this book.

The author had the additional opportunity to check up the efficiency of suggested methods for theoretical investigations in contensive problems of synthesizing precision automatic control complex for the relative motion of aerospace planes at their horizontal start and landing on moving ekranoplanes [86, 89]. The concept for construction of this perspective space transport system is being developed in cooperation with professor N. Tomita and other colleagues from Tokyo. The communication between them provided new ideas for improving the material in the book.

On every stages of this book preparation the author was inspired by support of many other friends and colleagues, who are greatly faithful to science and who give all the strength they have to it. Many thanks to all of them.

St. Petersburg, December 2003 Alexander Nebylov

# Contents

# Chapter 1

## Main concepts and definitions

### §1.1 Dynamic systems

**The dynamic system concept.** One of the major results of the theoretical cognitive process is the possibility to evaluate the consequences of different human or natural force actions on those or other complicated objects or systems, capable to accept such actions. It is impossible to discover correct solutions when constructing engineering devices, technological processes, socioeconomic mechanisms, and elements of human environment or when excluding it in other areas of intellectual activity.

Real systems of theoretical research are represented, such as models having a certain formal description, which are more often mathematical. The legitimacy of model choice is proved by the adequacy of real systems and model responses to excitations of an identical kind.

The mathematical model of a system enables one to find the relationship between some input $g$ (setting action) and output $y$ (response). The external actions and responses can be described thus by functions of continuous time $t$ or discrete time $n$ and look like $y(t) = y(t,g)$ or $y(nT) = y(nT,g)$, where $g = g(t)$ or $g = g(nT)$; $T$ is the discretization period, $n = 0, 1, 2, \ldots$ (Fig 1.1).

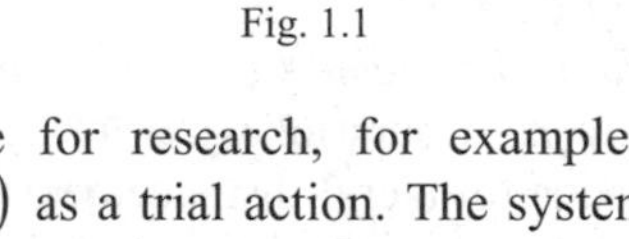

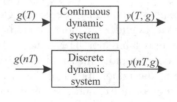

Fig. 1.1

It is convenient to use for research, for example, the unit step-function $g(t) = 1(t)$ or $g(nT) = 1(nT)$ as a trial action. The system reaction to such a step generally is not a step and is described by a smoother function. The form of such function $y(t)$ or $y(nT)$ characterizes the dynamic properties of a system. The system possessing such dynamic properties is called the dynamic system.

Thus, input and output actions in dynamic system are functions of time, and the current value of output is determined not only by current, but also by previous values of input, i.e. the system has some "memory" persistence. The mathematical model of dynamic systems is inhomogeneously differential or a difference equation, whose left member is written concerning a response, and the right member includes the external action.

One of the possible definitions for a dynamic system is: *the structure, into which something (substance, energy or information) is entered in the defined moments and from which outputs something in any instant. It is used to describe a cause-effect correlation from the past into the future.*

**Examples of dynamic systems.** Give examples concerning the area of engineering systems (1 — 4), physics-biological systems (5 — 8), socioeconomic systems (9 and 10).

*Example 1. An electromechanical servosystem of rotation angle reproduction.* Input $g(t)$ is the adjustment of a new value for the rotation angle at the reference-input element. Output $y(t)$ is a rather smooth turn of the executive axis of control plant on a required angle. A control plant can be a physical object with large mass and moment of inertia: the large-sized directional antenna, telescope, launcher, rudder of a sea vessel, special purpose robotics handling actuator, etc. The final value of a rotation moment, explicated by an executive electric motor concerns the number of factors, defining the dynamic properties of a system, which restricts the possible angular acceleration of the executive axis with the defined reduced moment of inertia.

*Example 2. A temperature stabilization system on a space orbital station bay.* When the station passes a planet shadow, then the step decrease of the radiant energy stream from a star $g(t)$ should be balanced by a connection between heat-generating devices. However, their deduction of required thermal power is possible only under the smoother law $y(t)$. Considering the stabilization of temperature in different parts of a bay also occurs only gradually.

*Example 3. A system for the automatic tuning of radiation frequency in the jammer of radio interferences against missile control systems.* At a step variation of carrier frequency of a suppressed transmitter signal $g(t)$, the frequency of a generated radio interference varies under the more "smooth" law $y(t)$ and can coincide with the signal frequency only a bit later. The value of this piece of time is the important characteristic of counter radio measurement efficiency. It is basically restricted from below due to the impossibility of the instant estimation of a signal frequency with the presence of noise, and also by limited operation speed of the radio interference generator device.

*Example 4. A digital control system of a plane when driving in a landing final approach.* Perturbation is a hard burst of wind in the direction, perpendicular to the runway axis, which leads to lateral displacement of the plane with values of discrete reference input $g(nT)$, where $T \approx 10^{-1}$sec. The output is a deviation of the steering mechanisms which causes the lateral controlling force and gradual inverse lateral displacement of the plane with values of discrete reading $y(nT)$.

*Example 5. A mans blood pressure stabilization system.* This action consists of the patient's reception of the defined dose of a medicinal preparation $g(t)$. The response is a smooth drop down of blood pressure $y(t)$ with the onset of drug action and then in it's ending process — again a gradual rise of pressure. A correct choice of dosage and periodicity of medicine maximize the treatment efficiency.

*Example 6. A monitoring system for spreading infectious diseases.* This input action consists of the vaccination of a population in a given area, which ensures an increase of immunity to a disease agent, or other protective procedures. The result is the decrease of patients at the appropriate choice of action.

*Example 7. A fish stores reproduction system in some fishing craft area.* The action is a scheduled dropdown of annual catch $g(nT)$, where $T = 1$ year, and the transfer of fish catching fleet from one area to another. A reaction is gradual increase of fish stores $y(nT)$. The correct mathematical model of a system allows one to reasonably plan the results and to produce the additional fish guarding procedures in time.

*Example 8. A system for thermal balance maintenance of the Earth.* This anthropological action consists of sharp growth of mankind's energy needs, of which the significant part is produced at the expense of burning different kinds of chemical fuel. It leads, in particular, to the increase of carbonic gas percentage in the Earth's atmosphere $g(t)$ and the "greenhouse effect" development that changes the balance of incidents on the Earth and emission by the Earth's radiant energy. This response consists in the increase of the annual temperature $y(t)$ in characteristic geographical areas. The authentic mathematical model of this system has not yet been created.

*Example 9. A governing system for branches of national industry (agriculture, transport, education, public health services).* This input action is the additional tool or recourse allocated for development of manufacture in the current year $g(nT)$, where $T = 1$ year. This response consists of increasing the amount or (and) production quality $y(nT)$, worked out during the year. The possible criteria of assigned resource distribution are the minimization of time, for which the response rises up to a target level, and the maximization of the response level which is guaranteed to be achievable during the preset time, etc.

*Example 10. Consumer demands prediction system for a given group of goods.* The input is a variation of the goods price $g(nT)$, where $T = 1$ day. The output $y(nT)$ is the volume of sales estimation. Possible optimization criterion for function $g(nT)$ consists of the maximization of the expected producer's profit.

**Additional remarks to the dynamic system concept**. The same real system depending on the research problem definition can be considered either as dynamic or as not dynamic. For example, the radar detector of a legitimate signal in the noise consisting of the matched filter and the threshold unit [34, 56] is a dynamic system, if the time lag between the occurrence of signal and threshold unit operation moment is investigated. If the detector operation speed is not being controlled, but only the probability performances of correct detection and false alert are of interest, then it is necessary to consider the detector as a non-dynamic system.

**Classification of dynamic systems.** The scheme of dynamic systems classified on character and on correlation of processes, taking place in them, is shown in Fig.1.2.

The linear dynamic system unlike the nonlinear one is described by a simple equation (differential or difference), as well as any part of a linear system, i.e. any dynamic unit, included in it. Dependence of a steady state output value from the fixed value of input $x_2 = F(x_1)$ is called a static characteristic, for a linear unit or for a whole linear system this is represented by straight line (Fig.1.3*a*).

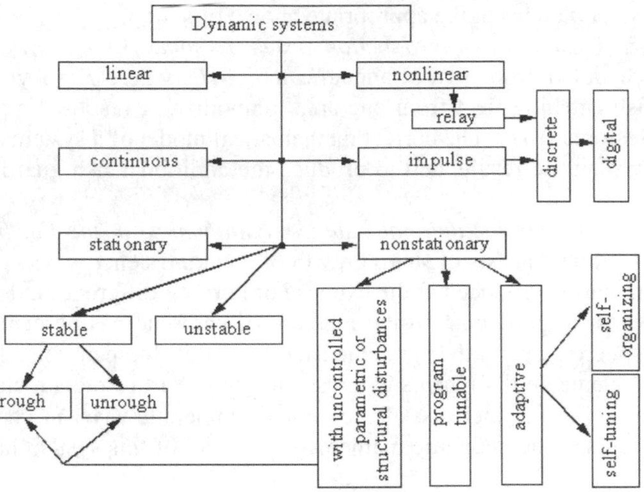


Fig. 1.2

The concept of a linear system actually is related to the accepted mathematical model, instead of real systems, any of which is certainly non-linear. However, with ranges of processes restricted by values, taking place in a system, the linear model in many cases represents quite adequately the properties, really shown by a system. First of all it is valid for the systems constituents of which have the restricted-linear static characteristics (Fig. 1.3*b*).

Remarkable, is that the impossibility to accept the linear model systems causes the essential complication of investigating its dynamic properties. Some of the most universal analytical methods of such investigation are coupled again to the substitution of nonlinear parts of a system by "equivalent" (in this or that sense) linear units. The methods of harmonic linearization, static linearization and many others follow this principle (see [9, 74, 80]).

Among the essentially nonlinear systems there are relay systems, in which the quantization of some process on a level takes place. Such systems include a unit or some units with a relay static characteristic (Fig.1.3 *c*, *d*).

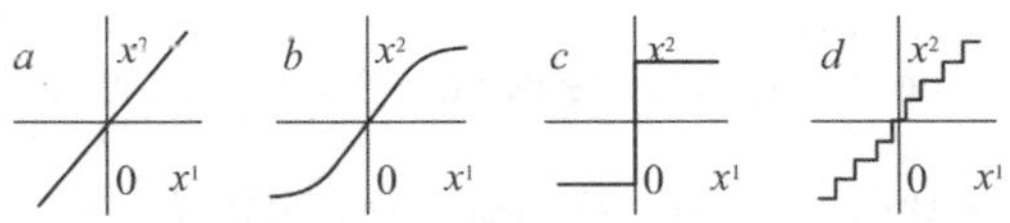


Fig. 1.3

Independent of whether the system is linear or non-linear, it can be continuous or pulsed (sampled). The difference of pulsed systems from continuous ones is the available quantizations of any process or processes in time.

The pulsed and relay systems are joined together by one concept of discrete systems. There is a quantization of processes in time or (and) on a level in them.

A system, whose properties do not vary in time, is named stationary (really it is often possible to speak only about quasi-stationary). When the system proprieties vary essentially in time then it is named non-stationary.

In most cases the nonstationarity is coupled to unmonitored parametric or structural perturbations that cause the casual modifications of parameters or form of the equation describing a system, not having purposeful character.

In some highly organized dynamic systems the purposeful variation of dynamic properties for more successful system operation takes place under varying the ambient conditions.

If the variation of ambient conditions is applicable under the determined law, then the indicated positive effect is accessible in the program-tunable systems, whose properties vary under the defined fixed program. An example is the seasonal reorganization in photochemical synthesis systems for over-year plants.

If the law of ambient conditions variation beforehand is unknown, then the influence of these modifications on relative successfulness of system operation can be removed in adaptive systems, where the reorganization is carried out depending on continuously analyzable ambient conditions. An example is a radar servomechanism with adaptation depending on a power or other properties of radio interferences [34].

An adaptive system with tunable parameters is called a self-adapting system and the system with tunable structure — a self-organizing one.

The stationary system depending on those or other dynamic properties can be referred either to stable, or unstable systems (the intermediate state corresponds to the stability limit). These concepts will be illustrated further in §1.2. Remarkable is that the defined fixed status of non-stationary system can also be stable or unstable. If the stable system status (unstable status theoretically can also be taken as a basis, however it would not be practical) is kept in the presence of small parametric perturbations then such a system is named rough, otherwise — not rough.

## §1.2. Stable and rough systems

**Stability**. The concept of stability is applicable to any dynamic system.

The system is called stable, if it recovers back to an initial state after removal of perturbation. Perturbation plays the role of any external action. Thus the initial undisturbed state can be implied with not only static status with constant inputs and outputs, but also the status described by variation of input and output values in the interacted laws.

The stability of linear system is equal to the property exclusive of the system that does not depends on excitations, applied to a system or having been applied to it earlier. A convenient test input action $g(t)$ when estimating a linear system stability is the short rectangular pulse with a final square (in a limit — $\delta$-function).

If the response of a system on such input $y(t)$ at $t \to \infty$ tends to zero, the system is stable. If the response is restricted on magnitude, the system is on the limit of stability. If the response is not restricted, the system is unstable. It is illustrated by the graphs in Fig.1.4 *a, b, and c* accordingly.

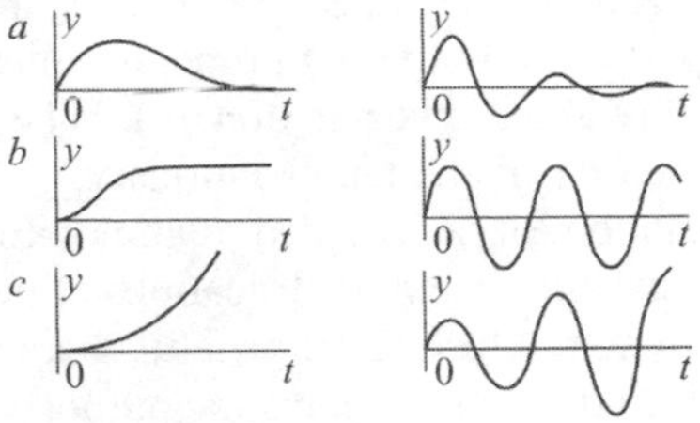


Fig. 1.4

The stability of nonlinear system can depend on external actions and is defined in a special way with the account of their value.

The analytical stability conditions of any dynamic system can be formulated only after the representation of its mathematical model is provided. These conditions look like inequalities written for parameters of a system.

**Stability regions**. Geometrical maps of stability conditions contain the stability regions constructed in a plane (generally — in multidimensional space) of system parameters.

The possible kind of stability regions when analyzing two parameters $A$ and $B$ is shown in Fig.1.5.

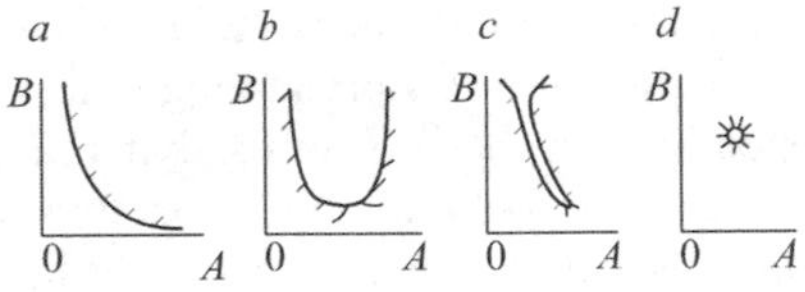


Fig. 1.5

**Concept of system roughness**. If the stability region has significant or even indefinitely large width in all directions (Fig.1.5*a*, *b*), then the system can have roughness property or else is rough. In a rough system with nominal values of parameters, lying far enough away from stability limit, then the stability is kept at possible deviations of parameter's values away from nominal ones.

If the stability area has a small (in a limit — infinitesimal) width in some or in all directions (Fig.1.5*c*, *d*), the system is not rough.

If the stability area is completely absent from a plane (generally — in multidimensional space) of all system parameters, then the system is called structurally unstable.

Naturally, in a society and in engineering both unstable and non-rough dynamic systems, as a rule, are impractical, nonviable and have no working capacity.

## §1.3. Influence of external action properties on indexes of system successful functioning

**Multidimensional and multicoupled systems.** In real complicated dynamic systems there are usually some inputs (input actions) and some outputs (responses). For each possible pair of input-output it is possible to define some dynamic properties, which completely characterize such a system, called multidimensional. For example, the linear multidimensional dynamic system is completely characterized by a matrix transfer function $H$, linking a matrix-column of action representations $G = [g_1 g_2 \ldots g_m]^{\mathrm{T}}$ and matrix-column of responses representations $Y = [y_1 y_2 \ldots y_m]^{\mathrm{T}}$ by the relation $Y = HG$.

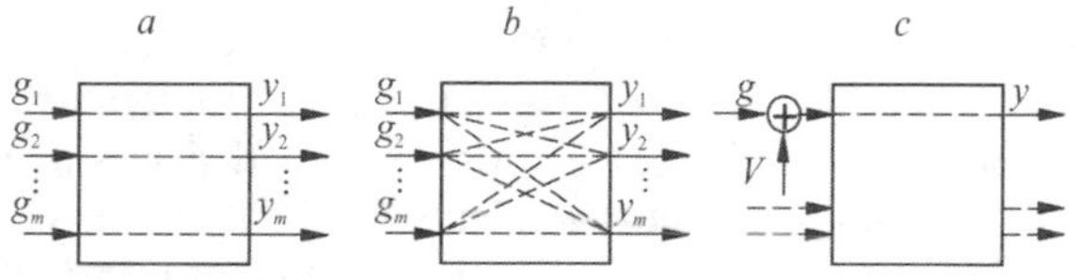


Fig. 1.6

In special cases the multidimensional system can have one input and some outputs (then $H$ is a functional matrix-column) or some inputs and one output (then $H$ is a functional row matrix).

If each of the outputs in a multi-dimensional system depends only on one of the inputs (Fig.1.6*a*), then the system can be represented as several off-line channels with no cross-feed (the functional rectangular matrix $H$ is diagonal in this case). Otherwise (Fig.1.6*b*) the system is multicoupled.

**Types of external actions.** In the specific problems of systems research one input and one output are often discriminated, and the systems are multicoupled, for example, $g = g_1$ and $y = y_1$ in Fig.1.6*b*. The interaction between input and an output represents the main interest in the given concrete problem. Thus the dependence of response on other input actions only spoils the determination of the main investigated dependence. In this case it is necessary to consider such actions in a concrete considered problem such as noise or disturbance.

In a linear multicoupled system, where the principle of superposition is fulfilled, the reaction from a disturbance can be immediately added to a reaction from the main action. If the dynamic properties of transmission channels in the main action and the disturbances coincide then it is possible to consider, that the action and the disturbance are applied to one input or are reduced in one input as a result of equivalent transformations of the system model (Fig.1*b*, *c*). It allows for the consideration of more simple research of the problem of a single-channel system with an additive mixture of the fundamental action and interference on input instead of the research of the problem of a multicoupled system.

The reaction of interference in a nonlinear multicoupled system can lie in a rescaling of an output reaction from the fundamental action. In this case the interference is called multiplicative. The same interference can have both additive and multiplicative components in practice.

It is possible to allocate the other characteristic case, when the interference action causes the value of some parameter to vary in the mathematical model of the dynamic system. Such interference is called parametric or parametric perturbation. There is a concept of structural perturbation leading to a variation of the kind mathematical formulas, describing a system, i.e. to a modification of system structure.

The example of parametric perturbation in a plane control system can be a variation of rudder efficiency owing to a variation of the planes' aerial velocity or air density, the example of structural variation or the jamming of rudder.

**Factors of successful system operation.** Stability and roughness properties still do not ensure the successful operation of dynamic system or, as it is usually spoken of in an engineering system, it's working capacity. These properties create only the necessary background for this purpose. The system operation is considered successful only when the values of major numerical factors describing it lye in the permissible limits. The indicated factors are included in some formal successful condition of operation permitting selection of the subset of successfully operating variants of dynamic systems from a set of actual or hypothetical ones.

The choice of particular quantitative factors for successful system operation depends on many factors and is not always obvious. If the system is natural and is not created by men, then the fact of such system existence usually testifies the system efficiency and acceptable values of all quantitative factors. However, someone can evaluate even such a dynamic system by the subjective criteria. For example, people have individual conceptions about an ideal climate or weather characterized by air temperature, wind, rainfall amount and other parameters. Someone concludes a measure of atmospheric dynamic processes behavior depending on actual values deviations of such parameters from the “ideal”. But the successful operation conditions of engineering dynamic systems must be formulated with considerably less voluntarism.

Consider, that some factor of successful operation $I$ is given in one way or another, and it is evaluated depending on two factors:

- Dynamic properties of a system described by some operator or a set of operators $H$;
- Regularities of external actions variation described by some set of characteristics $S$.

The example of an operator $H$ in the case of a linear system can be its transfer function. Examples of characteristics $S$ are the spectral densities, and at the determined approach particular analytical expressions or graphs of external actions.

The factor of successful functioning generally can be a vector in $m$-dimensional space $I = (I_1, I_2, \dots, I_m)$. Its components are the scalar values $\{I_i\}_1^m$, comprehensively describing, for example, the possible deviations of system response from the values, which would correspond to its "ideal" (in the sense of the adopted criterion) operation at different external actions. In the case of servo-mechanism the components of vector $I$ can be, for example, its maximal error, root-mean-square error, constant component of error, decay time of step response, maximal relative value of system overshoot in such step response and other similar characteristics.

Knowledge of factor $I = I\ (H,\ S)$ allows finding out in each particular case, whether the operation of dynamic system is successful. For this purpose the region of satisfactory values for this index should be allocated in $m$-dimensional space which will be denoted as $I_{sat}$.

The operation of a system with the defined dynamic properties $H$ at external actions with characteristics $S$ is successful, when it satisfies the condition

$$I(H,S) \subset I_{sat}. \tag{1.1}$$

Otherwise the system operates poorly or, when speaking with reference to engineering systems, has no working capacity.

Underline that the condition (1.1) is written at quite defined and univocally known characteristics $H$ and $S$.

## §1.4. Robust systems

**Robustness concept.** Evaluating the successfulness of system operation on the base of (1.1), it is necessary to take into account, that due to the parametric and structural perturbations the dynamic properties of system can vary or have a scatter for different copies of a system. Only the class of possible dynamic properties of system $M_H$ is supposedly known, i.e. for each particular system the accurate kind of operator $H$ is unknown, but the following condition is certainly satisfied

$$H \subset M_H. \tag{1.2}$$

For conditions of system operation and appropriate characteristics of external actions $S$ the complete determinacy is not usually proper. A particular kind of these characteristics is considered unknown, but some of their class $M_S$ is given, and the following condition is necessarily satisfied

$$S \subset M_S. \tag{1.3}$$

Now evaluate the realization of condition (1.1) at all admissible characteristics $H$ and $S$, corresponding to conditions (1.2) and (1.3), which can then be mathematically written as:

$$\underset{\substack{H \subset M_H \\ S \subset M_S}}{I(H,S) \subset I_{sat}}. \tag{1.4}$$

If condition (1.4) is fulfilled, then the system has a property of robustness and is called robust. If condition (1.4) is not fulfilled, the system is not robust, but still can be stable and rough.

The robust property means saving the beforehand assigned system operation performance at a high enough level. It must be fulfilled at all possible characteristics of external actions within the framework of some of their preset class at the account of an admissible variation of system dynamic properties due to parametric and structural perturbations.

For simplification of condition (1.4) entry, general input actions, disturbances and also parametric and structural perturbations will be referred further to external actions (its properties are described by a set of characteristics $S$). Thus it will be considered that the operator $H$ characterizes the nominal dynamic system properties, defined at absence of parametric and structural perturbations. Then condition (1.4) looks like

$$I(H,S) \underset{S \subset M_S}{\subset} I_{sat}, \tag{1.5}$$

**The most unfavorable properties of actions.** Suppose some formal rule is chosen, permitting us to define the kind of different successfully enough operating variants of dynamic system, according to the condition (1.5), operating more successful than others. Actually it means that the vectorial successfulness factor of operation $I = (I_1, I_2, \dots, I_m)$ is replaced by scalar or is converted into a scalar factor. For example, it can look like

$$I = \sum_{i=1}^{m} \alpha_i I_i , \tag{1.6}$$

where $\{\alpha_i\}_1^m \in [0, \infty)$ are weight coefficients. In that specific case all weight coefficients in (1.6), except only one, can be equal to zero.

In that particular case it will be considered, that the smaller value of index $I$ corresponds to the more successful system operation. Still factor $I$ is a function of nominal system dynamic properties $H$ and properties of external actions $S$ (i.e. properties of the fundamental input action, disturbance, parametric and structural perturbations).

It is possible to find such properties of actions $S = S_{mu}$ within the framework of their preset class $M \subset M_S$, at which the system with the defined dynamic properties $H$ will operate least successfully and the value of the factor $I$ will reach maximum

$$I(H, S_{mu}) = \max_{S \in M_S} I(H,S).$$

The actions with the indicated properties $S_{mu}$ are considered the most unfavorable.

It is possible to find the optimal dynamic properties of a system $H_{\text{opt}}$, which provides the most successful operation at the defined properties of external actions $S$, i.e. the minimal value of the factor

$$I(H_{\text{opt}}, S) = \min_{H} I(H,S). \tag{1.7}$$

**Minimax robust systems.** Set the problem of dynamic system properties optimization by criterion (1.7), under the requirement that the external actions necessarily have the most unfavorable properties $S = S_{mu}$ at any selected operator of system $H$. Such a task can be treated from the positions of game theory with a two player game. One of the players gives orders for choice of external actions properties and tends to maximize the factor $I$, and another one gives orders for choice of system properties with the purpose to minimize this factor. As a result of solution the minimal guaranteed value of successfulness index of system operation can be found as:

$$I_{\min} = \min_{H} \max_{S \in M_S} I(H,S) = \min_{H} I(H,S_{mu}) = I(H_{\text{opt}},S_{mu}), \tag{1.8}$$

with the optimum system operator $H_{\text{opt}}$, at which such value can be reached.

An approach to properties of the system optimization problem described by the formula (1.8) is named minimax. It provides reaching a level of the operation successfulness not worse then $I_{\min}$ at any property of actions from their preset class $S \in M_S$, and the indicated guaranteed value $I_{\min}$ is less, than in any non-optimal system with an operator $H \neq H_{\text{opt}}$.

If the value $I_{\min}$ expressed with the formula (1.8) concerns a region (interval) of acceptable values of operation successfulness index $I_{\min} \subset I_{sat}$, i.e. the condition (1.5) is satisfied. The system with an operator $H_{\text{opt}}$ is then named the minimax robust system [44].

In the case, when the fulfillment of the condition (1.5) is not controlled (more accurately, is controlled implicitly), a system is called simply minimax or optimal in minimax sense.

**Adaptive systems as an alternative to robust ones.** Abilities of robust systems to keep a required level of operation successfulness in a case of operation conditions indeterminacy cannot be unlimited. If the admissible class of external actions is too wide, then the same system cannot satisfactorily operate in all possible situations, at any dynamic properties it could have. In this case the positive effect of operation can be guaranteed only with the presence of adaptation in a system.

Unlike a robust system, in an adaptive system the independent automatic reorganization of dynamic properties after changing external actions (useful action, disturbance, and parametric and structural perturbations) is organized in someway with the purpose of successful operation provision in each particular situation. It can be made, for example, when allowing a system the possibility to analyze the properties of external actions and its tutoring to correct behavior depending on those or different results of such analysis.

In adaptive systems the current (but not just a priori) information on properties of external actions is used. Then the influence of a priori indeterminacy of these properties on system operation efficiency after completion of adaptation process basically can be completely eliminated. If the dynamic properties of a system become optimal for each particular situation in an ideal case, then it will provide the best of possible operation efficiency, which is regarded as the potential efficiency.

In a minimax robust system the potential efficiency can be reached only at some properties of actions. When observing all of their admissible class $M_S$, the guaranteed efficiency should be a bit less then potential, but not below the acceptable.

Thus, the robust systems lose to adaptive with respect to possibilities for reaching the best values of successfulness operation factor $I(H, S)$. The advantage of robust systems is the easiness of their construction.

The large reliability is peculiar to more simple system. Then so-called generalized efficiency, defined according to possible violations in a system operation because of fails in its elements, at robust system can appear higher than in adaptive system. The operation successfulness of adaptive systems decreases also because during the time necessary for adaptation process, the properties of external actions can vary.

The correct account of advantages and lacks of robust and adaptive systems allows us to prefer one of them in each particular situation, however without generalization of the conclusions on all varieties of dynamic systems. The variant of shared use of robustness and adaptation principles in the same system deserves attention also. Each of the enumerated approaches promotes solving the problem of successful dynamic systems operation in conditions of the priori indeterminacy and has the right for existence and development.

**The modern status of robustness problem**. The number of scientific papers and monographs published in the world on robust methods of information processing and control reaches many hundreds and permanently increases, although no longer so fast, as in the beginning of 1990. They are still mostly devoted to a robust stability of linear and nonlinear dynamic systems, and also to robust static identification, i.e. parameters and characteristics estimation for systems and processes.

The problems of parameters identification in the accepted model of a dynamic plant on noisy observations for input and output signals or for other required data in priori indeterminacy conditions are standard when researching the adaptation algorithms of automatic control systems, processing algorithms of medical, biological, economic, sociological information. Many authors develop the robust approach to solve such problems in different variants. The fundamental results on optimization of algorithms and ensuring the quality of statistic identification were obtained by Y.Z. Tsypkin and described in the monograph [84], where a good review of other publications on this theme is given also.

For close problems of identifying casual quantity distribution parameters, especially in the case of one or two unknown parameters, the robust methods have been well developed in mathematical statistics. A canonical example, on which the first robust procedures were completed, is an estimation of the shift parameter and dispersion [88, 35]. Unlike the classical theory of statistic estimations, the class of possible distributions is set but without a particular kind of a priori distribution. The estimations obtained in this way, as a rule the nonlinear one's, are considered robust on distribution, or *P*-robust (unlike *K*-robust estimations obtained by a statistical analysis of casual processes from the deficient information on correlation properties). Many of the obtained results should be treated as developments of the statistical solutions theory, instead of classical theory of estimations. They are tightly coupled with Bayesian approach; in particular, they can suppose the minimax solution determination.

In a smaller measure the robust approach has covered the filtering problems so far, i.e. the selections of current (or displaced on a time axis, or somehow converted by another known way) casual process values from samples corrupted by a disturbance. At the same time control accuracy research with the presence of disturbing actions is coupled with filtering problem analysis, but ensuring accuracy is provided by applications of robust filtering algorithms.

The synthesis of dynamic filtering algorithms basically can be considered as one of the problems of static solutions theory [56, 75]. Multidimensional distribution functions of investigated process and disturbance figuring in this theory provide information about correlation properties, which also play a leading role in filtering problems. However, it is possible to finish a solution up to constructive results only with Gaussian or Markov distribution laws approximations.

For a case of an additive mixture of useful action and disturbance, the Gaussian model optimizes this by square-law criterion solution of filtering problem linear, and leads it to Kalman or, in a stationary mode, to the Wiener filter. But when researching the linear systems on the correlation theory basis, only the spectral-correlation but not the probable properties of actions are essential. The robust approach to filtering should be coupled by providing some admissible areas in the space of spectral densities or correlation functions assignment, instead of probability densities, however the developed means of the static solutions theory badly provides such possibility.

All of the above explain the reason of relative dissociation of methods in solving dynamic filtering problems and identification problems in priori indeterminacy conditions. The majority of the obtained results on robust identification and on robust point estimation are not immediately usable in the filtering of processes with poor information about dispersion dynamics. These problems require special research methods.

There are not many publications on ensuring of control accuracy, including ones on the basis of a robust dynamic filtering use. It is possible to historically allocate one of the first groups of publications regarding research of minimax filtering with noncasual perturbations. The monographs by N.N. Krassovski [49, 50], A.B. Kurzhanski [55], the paper by D.S. Irger [37], and also later monograph by V.N. Fomine [27] are significant landmarks here. The perturbation is selected from a totality of solutions for differential equation with known coefficients. The defined restrictions are overlapped on the absolute value of its right member. Apparently, for the first time the problem of most unfavorable perturbation choice on the basis of game approach was posed in the publication by U. Grenander [33] in 1963 and is well described in the book [30].

It is essential, that the minimax approach in dynamic filtering problems is more justifiably represented, than in identification problems, because priori information usually is more extensive here. The minimax approach also causes pessimistic estimation of situation at identification, when only one numerical characteristic defining the class of actions is set (it is usually maximal value or dispersion). The necessity to account several numerical action characteristics causes the sharp increase of mathematical difficulties at a particular task solution. They often become insuperable when using the methods explained in mentioned publications. The practical examples of solutions, dealing with constructive result, as a rule are limited by a case of one numerical characteristic representation, and these solutions can not belong to a class of filters with fractional rational transfer functions [44, 32, 70].

Considering an ensuring of control accuracy problem in the historical plan, it is necessary to mention the significant publication by B.V. Bulgakov [10] on the strict analysis of perturbations accumulation in a linear dynamic system. His work has generateda series of publications [31, 87], in which the procedure of maximal dynamic controlling error estimation was developed. The maximal error must certainly be considered as a major factor of control quality, in spite of the fact that its research is coupled to some analytical difficulties (see the chapter 6).

An attempt to generalize the ensuring of accuracy problem definition and to ensure the error restriction at determination and correction of dynamic system phase coordinates, has been made in the book by B.T. Bahshiyan, R.R. Nazirov and P.E. Elyasberg [4]. There the distribution function of error probability for the initial conditions are considered as unknown, and only some sets which they can belong to are given. The problem is formulated in a general view, but basically the identification problems are actually considered; in particular, originating at an estimation of space vehicles orbits parameters.

In common problem definition on a guaranteed estimation of dynamic systems phase state are researched in papers by F.L. Chernousko [17] and in his monograph [18], completely devoted to an ellipsoids method development. The linear and nonlinear controlled dynamic systems with continuous or discrete time are considered. A "Set of attainability" is constructed in the space of system phase coordinates. It is important to estimate each particular realization of possible deviations, instead of average characteristics. One of phase coordinates can be, in particular, the control error. The possibility to treat the mode of estimation on the basis of an ellipsoid method as guaranteed analog of a Kalman filter is indicated when lacking the reliable information about perturbations. However the boundaries of these perturbations are know. At the same time the significant calculation difficulties of this mode realization are marked.

there is also a group of publications devoted to special Kalman robust filter research. The paper by A.A. Ershov and R.S. Lipzer [26] has been the most successful. Robustness to action distribution function, instead of its correlation properties, is analyzed in them. It is shown that there is a nonlinear filter which posses some advantage against the classical linear Kalman filter at the Gaussian distribution with "pollution".

One of the first research problems for linear robust dynamic filtering in a sense of signal recovery from its additive mixture with a disturbance in the frequency domain at the deficient of a priori information is covered in the paper by A. Kassam and T.L. Lim [43]. Two particular representation variants of the admissible class of action spectral densities are considered, and the accessible guaranteed filtering accuracies are analyzed. However, this analysis is rather rough and has no requirements for the system physical feasibility condition. The first variant represents the regular Huber model with "pollution", but applied to spectral densities, instead of probability distribution functions. Such a model has enabled researchers to make a conclusion about sharp degradation of Wiener filter accuracy with the occurrence of "pollution" in action spectral densities with a restricted dispersion. But this model has not received the propagation as a robust filters synthesis instrument. The second variant differs by the representation of action spectral densities range which is bounded from above and from below by curves with known analytical expressions.

The indicated mode of action properties description is used in a series of other works. The synthesis of filter with the best-guaranteed accuracy in such statement of problem solution is called bounding filters. The procedure of synthesis thus includes the determination of spectra laying in preset boundaries and is most unfavorable for a filtering realization. Sometimes the boundaries are given for parameters that are included in common expressions for spectral densities, instead of boundaries for spectral densities themselves. A shortage of all these approaches is the difficulty when assigning particular boundaries in application problems.

The bounding filters are treated in many papers as a particular and the most developed type of linear robust filters. The publications on some other directions of linear robust filtering in stationary statements are too abstract and practically do not contain any new constructive outcomes, which would be effective in informal problems. However, it is necessary to signify the paper by H. Poor [70], distinguished by a good procedure and expound width.

When analyzing the problem of ensuring control accuracy, it is necessary to mark a significant role in its occurrence and development of classical frequency domain methods for dynamic systems research. The means of frequency transfer functions and logarithmic frequency characteristics, the start of which wide use in Russia in the fifties — sixties of the 20 century, is coupled first of all with works by A.V. Mikhailov, V.V. Solodovnikov, E.P. Popov and V.A. Besekersky and other scientists. These works have allowed the formulation of rather simple and obvious conditions of dynamic error regulating limitation. Forbidden areas of the Bode diagram, constructed on the basis of equivalent harmonic action reviewing [6, 7, 9], have already been widely applied in engineering practice for a long time and have proved high efficiency. However they are closer to heuristic, than theoretic, and do not ensure the strictness of mathematical results. But such strictness is less important for the designer, than the possibility to get useful landmarks for a choice of constructive solution on account of all showed system requirements rather fast and simple. On the other hand, modern conditions of computerization design and revising many other criteria of design processes organization force one to adapt and develop useful research methods adequately.

It is shown in the book [8] that the constructive results of harmonic action model application in ensuring control accuracy problems can be strictly justified, specified and generalized on the basis of mathematical means of the problem of moments and nonlinear programming methods. Thus the frequency domain method of robust dynamic systems researching is formulated. It combines the reliability of used priori information, ease and obviousness of an equivalent harmonic action with reasonable strictness and community of methods for calculus and programming. The significant part of the present book is devoted to the development of such a method.

## §1.5. Low-sensitivity systems

**Parametric classes of actions characteristics.** A priori indeterminacy in properties of external actions applied to dynamic systems increases with the enlargement of permissible class (or classes depending on numbers of actions) of their characteristics $S \subset M_S$. Such classes of characteristics are divided into two groups due to representation mode: parametric and nonparametric.

The parametric class of characteristics is set by some common analytical expression $S = S(\alpha_1, \alpha_2, \ldots, \alpha_l)$, correct for each of the possible particular characteristics, but at different values of parameters $\{\alpha_i\}_1^l$. In the elementary case $l = 1$ those parameters can accept the values restricted by inequalities $\alpha_{i\min} \le \alpha \le \alpha_{i\max}$, $i = 1, 2, \ldots, l$, at known finite or infinite values $\{\alpha_{i\min}\}_1^l$ and $\{\alpha_{i\max}\}_1^l$. It is assumed that the values of function $S$ should also depend on some independent variable $\vartheta$, which is the argument in the characteristic, described by that function, i.e. the expression is actually considered as

$$S(\vartheta) = S(\vartheta, \alpha_1, \alpha_2, \ldots, \alpha_l). \tag{1.9}$$

Here are three examples for parametric classes of characteristics.

1. *Class of exponential correlation functions*

$$R(\tau) = D\,exp(-\mu|\tau|)$$
$$(\vartheta = \tau,\ l = 2,\ \alpha_1 = D,\ \alpha_2 = \mu).$$

2. *Class of rectangular spectral densities*

$$S(\omega) = \begin{cases} S_* & \text{at } |\omega| \le \beta, \\ 0 & \text{at } |\omega| > \beta \end{cases}$$
$$(\vartheta = \omega,\ l = 2,\ \alpha_1 = S_*,\ \alpha_2 = \beta).$$

3. *The polynomial input actions of the third order class*

$$g(t) = a_0 + a_1 t + a_2 t^2 + a_3 t^3$$

(the characteristic $S$ is the action itself, $\vartheta = t,\ l = 4,\ \alpha_1 = a_0, \alpha_2 = a_1, \alpha_3 = a_2,$ $\alpha_4 = a_3$).

**Nonparametric classes of actions characteristics.** When giving the nonparametric class of characteristics there is no common analytical expression, correct for any characteristic, which belongs to this class. Hence, each particular characteristic can not be coupled with the set of finite number of parameter values, or else, with some point in finite-dimensional space of parameters. Then it is impossible to put in univocal correspondence of some area in finite-dimensional space of parameters, into a whole class of characteristics. Any class of characteristics, whose representation mode satisfies to such condition, can be named as nonparametric.

It is possible to give the following examples for nonparametric models of action spectral densities [44].

1. *Model of* $\varepsilon$*-pollution.* The spectral density $S(\omega)$ belongs to a class of functions, represented as a weighed sum of known function $S_0(\omega)$ and unknown "polluting" function $S_{uk}(\omega)$:

$$S(\omega) = (1-\varepsilon)S_0(\omega) + \varepsilon S_{uk}(\omega). \tag{1.10}$$

Here $\varepsilon$ is pollution coefficient, $0 < \varepsilon < 1$. For function $S_{uk}(\omega)$ (as well as for $S_0(\omega)$ the integrated limitation is usually set

$$\frac{1}{\pi}\int_0^\infty S_{uk}(\omega)\, d\omega \le \sigma^2,$$

by means of which the finiteness of action r.-m.-s. value $\sigma$ is taken into account.

Noticeable is that the model (1.10) is similar by form to the widely used non-parametric probability density model suggested by Huber [35].

2. $\varepsilon$*-neighborhood model.* The possible deviations of actual spectral density $S(\omega)$ from initial $S_0(\omega)$ are restricted only by an integral relation

$$\frac{1}{\pi}\int_0^\infty \left|S(\omega) - S_0(\omega)\right| d\omega \le \varepsilon\sigma^2, \tag{1.11}$$

at

$$\frac{1}{\pi}\int_0^\infty S(\omega) d\omega = \frac{1}{\pi}\int_0^\infty S_0(\omega)\, d\omega = \sigma^2.$$

Here the values e and $\sigma$ play the same role, as in the e-pollution model.

3. *p-point model.* The class of spectral densities $S(\omega)$ is determined by expression

$$\int_{\Omega_i} S(\omega) d\omega = \int_{\Omega_i} S_0(\omega)\, d\omega,\ i = 1, 2, ..., n. \tag{1.12}$$

Here, as before, $S_0(\omega)$ is the initial spectrum and the $\Omega_1, \Omega_2, \ldots, \Omega_n$ set are a partition of frequency axis on *n* regions. Practically, the relation (1.12) sets the action power values in each frequency interval for chosen *n* intervals.

4. *Band-pass model.* The spectral density $S(\omega)$ lies in some band between the upper and the lower preset functions

$$L(\omega) \le S(\omega) \le U(\omega),\ -\infty < \omega < \infty, \tag{1.13}$$

and still

$$\frac{1}{\pi}\int_0^\infty S(\omega)\, d\omega \le \sigma^2.$$

Notice that model (1.10) can be considered as a special case of (1.13) at $L(\omega) = (1-\varepsilon)S_0(\omega)$ and $U(\omega) \to \infty$.

5. *Model with restricted generalized moments.* The class of spectral densities $S(\omega)$ is set by the population of integral limitations defined as

$$\frac{1}{\pi}\int_{0}^{\infty} u_i(\omega)S(\omega)\,d\omega \le M_i,\ i=1,2,\ldots,N, \tag{1.14}$$

where $u_i(\omega)$ is known as basis functions, $M_i$ is known values. The left parts of inequalities (1.14) in mathematics are known as the generalized moments of function $S(\omega)$ concerning the functions $u_i(\omega)/\pi$. The physical treatment of inequalities (1.14) can be done as follows. At passage of action through $N$-connected in parallel linear filters with gain plots of a kind $\sqrt{u_i(\omega)}$, the powers of signals on these filters outputs should be restricted by the values $M_i$, $i = 1, 2, \ldots, N$.

Mark that $p$-point class (1.12) is a special case of (1.14) with

$$u_i(\omega)=\begin{cases}1 & \text{at} \quad \omega\in\Omega_i,\\ 0 & \text{at} \quad \omega\notin\Omega_i.\end{cases}$$

The maximal practical importance at investigation of automatic control systems accuracy has another special case of (1.14), when the basis functions $u_i(\omega)$ look like $u_i(\omega)=\omega^{2i}$. Generalized moments of spectral densities in this case make sense at dispersions of $i$-th derivatives of centered action (see chapter 5).

The nonparametric class of actions can be preset even immediately, without use of spectral or correlation models. The elementary example is the limitation of the absolute value of action or its derivative. This nonparametric action model differs by its large reliability and is widely used in practice (see chapter 6).

**The characteristics of parametric and structural perturbations.** The concepts of parametric and nonparametric classes of external actions require the additional explanation of their extension for such actions as parametric and structural perturbations. More often particular, "mechanisms" of such external processes or circumstances influence on the systems dynamic properties, not controlled by a system, is rather complicated and has no adequate model. Thereby it is necessary to consider the parametric and the structural perturbations as not a first cause but the result of the indicated external factors operation consisting in instability or scattering of system dynamic characteristics.

Note that perturbation separation on parametric and structure is conditional. Actually any structural perturbations can be presented as strong parametric perturbations. For example, the nullification perturbation or several parameters of system transfer function causes a variation of its kind, reflecting the system structure. Allowing this, the "parametric perturbations" term will be used further.

The parametric perturbations can be referred either to parametric, or to nonparametric actions classes depending on the character of their variation and on a type of operator $H$, accepted as the system dynamic properties characteristic.

The transfer function depends on several parameters of a system as an operator $H$. The result of parametric perturbations is the deviation of these parameters values from some nominal values. If such deviation goes rather slowly and within the considered piece of time, the parameter values are considered as quasi-constant, then unique perturbation characteristic will be its value that allows reference to such perturbation and to a parametric class of actions.

If the variation of parameter values increases rather fast and it is commensurable with the variation of other external actions on a velocity, then the parametric perturbation should be described by time function (determined or casual). Depending on such description types the perturbation can be referred to either as parametric, or as a nonparametric class of actions.

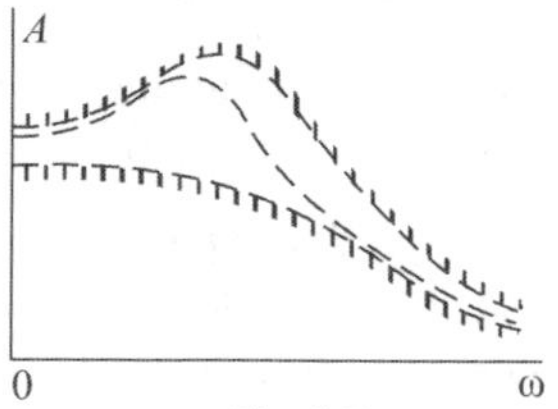


Fig. 1.7

Any dynamic system characteristic permitted to find an estimation of an operation efficiency index value $I(H, S)$ at defined external actions properties $S$, can be considered as an operator $H$. For example, experimentally found gain plot of a system sometimes can play this role. The parametric perturbations result appears as some range of possible values of ordinates to each value of abscissa in the gain plot. Graphically, some corridor on a gain plot (Fig.1.7) maps this circumstance within the framework of which the particular arbitrary shape graphs can lay. The shape of this class of system gain plot, having parametric perturbations influence, is nonparametric.

So, a priori indeterminacy in external action properties consists of the fact that the characteristics of all or part of these actions are particularly unknown, but necessarily belong to some preset class being parametric or nonparametric. The possible variants describing the dynamic system external action properties are shown in the classification scheme shown in Fig.1.8.

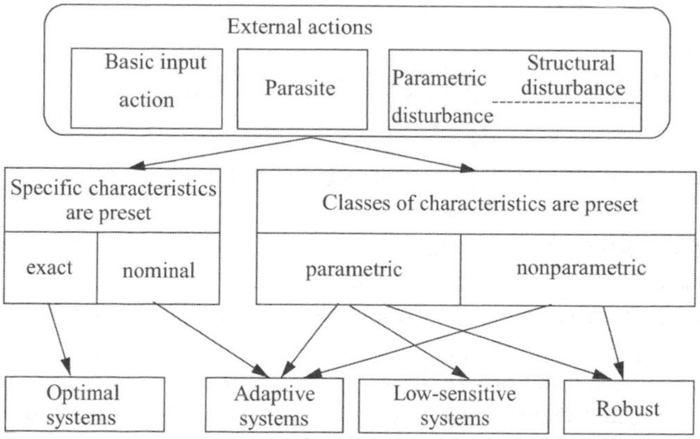


Fig. 1.8

The different methods of dynamic system operation efficiency investigation can be used depending on external actions type.

**The sensitivity functions.** If the external actions characteristics belong to the parametric class and common analytical expressions for these characteristics of the kind (1.9) are known, then the index (functional) of a system operation successfulness $I(H, S)$ is possible to present as a function of finite number of parameters $\{\alpha_i\}_1^l$

$$I = I(H, \alpha_1, \alpha_2, \dots, \alpha_l).$$

As a reminder, the ranges of possible parameters values must be preset by inequalities $\alpha_{i\min} \le \alpha_i \le \alpha_{i\max}$, $i = 1, 2, \dots, l$, in other words, the necessity to select some region in $l$-dimensional Euclidean space of parameters. Each of its points fulfills one of its possible particular $S$ characteristics.

Some point with coordinates $\alpha_i = \alpha_{i0}$, $i = 1, 2, \dots, l$ is fixed inside the permitted region of parameters space. It accords to a so-called base set of parameter $\{\alpha_{i0}\}_1^l$ values; these values can be considered as nominal. Thus the base value of efficiency index will be obtained as

$$I_0 = I(H, \alpha_{10}, \alpha_{20}, \dots, \alpha_{l0}).$$

Assuming this value allows us to consider the operation of a system successful, i.e. $I_0 \in I_{sat}$, let some increments to parameters values are set as

$$\alpha_i = \alpha_{i0} + \Delta\alpha_i .$$

If the function $I$ is differentiable on parameters $\{\alpha_i\}_1^l$, then for calculation of its new value it is possible to use the expansion of this function in a Taylor series in relation to a point $\{\alpha_{i0}\}_1^l$. Considering the parameter increments as small enough and the rejection of all nonlinear expansion terms admissible, it is possible to write the following expression:

$$I(H, \alpha_1, \alpha_2, \dots, \alpha_l) \approx I(H, \alpha_{10}, \alpha_{20}, \dots, \alpha_{l0}) + \sum_{i=0}^{l} \left.\frac{\partial I}{\partial \alpha_i}\right|_{\alpha_i = \alpha_{i0}} \cdot \Delta\alpha_i \qquad (1.15)$$

The partial derivatives $\left.\frac{\partial I}{\partial \alpha_i}\right|_{\alpha_i = \alpha_{i0}}$ included in this expression are called the functions of sensitivity for systems of a successful operation index for the appropriate parameters. At small values of sensitivity functions a system is considered low sensitive, at zero value — insensitive or invariant for the appropriate parameter.

Use of sensitivity functions is the fundamental method in investigating the influence of parameter increments on a system operation and at synthesis of low-sensitivity systems within the framework of so-called theory of sensitivity [42,

77]. The indispersion theory [74] is conceptually close to a sensitivity theory at least in the sense of an investigation problem statement.

**Comparison of low-sensitivity and robust systems.** The low-sensitivity and robust systems concepts are close, but do not coincide.

The dynamic system can be referred to as low-sensitivity only on the basis of sensitivity functions estimation and without accounting for acceptable ranges of parameter variation. The robustness concept is coupled with holding the successfulness factor of system operation $I$ in admissible limits, and takes into account not only the value of sensitivity functions, but also the possible parameters variation.

In addition, the small amount of sensitivity of a system can be set only in relation to properties of external actions, inhering to a parametric class, and at obligatory fixation of a base set of parameter values. Robustness is defined at any actions characteristics class, including first of all their nonparametric classes.

However, all these differences do not eliminate the possibility of using the sensitivity theory and an indispersion theory to investigate some robust systems.

The fundamental approaches to the dynamic systems synthesis at different variants of the information representation on properties of actions are illustrated by the scheme in Fig.1.8.

*Questions*

1. What is a dynamic system? What is the difference between dynamic and non-dynamic systems?
2. Give the original examples of dynamic systems.
3. Name the indications, of which the dynamic systems are subdivided into linear and nonlinear, stationary and non-stationary.
4. Are the adaptive and non-stationary systems concepts equivalent?
5. What kind of system is classified rough? Illustrate the dynamic system roughness concept.
6. How is the stability property in linear and nonlinear systems exhibited? What is the stability area?
7. What kind of dynamic systems are called multi-dimensional and multi-connected?
8. Name all kinds of possible external actions in dynamic systems.
9. How can the dynamic system characteristic be estimated and what does it depend on?
10. What does the dynamic system robustness property mean?
11. What is the minimax robust dynamic system?
12. Compare the robust and adaptive systems on attainable operation successfulness level.
13. Give examples of setting the parametric classes of external actions characteristics.
14. Give examples of setting the nonparametric classes of external actions characteristics.
15. What is the difference between low-sensitivity and robust systems concepts?

# Chapter 2

## Formalization of dynamic systems investigation problems

### §2.1. Mathematical description of dynamic systems

**Calculus and operational calculus.** Isaac Newton, the great Renaissance thinker, made the first mathematical model of a dynamic system. In "Mathematical Beginnings of Natural Philosophy" (1687) he described the free motion of celestial objects with the use of differential equations. Practical use of calculus began at that time. Newton shares the founders' laurels with his great contemporary Leibnitz. But the dynamic system described by Newton was uncontrollable, and its motion could not be modified by man's will. However, the transition to controlled dynamic systems connected with interposition of the inhomogeneous differential equations, was a quite natural step and was soon carried out.

The main merit of Leibnitz' approach was the interpositioning and use of the concepts of variable $x = x(t)$ and differential. This allowed for the definition of the derivate and integration operators:

$$\dot{x} = \frac{dx}{dt} = \lim_{\Delta t \to 0} \frac{\Delta x}{\Delta t}, \quad \int_0^t x(t)\,dt = \lim_{\substack{n \to \infty \\ \Delta t_i \to 0}} \sum_{i=1}^{n} x(t_i)\,\Delta t_i \,,$$

which are mutually inverse.

The indicated operations underlie differential and integrated calculus, generating a calculus in aggregate.

Elementary algebra was rather well developed at the time of calculus' origin. Therefore attempts were actively undertaken to algebralize a calculus, to avoid limit passages requiring execution of derivation and integration operations, and to replace them by their symbolic operations according to the rules of algebra [89]. Papers by Leibnitz, and also by Euler, Lagrange, Laplace, Fourier, Cauchy and other outstanding mathematicians of the XVIII-XIX centuries, created symbolic or operational calculus. English engineer Heavyside widely applied it for the first time.

Operational calculus is based on the introduction of images $X(s)$ of functions $x(t)$, that are a multiplication of an image by a complex variable $s$ corresponding to the derivation of origin, i.e. $\dot{x}(t) \fallingdotseq sX(s)$. The division of an image by $s$ corresponds to the integration operation of the origin. The rules for other standard operations also then became very simple.

The image evaluation is given on the basis of an integrated Laplace transform

$$X(s) = L\{x(t)\} = \int_0^{\infty} x(t)\, e^{-st} dt,$$

for which detailed tables are tabulated. Assume, that $x(t) = 0$ at $t < 0$.

Thus, the operational calculus allows for a transfer from the differential equations written concerning the origin, to the appropriate algebraic equations concerning the images. Having solved these equations, it is possible to return to the origins, having executed a Laplace inverse transform $L^{-1}\{X(s)\} = x(t)$.

In impulse and digital dynamic systems, in particular, the processes are described by lattice time functions of a kind $x[k] = x(t)\big|_{t=kT}$, where $T$ is the discretization period, for which the image can be found with use of a $z$-conversion

$$Z\{x[k]\} = \sum_{k=0}^{\infty} x[k]\, z^{-k}.$$

Here the complex variable $z^{-1}$ is a lag operator for one period:

$$Z\{x[k-1]\} = z^{-1} Z\{x[k]\}.$$

The last equality expresses the time lag theorem; generally

$$Z\{x[k-1]\} = z^{-i} Z\{x[k]\}.$$

Accordingly $z = e^{sT}$ is usually established between $p$ and $z$ operators. Thus the function

$$X_*(s) = \sum_{k=0}^{\infty} x[k]\, e^{-ksT},$$

into which $z$-image is transformed, is a Laplace transform for a process such as "comb" of δ-impulses

$$x_*(s) = \sum_{k=0}^{\infty} x[k]\delta(t - kT).$$

This is named the discrete Laplace transform.

**Transfer functions**. Writing the differential equation for the continuous linear dynamic system with constant parameters

$$\sum_{i=0}^{n} a_{n-i} y^{(i)}(t) = \sum_{j=0}^{m} b_{m-j} g^{(j)}(t), \tag{2.1}$$

where $g(t)$ and $y(t)$ are external actions, and appropriate system response respectively, $m \le n$.

Applying the Laplace transform for the left and right members of this equation, with the $i$-th derivative image of some function $x(t)$ at zero initial conditions, makes $L\{x^{(i)}(t)\} = s^i L\{x(t)\} = s^i X(s)$. Then, using the Laplace transform linearity property the following is obtained:

$$\sum_{i=0}^{n} a_{n-i} s^i Y(s) = \sum_{j=0}^{m} b_{m-j} p^j G(s)$$

or

$$Y(s) \sum_{i=0}^{n} a_{n-i} s^i = G(s) \sum_{j=0}^{m} b_{m-j} s^j .$$

Having designated $A_n(s) = \sum_{i=0}^{n} a_{n-i} s^i$ and $B_m(s) = \sum_{j=0}^{m} b_{m-j} s^j$, the expression for a system response image is written as

$$Y(s) = H(s)\ G(s), \tag{2.2}$$

where

$$H(s) = B_m(s) / A_n(s). \tag{2.3}$$

Function $H(s)$ is called the dynamic system transfer function. It links the Laplace image of a system response to the external action that has caused this response. It is visible from (2.3), that the transfer function is a fractional rational function of argument $s$.

After the determination of the system response image in formula (2.2) it is possible to apply to the origin by a Laplace reconversion $y(t) = L^{-1}\{Y(s)\}$.

If a transfer function $H(s)$ is considered as the image and it is passed to origin $w(t) = L^{-1}\{H(s)\}$, then the system response to the action $g(t)$ can be expressed by convolution

$$y(t) = \int_0^t w(t-\tau) g(\tau)\ d\tau . \tag{2.4}$$

Function $w(t)$ included in expression (2.4) is the pulse response of a system. It can be defined as a system response of a $\delta$-function, whose image is equal to one. Thus, the transfer function $H(p)$ and pulse response $w(t)$ are connected among themselves by direct and inverse Laplace[1] transforms.

Note, that the condition $w(t) = 0$ at $t < 0$ should be satisfied in the physically realizable system. Due to this, the upper limit of integration in (2.4) can be seen as indefinitely large.

**Difference equations and discrete transfer functions.** For lattice time functions used in describing processes in impulse systems, it is possible to define the finite differences of the different orders. The first reciprocal difference of a lattice

---

[1] Following a reasonable rule to designate an origin and its image by one letter (according to lower case and capital), it should be necessary to designate a pulse response through $h(t)$. However, it is accepted to denote a step response characteristic [9] in this way. Another possibility is to avoid this problem consists of denoting a transfer function through $W(s)$. However so it is convenient to denote a transfer function of the open loop circuit of a system. Therefore by way of exception it is necessary to accept an entry $L^{-1}\{H(s)\} = w(t)$.

function $x[k]$ is an increment of its value during one discretization period $\nabla x[k] = x[k] - x[k-1]$. The second reciprocal difference is an increment of the first difference $\nabla^2 x[k] = \nabla x[k] - \nabla x[k-1] = x[k] - 2x[k-1] + x[k-2]$. Generally, the $i$-th reciprocal difference is an increment of the $(i-1)$-th difference $\nabla^i x[k] = \nabla^{i-1} x[k] - \nabla^{i-1} x[k-1]$. Direct differences are used less often, as their determination requires knowledge of the future values of lattice function.

The relations noted for lattice functions and their differences are called difference equations. The difference equations are used as a mathematical exposition for impulse dynamic systems; i.e. they play the same role as the differential equations for continuous systems.

The linear impulse dynamic time-invariant system is described by this kind of difference equation

$$\sum_{i=0}^{n} a_{n-i} \nabla^i y[k] = \sum_{j=0}^{m} b_{m-j} \nabla^j g[k], \tag{2.5}$$

where $g[k]$ and $y_0[k]$ are the lattice functions of system input and output. The relation between the orders $n$ and $m$ in the left and right parts of the equation (2.5), unlike the differential equation (2.1), can be arbitrary.

If finite differences in the left and right members of the equation (2.5) are expressed immediately through values of lattice functions $g[k]$ and $y[k]$, the following equation is then the accepted format

$$\sum_{i=0}^{n} \alpha_{n-i} y[k-i] = \sum_{j=0}^{m} \beta_{m-j} g[k-j], \tag{2.6}$$

This is also usually called a difference equation. Here the factors $\alpha_i$ and $\beta_j$ are uniquely expressed through factors $a_i$ and $b_j$ in the equations (2.5).

Note, that the equations (2.6) are easily convertible to a recurrence equation

$$y[k] = \left( \sum_{j=0}^{m} \beta_{m-j} g[k-j] - \sum_{i=1}^{n} \alpha_{n-i} y[k-i] \right) \alpha_n^{-1}, \tag{2.7}$$

permitting sequential calculation, step by step, the values of the output lattice function $y[0]$, $y[1]$, $y[2]$, ... on known values of the input lattice function and at the known initial conditions, i.e. with the set of values $y[-1]$, $y[-2]$, ..., $y[-n]$.

Having subjected the left and right members of the equation (2.6) to $z$-transformation with allowance for the linearity property and time lag theorem, the following is obtained

$$\sum_{i=0}^{n} \alpha_{n-i} z^{-i} Z\{y[k]\} = \sum_{j=0}^{m} \beta_{m-j} z^{-j} Z\{g[k]\},$$

from here

$$S(\vartheta) = S(\vartheta, \alpha_1, \alpha_2, ..., \alpha_l), \tag{2.8}$$

where

$$H(z) = \sum_{j=0}^{m} \beta_{m-j} z^{-j} \Bigg/ \sum_{i=0}^{n} \alpha_{n-i} z^{-i} . \tag{2.9}$$

The $H(z)$ function expressed as the formula (2.9) is called the discrete transfer function of an impulse system. According to (2.8), it links the $z$-images of the input and output lattice functions.

The passage from $z$-image to the origin can be executed using a table or by expansion of the image as a Laurent series [7, 22, 60]. Other methods are rarely used.

The origin for a discrete transfer function is the lattice pulse response of an impulse system $w[k] = Z^{-1}\{H(z)\}$. It has the sense of a system response on an input step impulse lattice function

$$g[k] = \delta_0[k] = \begin{cases} 1 & \text{at } k = 0, \\ 0 & \text{at } k \neq 0 \end{cases}$$

with zero initial conditions.

The lattice impulse pulse response allows an impulse system response to be found in the arbitrary action $g\,[k]$ using the convolution formula

$$y[k] = \sum_{\nu=0}^{n} w[\nu] g[n-\nu] .$$

**Frequency-domain characteristics**. Determination of a system response $y(t)$ or $y[k]$ is simpler, if the action is harmonic. Writing the continuous harmonic input action in the complex form as:

$$g(t) = g_M e^{j(\omega t + \varphi)} = \tilde{g}_M e^{j\omega t} ,$$

where $g_M$ is an amplitude, $\omega$ is a circular frequency, $\varphi$ is an initial oscillations phase, and $\tilde{g}_M = g_M e^{j\varphi}$ is the complex amplitude of action, if the system is linear, then the response is described by a potential function with the same circular frequency, but with other amplitude $y_M$ and phase shift $\psi$:

$$y(t) = y_M e^{j(\omega t + \varphi + \psi)} = \tilde{y}_M e^{j\omega t} .$$

Here $\tilde{y}_M = y_M e^{j(\varphi + \psi)}$ is the complex amplitude of response.

A relation can express the complex amplitude of response through the complex amplitude of action as

$$\tilde{y}_M = H(j\omega) \tilde{g}_M , \tag{2.10}$$

where $H(j\omega)$ is the complex circular frequency function depending only on dynamic properties of a linear system, which is called the frequency transfer function.

It is easy to show that the frequency transfer function is uniquely related with a transfer function of a system $H(s)$ and can be obtained by substitution

$$H(j\omega) = H(s)\Big|_{s=j\omega}.$$

For calculations using formula (2.10) the frequency transfer function should be represented as

$$H(j\omega) = A(\omega)\ e^{j\psi(\omega)},$$

where $A(\omega) = |H(j\omega)|$ is a module of the function $H(j\omega)$, defining the gain plot of a system, $\psi(\omega) = \arg H(j\omega)$ is argument of function $H(j\omega)$, defining the phase plot of a system.

Graphic constructions of gain and phase plots are often given on a logarithmic scale. Two indicated characteristics and other [9] frequency characteristics are widely used for the investigation of dynamic systems by frequency domain methods.

Frequency investigation methods are also applicable to impulse systems. If $H(z)$ is a discrete transfer function for the linear impulse system, then the appropriate frequency transfer function looks like

$$H\left(e^{j\omega T}\right) = H(z)\Big|_{z=e^{j\omega T}}.$$

Its module characterizes the ratio of harmonic lattice function amplitudes on the input and output of an impulse system, and its argument — phase shift between these harmonic lattice functions.

It is convenient to use absolute pseudo-frequency $\lambda = \frac{2}{T}\mathrm{tg}\frac{\omega T}{2}$ instead of circular frequency $\omega$ when writing the frequency transfer function of an impulse system as:

$$H^*(j\lambda) = H(z)\Big|_{z=\frac{1+j\lambda T/2}{1-j\lambda T/2}}.$$

Thus the function $H^*(j\lambda)$ unlike the function $H\left(e^{j\omega T}\right)$ has a fractional rational nature and is not periodic.

## §2.2. Basic structures of control systems

**Dynamic units and block diagrams of a system**. The transfer function of a dynamic system describes the correlation between action and response, but does not carry the information about the inner pattern of the system, considered as "a black box". At the same time the dynamic system can be usually submitted as a set of more simple dynamic units[2], joined in a certain manner. A connecting circuit of dynamic units, included in a system, is called the block diagram. The separation of a system into units when constructing its mathematical model is usually stipulated by the heterogeneous character of elements that are included in the structure of real complicated systems. The procedure of determination of the differential (or

[2] The dynamic unit is part of the system described by differential equation of a defined kind. The standard dynamic units appropriate to differential equations not higher then second order are described in standard textbooks on automatic control.

difference) equation for each of these elements can be rather specific. Therefore, mathematical models of separate elements must be found first. Then they are used for assembling the resulting model of the whole system according to how dynamic units are connected.

Another situation arises at the system synthesis stage according to any criterion that clarify the transfer function of a system when its operation is best or good enough. Thc solution of the synthesis problem is common in relation to those systems, whose transfer function can be reasonably changed. Including dynamic units into the system structure with specially selected properties in addition to already available units usually makes such variation or correction. These units are appropriate to the so-called invariable part of the system. When the system is realized, correction not only is desirable to the transfer function $H(s)$, found as a result of synthesis, but also the system inner pattern is essential. Its required dynamic properties vary depending on the system block diagram and the placement of correcting unit connection.

Note that the synthesis and correction problem not only relates to engineering systems of automatic control, but also to some socio economic, ecological and other dynamic systems. However, it is usually necessary to take into account the much greater number of factors and circumstances both at the criterion choice stage and at the estimation of action properties stage in this case. This complicates the synthesis.

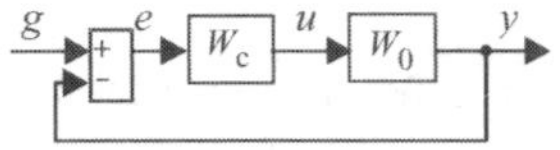


Fig. 2.1

Consider the possible variants of a dynamic system block diagram used for an automatic control system. Let the fundamental units of a system correspond to the control plant with transfer function $W_0(s)$ (or discrete transfer function $W_0^d(z)$) and to a control device with a transfer function $W_c(s)$ (or discrete transfer function $W_c^d(z)$). The control action $u$ is applied to an input of the controlled plant and its output is characterized by controlled variable $y$. Its desirable valuc then coincides with the reference input action $g$.

**Feedback systems**. The closed loop circuit shown in Fig.2.1 is the most widespread block diagram used in automatic control systems since the time of the first industrial regulators designed by J. Watt and I. Polzunov. The control by feedback principle or on deviation is realized in such a system. Occurrence of undesirable mismatching (control error) $e = g - y$ is detected. Its value is metered and devices eliminating the deviation by corresponding variation of the controlled variable $y$ value are actuated.

Note that use of the feedback when constructing a dynamic system is often predetermined by the fact that it is easier to meter the mismatching $e$, than to meter the input action $g$ directly. A typical example is in radar servomechanisms used as transmitters for mismatching. Here the angular, temporal, frequency or phase discriminators are rather easy to deal with for feasibility and posses high noise immunity.

However, the feedback principle is flawed. The point that matters is that the tools eliminating the harmful effect (deviation $e$) work after the harmful effect is already detected, and, hence, has already taken place. Therefore, the countermeasures to deviation can only be performed when the deviation has already arisen. It does not allow maintenance of the identical equality of system output and input $y = g$ with the appropriate choice of transfer function $W_c$ and, hence, to ensure the unit transfer function of closed loop system $H \equiv 1$. The transfer function of a system with the block diagram shown in Fig.2.1, is expressed as

$$H = \frac{W}{1+W}, \tag{2.11}$$

where $W = W_c W_0$ is a transfer function of system open loop. Hence, the function $H$ cannot be a unit at finite values of $W$. The inverse to the relation (2.11) looks like

$$W = \frac{H}{1-H}. \tag{2.12}$$

**Direct control systems.** The French engineer Poncelet suggested another principle of automatic system engineering in 1829. This is called the prediction principle, or principle of compensation. The block diagram of the direct control system realizing the indicated principle is shown in Fig.2.2.

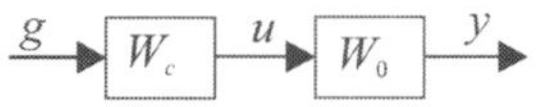


Fig. 2.2

The idea of Poncelet consists of the following statement: it is necessary to put the disturbance on the control device immediately and put the disturbance on the control plant with time lag. The control device will then have time to work out the control action, completely compensating the casual disturbance. The Poncelet idea could not be realized in practice because it is technically impossible to ensure the time lag for a force disturbance. The compensation principle, underlying this sentence, received recognition and propagation later.

The engineering execution of the compensation principle has appeared possible not in the way planned by Poncelet but in the way indicated by N. Wiener[3]. It is necessary to realize not only time lag of force disturbance but also its prediction according to Wieners' idea. The controller operating on predicted disturbance values can utilize compensation techniques to compensate for the consequences of the disturbance beforehand. Wiener correctly pointed out, that the prediction of signals and, hence, compensation of disturbances can be only statistical. The statistical theory of prediction and filtering of signals, founded by N. Wiener, A.J. Hinchin[4] and A.N. Kolmogorov[5], has led to the idea of statistical compensation of disturbances.

---

[3] Wiener N. The extrapolation, interpolation and smoothing of stationary time series. — New York: Wiley, 1949.

[4] Khinchin A.Ja. The theory of stationary equilibrium processes correlation.//Achievements of Math Sciences. — 1938.— B. 5.

However, use of the compensation principle in the pure state is hampered by the fact that even small interactions between the control device and properties of the controlled plant can cause the qualitative modification of system properties, in violation of its roughness (see §1.2)

**Combined systems**. Combination of the feedback and compensation methods is represented quite naturally, taking into account the shortfalls peculiar to feedback systems realizing control on error deviation, and direct control systems.

The block diagram of the combined system is shown in Fig.2.3. This system contains two control devices with transfer functions $W_{c1}$ and $W_{c2}$ and is expressed by the equations

$$Y = W_0 U, \; U = W_{c1} G + W_{c2} E, \; E = G - Y. \tag{2.13}$$

The elimination of intermediate variables gives the expression:

$$H = \frac{\left(W_{c1} + W_{c2}\right) W_0}{1 + W_{c2} W_0}. \tag{2.14}$$

for the transfer function of whole system $H = Y/G$.

At $W_{c1} = 1/W_0$ the condition $H = 1$ is satisfied. When considering a transfer function of closed loop system for an error:

$$H_e = 1 - H = \frac{1 - W_{c1} W_0}{1 + W_{c2} W_0},$$

then $H_e = 0$ in the indicated case. It allows a system to be called absolutely invariant; meaning that there is independence between an error $e$ on assigning action $g$ after the improvement of initial mismatch.

It is necessary to remember that the fundamental action $g$ and interference disturbances also are usually applied to a system. If the information on action $g$ in the combined system comes from one gauge, and information on a mismatching $e$ — from another one, and both gauges have errors, then such errors should be treated as two interference actions applied at different points in a system. It is impossible to ensure the indispersion of a system in relation to these interferences. However, their influence on system operation can be minimized by the appropriate choice of transfer function $W_{c2}$. With the lack of a priori information on properties of interferences it can be made on the basis of robust approach.

**Other variants of system structure**. The feedback system structure becomes complicated sometimes at the expense of inclusion of additional units into the system. The block diagram of a system when including a dynamic unit in a feedback circuit changes to that shown in Fig.2.4.

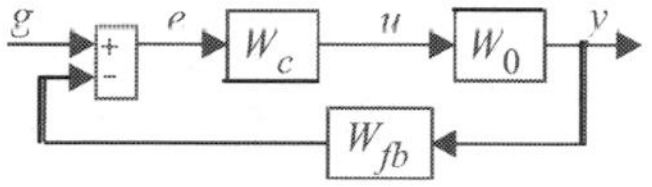


Fig. 2.4

[5] Kolmogorov A. N. Interpolating and extrapolation of stationary casual sequences.//Reports of the USSR Academy of Sciences: Math series, 1941, V.5, №1.

The conformable transfer function is

$$H = \frac{W_c W_0}{1 + W_c W_0 W_{fb}}, \tag{2.15}$$

where $W_{fb}$ is the transfer function for a circuit of non-unitary feedback.

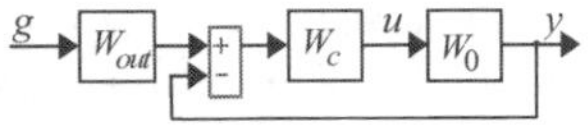


Fig. 2.5

With inclusion of an additional unit with a transfer function $W_{in}$ into the input circuit the block diagram shown in Fig.2.5 occurs. The conformable transfer function is

$$H = W_{in} \frac{W_c W_0}{1 + W_c W_0}. \tag{2.16}$$

If $W_{in} = k_{in}$, where $k_{in}$ is some factor, then a system with scaling of input signal is developed.

Freedom in choosing between transfer functions $W_{fb}$ and $W_{in}$ in expressions (2.15) and (2.16), in addition to freedom in selection of the transfer function of control device $W_c$ in the invariable transfer function of control plant $W_0$ expands the possibilities to realize the required dynamic properties of the system with the simplest tools.

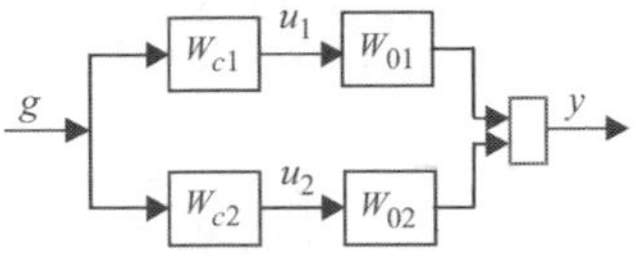


Fig 2.6

The structure of the direct control system can also be complicated, for example, at the expense of inclusion of a second bypass channel of direct control, as shown in Fig.2.6.

In this case two control actions $u_1$ and $u_2$ are applied to the controlled plant. It has different transfer functions $W_{01}$ and $W_{02}$ in relation to those actions. Plants that exhibit this form are planes (its course can be changed by a deviation of ailerons $u_1$ or direction rudder $u_2$), the national currency (the rate of which is controlled by the central bank by the volume of money variation $u_1$ and variation of monetary reserves $u_2$), and many others.

Note that the absolute indispersion for an error $e = g - y$ is possible in certain conditions in a double-channel system, as well as in a system of combined control. This follows the "principle of dual channel" by B.N. Petrov. According to this principle, the absolute indispersion is accessible only in systems not having less then two control channels.

The overview of possible structures of dynamic systems is not yet complete. The other variants of block diagrams shown in Fig.2.1 — 2.6 can also be realized. However, at synthesis of desirable dynamic properties of a system, i.e. at transfer function $H$ determination, ensuring the most successful system operation, usually the type of its block diagram is not so important. The required transfer function $H$ can be realized in systems with different structure, and the choice of this or that variant of structure is determined by factors, as a rule, that do not have direct relation to a synthesis of the function $H$ problem. Therefore, the transfer function $H = Y/G$, is also uniquely connected to it by the formula (2.11) transfer function of the open loop system with a unit feedback (see Fig.2.1) $W = Y/E$. This will be considered as the main characteristic of a closed loop linear dynamic system. It does not depend on the type of inner pattern, which the system has or could actually have.

## §2.3. Classification of investigation methods for dynamic systems

**Fundamental investigation stages**. Conditionally the following five main investigation stages of dynamic systems can be allocated:

1) Mathematical representation of a system;
2) Linearization of nonlinear units;
3) Mathematical representation of actions and other excitations;
4) Formalization of requirements to system dynamic;
5) System analysis and synthesis.

Modes of mathematical representations, i.e. the creation of a mathematical model of a system, are explained in §2.1. Note only, that the fundamental sense of this investigation phase consists in determining a differential or difference equation connecting the defined input and defined output variables of a system. If in multi-dimensional systems some input-output couples are of interest, then the determination of several equations on the number of the indicated couples is required accordingly.

If the system model is linear, it can be completely described by a system transfer function $H$, which can be treated as an abbreviated form of a system equation. It relates to a frequency transfer function biunique connected with the transfer function in an equal measure.

The frequency transfer function as a complex function of frequency can be preset by two real functions of frequency on a number of complex plane coordinates. When using the polar frame, such real functions are gain plot $A = \mathrm{mod}\, H$ and phase plot $\Psi = \arg H$. When using the Cartesian frame they are real and imaginary frequency characteristics $U = \mathrm{Re}\, H$ and $V = \mathrm{Im}\, H$. They are related by the formulas

$$A = \sqrt{U^2 + V^2}, \Psi = \mathrm{arctg}(V/U).$$

The elimination gives a minimal-phase system [9], where the frequency transfer function, i.e. the complete model of linear system, can be recovered using any of the four indicated real functions of frequency based on the relations

$$U(\omega) = -\frac{1}{\pi}\int_{-\infty}^{\infty}\frac{V(\omega)}{u-\omega}du,\ V(\omega) = \frac{1}{\pi}\int_{-\infty}^{\infty}\frac{U(\omega)}{u-\omega}du, \tag{2.17}$$

$$\Psi(\omega) = \frac{1}{\pi}\int_{-\infty}^{\infty} \frac{dL}{d\lambda} \ln\mathrm{cth}\left|\frac{\lambda}{2}\right| d\lambda,$$

where $L(u) = \ln A(u)$, $\lambda = \ln(u/\omega)$, $u \in (-\infty, \infty)$ is a variable of integration. The formulas (2.17) describe the Hilbert transformation.

It is essential, that the system model as a rule becomes linear only on the second phase of investigation after the linearization of nonlinear units. If the linearization is not feasible for all system units, then the model remains nonlinear. This complicates its investigation.

The mathematical description of external actions is the third investigation stage, which is as important and necessary, as the first one. It is expedient to describe not only the direct actions themselves but also the nonparametric or parametric classes, which these actions should belong to. In this way the specificity of robust systems is exhibited, thus either statistical or non-statistical approaches can be used.

Formalization of dynamic system requirements consists of assigning the indications that permit estimation of the successfulness of its operation. Usually such indications look like restrictions on numerical factors describing a stability margin, speed of response and accuracy of a system. The similar indexes are well developed in automatic control theory. In particular, stability margin and speed of response can be defined practically on any dynamic characteristic of a system in the time or frequency domain.

When investigating a step response of a closed-loop system $h(t) = L^{-1}\{H(s)/s\}$ the speed of response is evaluated using step response damping time $t_{tr}$ down to a preset small level $\Delta$ (for example, $\Delta = 3\%$), and the stability margin is evaluated on the value of overshoot $\sigma = [\max h(t) - h(\infty)]/h(\infty)$ (Fig. 2.7), which is a system reaction at input $g(t) = 1(t)$ at zero initial conditions.

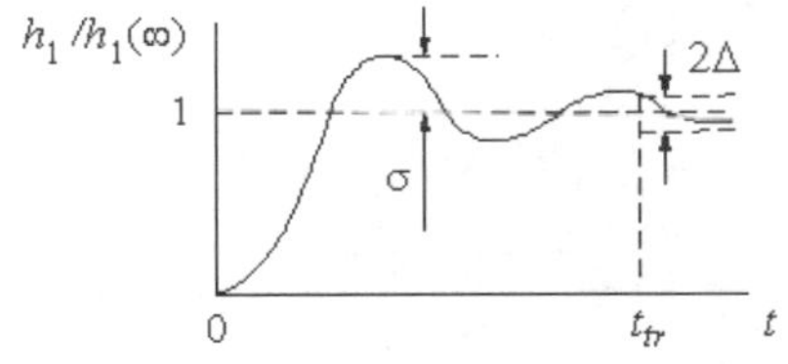


Fig. 2.7

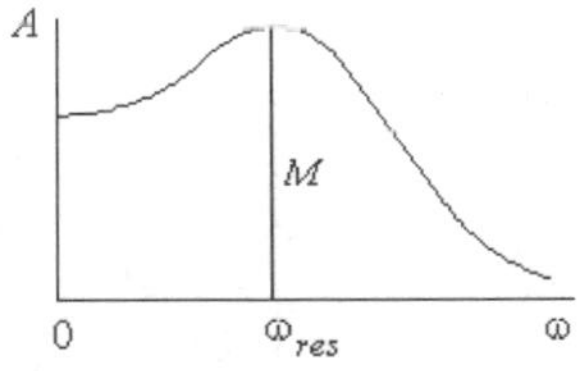


Fig. 2.8

By reviewing the gain plot of a closed loop system $A(\omega) = |H(j\omega)|$ the stability margin is determined on the index of oscillations $M = \max A(\omega)/A(0) = A(\omega_{res})/A(0)$, where $\omega_{res}$ is a resonance frequency of the closed loop system (Fig. 2.8), and transient period with a good system stability margin, estimated using the approximated formula:

$$t_{tr} = (1 \div 2)\frac{2\pi}{\omega_{res}}. \quad (2.18)$$

By reviewing the Bode diagram of the open loop system the stability margin is determined on amplitude and phase $\Delta L = -L(\omega_r), \mu = 180^\circ + (\omega_{-r})$, where $\omega_r$ is the resonance frequency of the open loop system and $\omega_{cut}$ is the cutoff frequency, i.e. $\Psi(\omega_r) = -180^\circ$, $L(\omega_{cut}) = 0$ (Fig. 2.9). For speed of response estimation, the formula (2.18) is applicable at substitution of the magnitude $\omega_r$ instead of $\omega_{cut}$.

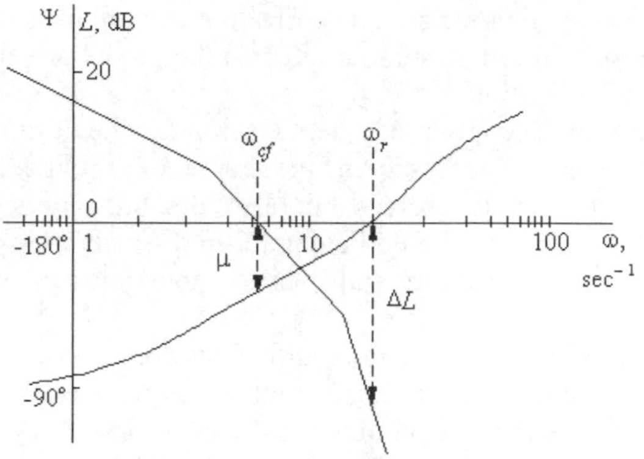


Fig. 2.9

The values of magnitudes σ, *M*, Δ*L*, μ, $t_{tr}$ and other stability margins and speed of response characteristics required for successful system operating depend on the destination of system. For example, they can make $\sigma \leq 30^\circ$, $M \leq 1.5$, $\Delta L \geq 6$, $\mu \geq 30^\circ$, $\Delta L \geq 6$ dB, $t_{tr} \leq 1$ sec for instrument servomechanism [6].

The methods for requirements formalization for a given system accuracy are well described in §2.4. Note that such requirements can be unequal for different modes of system operation.

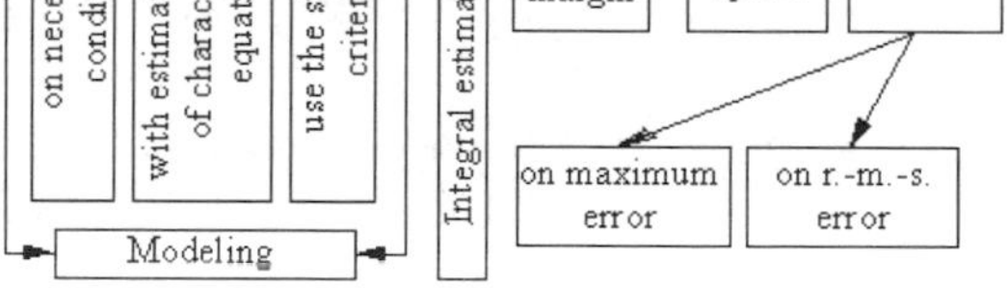

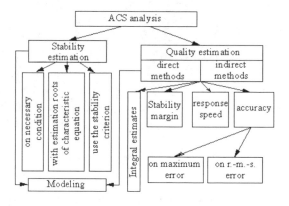

Fig. 2.10

The last and fundamental investigation stage of a system consists of its analysis and synthesis, and, as a rule, it is usual to speak about the triad analysis-synthesis-analysis. The preliminary analysis of common system operation regularities allows the synthesis problem to be set correctly. The solution of a synthesis problem gives the particular variant of the systems fitting to the formal requirements imposed to it (at optimal synthesis — the best variant according to the accepted optimality criterion). At final analysis of the synthesized system it is possible to investigate its criticality to deviations from the initial assumptions. Then it is possible to check the execution of additional, difficult to formulate requirements, and investigate the successfulness of system operation in nonconventional situations, etc.

The brief classification for methods of analysis and synthesis of dynamic systems within the frame of automatic control classical theory is given below.

**Analysis methods**. The fundamental types for analysis of automatic control systems (ACS) are given in Fig. 2.10. They allow estimation of stability and quality of the dynamic system.

The analysis begins from the estimation of stability, because if the system is unstable it must be considered as having no working ability and further analysis is senseless. The stability of a linear system is determined by attachment of its characteristic equation roots to a stability region. A stability region, in the case of a continuous system is the left half-plane of complex plane $s$ and in the case of an impulse system it is a unit radius circle in a complex plane $z$. If in a characteristic equation of an impulse system one passes to a $w$-transformation [7, 9, 22, 60] by substitution $z = (1 + w) / (1 - w)$, the stability region is transformed to the left half-plane of a complex plane $w$. Then the problems of impulse and continuous systems stability estimation do not differ formally and can be solved by identical methods.

Sometimes it is possible to reveal instability of a system only on an indispensable stability condition consisting of the degree of positiveness of all characteristic equation coefficients. If the indicated indispensable stability condition is fulfilled, then investigation of stability should be more thorough. Thus it is necessary to find the characteristic equation roots and ascertain their membership in stability region immediately, or to use one of the stability criteria.

The stability criterion gives an algorithm permitting estimation of the system stability without determination of its characteristic equation roots. The criteria are distinguished as algebraic, connected with the analysis of the relation between the characteristic equation coefficients, and frequency domain criteria connected with analysis of the frequency transfer functions of the closed or open loop system. An algebraic criterion by Hurwitz or Routh and the frequency criteria of Nyquist [9, 72, 80] are used most often. Note that the Routh criterion was developed first in 1877, but soon it was superseded to a second Hurwitz criterion, more convenient for manual calculation. However, the Routh criterion now has gained an importance, because the cyclical procedures are well applicable on computers.

After system stability estimation, it is necessary to perform the quality analysis, which includes the determination of the stability margin and speed of response, and also accuracy at the defined models of input actions. Sometimes integrated estimations of quality, formally describing all three indicated quality factors by one single-dimensional magnitude are used for comparison of quality for different system variants. For example, "improved square-law" integrated estimation [9] looks like

$$I_{sq} = \int_0^\infty \left[x^2(t) + T_*^2 \dot{x}^2(t)\right] dt\,,$$

where $x(t)$ is deviation of the output magnitude from the steady-state value at step input, and $T_*$ is the time constant eliminating the most acceptable velocity of the transient process in a system.

The quality estimation methods are subdivided into straight methods connected with construction of the step response curve of a system, and indirect methods, where such construction is not required. Frequency domain methods permitting estimation of the quality of those or other frequency characteristics of a system are the most widespread indirect methods. It is possible to perform simulation of a system on a computer permitting investigation of system response to input actions as one of the major direct methods

The accuracy of estimation methods is distinguished depending on the kind of mathematical description of the input actions (see §1.5) and on the accepted measure of control accuracy. Maximal error or r.-m.-s. error is usually used as a measure of control accuracy.

**Synthesis methods**. The synthesis of a system can be heuristic or mathematical (Fig.2.11). The problem of mathematical synthesis is completely formalized and reduced to investigation of some function by mathematical methods. Usually this investigation has the purpose of finding the extremum of a function that corresponds to determination of a system with the best operational quality in the sense of the accepted criterion.

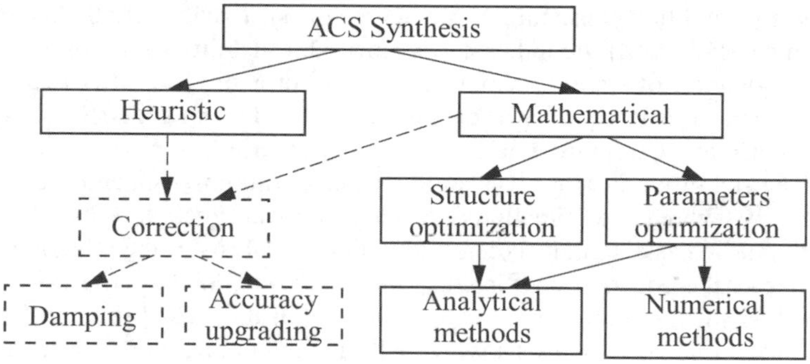


Fig. 2.11

System structure, i.e. the type of its transfer function, and values of all its parameters, or the parameters of a system preset beforehand, or selected structure can be optimized at mathematical synthesis. The structure optimization is a more complicated problem and can be carried out only by analytical methods. The parameter optimization is mathematically treated as the problem of determination of the extremes of several variables function, but not a functional. It can be solved not only analytically, but also by numerical methods, whose role has strongly increased with development of possibilities in computer technology.

The heuristic synthesis is not reduced to a solution of a mathematical problem. It consists of a choice of acceptable system variants on an experience basis, intuition or recommendations, which cannot be reduced to any precise formal algorithm. The priority of heuristic synthesis is largely possible on account of all population of requirements imposed to a system unlike in mathematical synthesis. However, its efficiency strongly depends on the qualification of the investigator.

Correction is a tool required (found as a result of synthesis) by dynamic property execution of a system and occupies a special place in automatic control systems synthesis. Including the correcting units into the system fulfills the correction. The fundamental purposes of correction are usually damping of a system and improving its accuracy.

Damping is the increase of system stability margin without variation of its gain parameter. Three basic modes of damping are known: with suppression of high frequencies, with rising of high frequencies and with suppression of medium frequencies [9, 72, 80]. The increase of control accuracy can be achieved by variation of system gain parameter, increase of the system astatism, and by using non-unit feedbacks, scaling processes and other modes.

## §2.4. Control precision factors and classes of input actions

**R.-m.-s. and maximal values of error**. The choice of precision factor is one of the major stages of dynamic system, research problem formalization. It is determined by the applicability of the system, available information on properties of actions and the abilities of the mathematical means being used. The set of indicated circumstances, as a rule, restricts choice to three possible variants for robust systems investigation: an upper-bound estimate of r.-m.-s. error, maximal error and practically maximal error.

The advantages and disadvantages of using the r.-m.-s. error as the precision factor are well-known [56, 81], and the specificity of robust systems is exhibited here in necessity to estimate the upper bound of r.-m.-s. error or, if it is possible, to determine its exact upper bound*[6].

The maximal error $e_M$, i.e. the upper bound in the interval of possible current error absolute values $e(t)$, represents the most adequate performance of system accuracy properties in many cases, but practically is rather rarely used in this sense. The problem is that maximal error is not a statistical characteristic because its representation superimposes rigid restriction on each separate execution of the process $e(t)$, however it can be determined on the curve of probability density of error. Nonstatistical characteristics of action, for example, the maximal values of its derivative should also be preset in order to determine the maximal error. The spectral densities of action giving the full information on action properties, within the framework of the correlation theory, are not necessary for this purpose. In this

[6] The exact upper bound of a r.-m.-s. error is achieved at the defined spectral density of action that is the most unfavorable in a given class of spectral densities. On the other hand, even to a strongest upper estimate $\overline{\sigma}_e$ it is impossible to correspond the defined spectral density of action, at which it is achieved. As $\overline{\sigma}_e \geq \sigma_{e``}$, the use of the magnitude $\overline{\sigma}_e$ as an index of system accuracy is justified only at impossibility or difficulty in determination of the exact boundary $\sigma_{eM}$.

case the criterion $e_M \to \min$ is not applied in classical problems of statistical linear filtering. Chapter 6 is devoted to methods of maximal error investigation.

However, the other problem is that the strict maximal error badly characterizes the quality of broadband interference suppression by a system. Usually such suppression makes it impossible to transit all spectral components of interference $\upsilon(t)$ between input and output of the system. Exceptions are the extra low frequency components of interference, which form the appropriate component of resulting error $e_\upsilon(t)$. The maximal value of component $e_\upsilon(t)$ coincides with the perturbation $\upsilon(t)$ maximal value or even exceeds it when the system has a resonance property. The restrictions of the perturbation derivative $\upsilon(t)$, from above do not change the situation, because even a disturbance $\upsilon(t) = \text{const}$ passes through the system up to its output. However, r.-m.-s. error $\sigma_{e\upsilon}$ from a broadband disturbing perturbation is much less than the magnitude of perturbation. This testifies to the smoothing properties of the system.

**Practical maximal error from perturbation.** The strong smoothing of interference perturbation causes the normalization of the error $e_\upsilon(t)$ distribution law by virtue of the central limiting theorem. Knowing the r.-m.-s. value $\sigma_{e\upsilon}$ or even its upper estimate enables practical calculation of the maximal value [8] of such error $e_{\upsilon p}$. Its overflow is possible only with given small probability *p*, i.e. for an arbitrary moment $t_*$ the event $|e_\upsilon(t_*)| > e_{\upsilon p}$ has a priori probability *p*. For the normally distributed centered error the following relation is fulfilled

$$p = 1 - 2\Phi(e_{\upsilon p}/\sigma_{e\upsilon}),$$

where $\Phi(x)$ is a probability integral. From here at $e_{\upsilon p}/\sigma_{e\upsilon} = 3$ it is possible to obtain $p = 3\cdot 10^{-3}$, at $e_{\upsilon p}/\sigma_{e\upsilon} = 4 \div 5$ — $p = 6\cdot 10^{-5} \div 6\cdot 10^{-7}$, i.e. at $e_{\upsilon p} \geq 3\sigma_{e\upsilon}$ the probability of event $|e_v(t_*)| > e_{vp}$ is very small and decreases sharply with increase of $e_{\upsilon p}$. It allows the value $e_{\upsilon p} = 3\sigma_{e\upsilon}$ to be accepted or, in especially crucial cases the value $e_{\upsilon p} = 5\sigma_{e\upsilon}$ to be used as the practically maximal value of error from broadband disturbing perturbation.

The practical maximal error can also be connected with probability of error to overflow the value $e_{\upsilon p}$ even once during the system operation. In cases that are applicable for the analytical analysis [39], the value of coefficient in the formula $e_{\upsilon p} = (3 \div 5)\ \sigma_{e\upsilon}$ is specified.

For maximal resulting error $e_M$ determined at presence of reference action and interference perturbations it is necessary to summarize the maximal dynamic error $e_{gM}$ (from reference action $g(t)$) and practical maximal error from interference $e_{\upsilon p}$, i.e. to accept the following: $e_M = e_{gM} + e_{\upsilon p}$. The number of addends in this formula increases as the number of excitations applied to a system increase.

Notice that the given concept of practical maximal value cannot be used for dynamic error, which can be distributed arbitrarily using the unknown law of reference action distribution.

**About the stationarity and ergodicity of casual input actions**. The investigation methods for basically linear (more correctly linearized) stationary dynamic systems are described in this book. This stipulates the expediency of use of statistical investigation of systems accuracy and the stationary ergodic models of actions. The determination of its characteristics is made without taking into account those executions or explicitly selected segments of executions, which correspond to obviously highlighted modes of system operation. It is supposed that, the stationarity property and the ergodicity are fulfilled only approximately for real actions, unlike its model. This is given in an example:

Scrutinize the process for vessel roll angle variation considering rolling motion on non-regular sea waves. This process is an input action for stabilization of the system for the sweep plane of an onboard radar antenna. The roll performance essentially depends on the vessel course angle in relation to the prevailing direction of sea waves under other constant conditions. The possible execution of the roll process is given in Fig. 2.12, where points $t_1$ and $t_2$ on the time axis correspond to the moments of vessel course modification. The nonstationarity of the process is obvious in the sense that its properties essentially differ at time intervals divided by points $t_1$ and $t_2$. However the conclusion about a stationarity of the considered process can be made, if the properties of the process are evaluated on average with a large number of executions, and if $t_1$, $W_{c2}$ similarly at points in each execution by a causal way are allocated on the time axis. Therefore at any instant on the average, the execution intensity and other roll indexes have constant values..

What model of action can be accepted at system investigation? Apparently, the necessity of non-stationary model application arises in the case, when it is necessary to construct a system using adaptation principles and to realize its optimization for processing each interval of input action with defined properties. If the system, on which the action is applied, is stationary and it has no self-tuning, purposeful program variation of parameters, then the stationary model of action is quite acceptable.

Fig. 2.12

Using the stationary model in the considered example actually means, that the roll process with an intensity, which is not varying only at separate time intervals, is substituted by isotropic in time process with some constant intensity. If this intensity is average on all possible executions, then the effect of such a process on a system is on the average equivalent to an operation of the real process shown in Fig. 2.12.

In practice, the requirement on control accuracy is often set to ensure an acceptable error value in the heaviest possible mode of system operations. Thus not average, but the most unfavorable values for statistical characteristics of action are of interest. In the considered example, this means accepting a stationary model of the roll process, of which its performance should be defined with reference to time intervals, where roll is most intensive, in particular, to a segment $t_1\, t_2$. Such an approach is characteristic for ensuring control accuracy problems.

It is possible to approach the ergodicity of process estimation in a similar way. For example, the intensity of the roll process depends not only on vessel course but also, naturally, on a number of sea conditions, which is possible to consider quasiconstant within the duration limits of one control process execution. The roll performance, found on separate executions for two different numbers of wave disturbances, will differ. In this sense the roll is not an ergodic causal process. However, when considering only the execution of the roll process, which are obtained at the most possible admissible number of sea condition for a synthesized system, it is possible to consider a process ergodic.

**Processes with stationary causal increments**. Input action in some cases is impossible to consider even approximately as a stationary causal process. However, its *K*-th derivative is stationary. Such properties are especially characteristic for reference actions. For example, if the acceleration of a moving object is a stationary casual process, then the coordinate of this controlled plant is a process with a stationary 2nd derivative. Similar actions concern the class of casual processes with stationary *K*-th increments, for which a theory has been developed [39].

The causal process $g(t)$ is named the process with stationary increments, if its increment during any fixed time slice $\Delta t$

$$\nabla g(t) = g(t) - g(t - \Delta t),$$

considered as a function of current time $t$, is the casual stationary process.

The division of process $\nabla g(t)$ values on a constant $\Delta t > 0$ cannot break its stationarity even if $g(t)$ converges to zero, when $\lim\limits_{\Delta t \to 0} [\nabla g(t)/\Delta t] = dg(t)/dt$. Therefore the process with fixed increments could formally be identified as a process with a stationary first derivative.

The more common concept of process with fixed *K*-th increments is introduced by reviewing the function

$$\nabla^K g(t) = \nabla^{K-1} g(t) - \nabla^{K-1} g(t - \Delta t) = \sum_{\nu=0}^{K} (-1)^\nu C_K^\nu g(t - \nu \Delta t),$$

that should be a stationary causal process. Here $C_K^\nu$ is the number of combinations of $K$ on $\nu$. The process with stationary *K*-th increments has a stationary *K*-th derivative.

It is obvious, that any stationary casual process is the process with stationary *K*-th increments, $K = 0, 1, 2, \ldots$ and its derivatives are stationary (if they exist). However, the converse statement is not fair.

Further account is needed according to stationary causal processes, and it is supposed, that if an action is not causal stationary, but has stationary $K$-th increments, then no action but its $K$-th derivative $g^{(K)}(t)$ is considered.

**About centrality of causal input actions.** It is expedient to consider all random input actions, applied to a system, as centered, i.e. having zero expectations, for the following reasons. If the expectation of action is known, the system is usually created in a way that some compensation signal eliminating influence of expectation on control accuracy is made. If the expectation is not known but the r.-m.-s. of action is known, then it is possible to accept a hypothesis of zero expectation, and put the dispersion equal to the mean quadrate. The physical treatment of assumption on zero expectation requires the spectral component of actions on frequency $\omega' = 0$ (the constant component) be transferred at frequency $\omega'' \to 0$. It cannot practically have an affect on results in research of system accuracy.

**Numerical characteristics bounding classes of input actions**. For input actions, except the common properties of stationarity, ergodicity and centrality, some minimum information about the differences in the dynamics of reference and interference excitations (or their separate component) should be known or found out. The statement on the filtration problem is senseless without it. Such information is usually richer for interference $\upsilon(t)$ and can be even full, when the spectral density $S_\upsilon(\omega)$ is considered known. The legitimacy of this assumption is connected with a small passband of the system in relation to a greater width of the disturbing perturbation spectrum.

Spectral density $S_g(\omega)$ for reference action $g(t)$ is usually unknown. First of all, in practice it forces use of a robust approach to the research problem of filtration accuracy. In §1.5 five versions for representation of possible spectral density classes of actions are described. Regarding the reference action it is possible to consider the last one as the most acceptable, connected with fixation of generalized moments of spectral density $S_g(\omega)$, concerning some system of basis functions $\{u_i(\omega)\}_{i=0}^{N} \in R_1^+$. Writing all these moments in one more common form as contrasted to (1.14) by the expression

$$M_i = \frac{1}{\pi} \int_0^\beta u_i(\omega) S_g(\omega)\, d\omega, \tag{2.19}$$

where $\beta \in (0, \infty)$ is some boundary frequency.

The exact values of dispersions in control resulting in error components are also evaluated using formulas of the kind (2.19). For deriving their good estimations it is desirable, that the functions $\{u_i(\omega)\}_0^N$ be close on the shape to quadrates of the appropriate gain plot of a system. However, for different versions of synthesized control system in those gain plots can strongly differ. Hence, it is expedient to choose the basic functions of elementary shape. Quadrates of gain plots must be approached by their linear combinations. Even power functions $u_i(\omega) = \omega^{2i}$ satisfy such conditions very well. This circumstance is successfully combined with the obvious physical sense of generalized moment's (2.19) at even power basis functions. They are the dispersions of the $i$-th derivatives of action:

$$D_i = \frac{1}{\pi} \int_0^{\infty} \omega^{2i} S_g(\omega)\, d\omega,\ i = \overline{K, N},\ 0 \le K \le N\,. \tag{2.20}$$

The boundary values of several dispersions $D_i$ found at the theoretical analysis of operational conditions of a system or by experimental researches are the typical set of a priori data about the reference action. Sometimes the dispersions of not exact but smoothed derivatives of action are experimentally measured:

$$D_{ri} = \frac{1}{\pi} \int_0^{\infty} \left( \frac{k_s \omega^2}{1 + \omega^2 T_s^2} \right)^i S_g(\omega)\, d\omega\,, \tag{2.21}$$

where $k_{ds}$ and $T_s$ are the parameters of differentiating-smoothing devices, or generalized moments of another kind.

It is also possible to use its generalized moments instead of spectral density of disturbances. In this case the basis functions appropriate to the gain plot of low-pass filters are convenient.

Maximal values of derivatives of reference action $\left\{ g_M^{(i)} \right\}_K^N$, ensuring the execution of inequalities

$$\left| g^{(i)}(t) \right| \le g_M^{(i)}, t \in (-\infty, \infty),\ i = \overline{K, N},\ 0 \le K \le N \tag{2.22}$$

must be used instead of application of dispersions at estimation of maximal values of control error.

It is shown in chapter 6, that the magnitudes $\left\{ g_M^{(i)} \right\}_K^N$ can be treated as generalized moments as a module of spectral density of the amplitude (instead of power) of the reference action.

Such qualitative characteristics of $S_g(\omega)$ and $S_\upsilon(\omega)$ curves such as unimodality, monotonicity, continuity and similar curves are useful for research in control accuracy.

It is possible to use the information on a real width of action spectrum in some cases and to accept the assumption on the lack of spectral component in region $|\omega| > \beta$, where B is some bound frequency. The theoretical spectral density of a true causal process cannot be a finite function. It would mean its exact extrapolation at any time slice is possible. However, there are no basic obstacles for use of such an action model for research of dynamic systems described by differential or difference equations of finite order. The only important thing is that the spectral components are the most influential for system accuracy, whichwere not discarded.

**Representing the action as a sum of additive components**. An effective way for increasing the details of action property description is its separation on additive component, for properties each of which the separate mathematical description is given. Such representation of action well expresses the real association of its law of variation on the several independent factors in many practical cases. Problems of similar signal filtration arise for radar tracking of mobile objects from the mobile carrier, at joint processing by a tracking drive, the stabilization errors and commands of program control, at integration of navigational transmitters, etc.

For example, at integration of positional sensor and accelerometer in systems of motion measurement, the error of accelerometer subjected to filtration is determined by shift of the scale zero, inaccuracy in stabilization of sensitivity axis in required direction, inaccuracy in compensation of gravity acceleration and other reasons. It allows justification of the model of accelerometer error as the sum of several components, for each of which (reduced to dimension of distance) the second derivative is restricted, and, for the majority, where the third derivative (velocity of acceleration) or first derivative (velocity) is also restricted [64]. The model of indicated action, outstanding by heightened reliability and entirety, is accepted.

An example of such model use is observed in § 8.2.

Sometimes it is possible to select the authentic models as the determined time functions or causal functions with a completely known spectrum for some additive component of causal action.

*Questions*

1. What is the transfer function of a system?
2. What is the correlation between transfer function and pulse response of a system?
3. What kind of mathematical mean is used for dynamic systems research?
4. In what way are the frequency characteristics of a system introduced?
5. What is the block diagram of a system?
6. What is the control system correction performed for? How is it performed?
7. Give the basic possible variants of system structure design.
8. Why is the feedback principle widely used for constructing control systems?
9. Describe the priorities and deficiencies for systems of direct control?
10. In what way can the absolute invariance in combined control systems be reached?
11. Why is the absolute invariance in single-channel systems of direct controlling impossible but possible in dual-channel systems?
12. Name the basic stages of dynamic system research.
13. How can stability of a dynamic system be estimated?
14. What factors can give the estimation of dynamic system quality?
15. Classify the methods of dynamic systems synthesis.
16. Name and characterize the possible factors of filtration accuracy.
17. In what cases can the stationary ergodic models of random input actions be used?
18. Give the examples of actions with stationary random increments.

# Chapter 3

# A priori information obtainment and analysis on input actions and its derivatives

## §3.1. Interaction of action and its derivatives characteristics

**Differentiability.** As in many applied problems of dynamic systems research a priori information on numerical characteristics for derivatives of action appears the most accessible. Some common regularities in similar numerical data are described below.

At derivation of casual process $g(t)$, the derivative of such function in the mean-square sense will be understood as the function $g^{(1)}(t)$, which satisfies to the condition:

$$\lim_{T_O \to \infty} \lim_{\Delta t \to 0} \frac{1}{T_O} \int_0^{T_0} \left[ \frac{g(t+\Delta t) - g(t)}{\Delta t} - g^{(1)}(t) \right]^2 dt = 0.$$

The concept of $i$-th derivative is introduced in the same way.

Derivatives of stationary casual process are centered stationary casual processes.

The full statistical description of process is equivalent to full statistical description of its existing derivatives in research within the framework of correlation theory. Correlation functions and spectral densities of $i$-th derivative are evaluated through the appropriate characteristics of process $g(t)$ with known [13] formulas

$$R_{g(i)}(\tau) = (-1)^i R_g^{2i}(\tau),\ S_{g(i)}(\omega) = \omega^{2i} S_g(\omega),\ i \le N. \qquad (3.1)$$

For derivative dispersions it gives the expression (2.20), hence the necessary and sufficient condition of $N$-fold differentiability of the process is decreasing of its spectral density on high frequencies faster than $\omega^{-(2N+1)}$.

**Correlation between dispersions of different orders derivatives.** The integral analog of the Cauchy-Bunyakovski inequality is used for required relations derivation, according to which it is written

$$\left[ \int_0^\infty \omega^{2i+2} S_g(\omega)\, d\omega \right]^2 \le \left[ \int_0^\infty \omega^{2i} S_g(\omega)\, d\omega \right] \left[ \int_0^\infty \omega^{2i+4} S_g(\omega)\, d\omega \right].$$

Here according to (2.20) the following inequalities are accepted:

$$D_{i+1}^2 \le D_i D_{i+2},\ i = \overline{0, N-2}. \tag{3.2}$$

Sequentially changing $i$ from zero up to $N-2$, (3.2) can be transformed to a kind

$$D_1/D_0 \le D_2/D_1 \le \ldots \le D_N/D_{N-1}\ . \tag{3.3}$$

Validity of (3.3) is proven by the Ljapunov theorem [23]. Accordingly $\sqrt[i]{D_i/D_0}$ is not a decreasing function of $i$ at $i = 1, 2, \ldots$, or, in the common case, $\sqrt[i]{D_i/D_K}$ is not a decreasing function of $i$ at $i = K+1, K+2, \ldots, N$ and $K = 0, 1, \ldots, N-2$, i. e..

$$\begin{aligned} &D_1/D_0 \le \sqrt{D_2/D_0} \le \ldots \le \sqrt[N]{D_N/D_0}, \\ &D_2/D_1 \le \sqrt{D_3/D_1} \le \ldots \le \sqrt[N-1]{D_N/D_1}, \\ &\ldots\ldots\ldots\ldots\ldots\ldots\ldots\ldots\ldots\ldots \\ &D_{N-1}/D_{N-2} \le \sqrt{D_N/D_{N-2}}. \end{aligned} \tag{3.4}$$

Combining (3.3) and (3.4), the following expression is fulfilled

$$\begin{aligned} &D_1/D_0 \le \sqrt{D_2/D_0} \le D_2/D_1 \le \sqrt{D_3/D_1} \le D_3/D_2 \le \ldots \\ &\ldots \le \sqrt{D_N/D_{N-2}} \le D_N/D_{N-1}. \end{aligned} \tag{3.5}$$

The clear physical treatment of developed inequalities can be given using the concepts of average square-law, average quarter-law and, generalizing it, average ($2i$)-th frequency in spectrum of action $\omega_{av2i} = \sqrt[2i]{D_i/D_0}$ , and also concepts of themean-square frequency in $i$-th derivative of action spectrum $\omega_{rmsi} = \sqrt{D_{i+1}/D_i}$ . They clear the fact, that the spectrum of each subsequent derivative is higher in frequency, than the previous one. Average quarter-law frequency of action does not exceed average square-law frequency of its derivative in particular.

Realization of inequalities (3.5) is the sufficient condition for the sequence of magnitudes $\{D_i\}_0^N$ and can be considered as the sequence of derivative dispersions in real-time action with some spectral density $S_g(\omega)$.

**Estimation of action spectrum width by derivatives dispersions.** Each of inequalities (3.2) — (3.5) is converted into the equality at harmonic action $g(t) = g_M \sin(\beta t + \psi)$ with casual initial phase $\psi$. Dispersions of derivatives allow $D_i = g_M^2 \beta^{2i}/2$ for such unlimited differentiable action. It has only one spectral line on $\beta$ frequency, in the region of positive frequencies and $\omega_{rms0} = \omega_{rms1} = \ldots = \beta$.

If a spectrum of action is richer with spectral lines or it is continuous, then indicated inequalities become strict. Hence, the relation between the right and left-hand sides of those inequalities may serve as an index of action spectrum width. According to (3.2), such indexes are represented as:

$$\gamma_i = D_i D_{i+2} / D_{i+1}^2 = \left(\omega_{rmsi+1} / \omega_{rmsi}\right)^2 . \tag{3.6}$$

At $i = 0$, it characterizes the action spectrum width, at $i > 0$ — action derivatives. It is accepted to be $\gamma_0 = \gamma_1 = \gamma_2 = ... = 1$ in the case of harmonic signal with $\delta$-shaped spectral density. If magnitude $\gamma_i$ insignificantly exceeds 1, it testifies that $i$-th derivative of action is a rather smooth function and has the properties close to harmonic. The great values of $\gamma_i$ point out, that the function $g^{(i)}(t)$ has strong irregularity.

The character of factors $\gamma_i$ relation and, in particular, of factor $\gamma_0 = D_0 D_2 / D_1^2$ from the shape of curve $S_g(\omega)$ is such, that variation of factor $\gamma_0$ testifies to the relative decrease in power of mid-frequency component in the spectrum of action [8, 56]. In the mid-frequency region, the curve $S_g(\omega)$ is as though pressed, and in the low-frequency and high-frequency region it rises. This conclusion concerns the curve $S_{g(i)}(\omega)$ at $i > 0$.

Examples of spectrum width estimation by dispersions of action derivatives are considered in the book [8] in pages 19-21. It allowed for the conclusion that the factor $\gamma_0$ for one of vertex curve of spectral density $S_g(\omega)$ has almost the same sense, as the excess coefficient [56] for "density of probability curve" $S_g(\omega)$. At $\gamma_0 > 3$ the curve $S_g(\omega)$ has a sharper top and a more slanting high-frequency part, than the Gauss curve. At $\gamma_0 < 3$ it has a less sharp top and a steeper high-frequency part.

**Recovery of a correlation function of action by its derivatives dispersions.** $R_g(\tau)$ is the correlation function of unlimited differentiable action. It can be decomposed into McLoren's series

$$R_g(\tau) = R_g(0) + R_g^{(1)}(0)\tau + \frac{1}{2!} R_g^{(2)}(0)\tau^2 + \ldots$$

Due to parity of the correlation function, its derivatives of even orders in a point $\tau = 0$ are equal to zero. With account of (3.1) and formulas $R_g^{(i)}(0) = D_i$, it results in

$$R_g(\tau) = \sum_{i=0}^{\infty} \frac{(-1)^i D_i}{(2i)!} \tau^{2i} . \tag{3.7}$$

Thus, the correlation function of stationary casual action can be decomposed into the ascending power series. Its coefficients are obtained from dispersions of action derivatives. The question is: in what conditions do the series (3.7) converge? In what conditions does the infinite sequence of derivatives dispersions determine the correlation function (spectral density) univalently? Such conditions can be imposed on dispersions $\{D_i\}_0^\infty$, or on spectral density of action. They look like [8]:

$$\lim_{N\to\infty}\sum_{i=1}^{N}(D_i/D_0)^{-1/(2i)}=\infty,\quad \int_0^\infty S_g(\omega)e^{\omega|\tau_M|}d\omega<\infty,$$

where $\tau_M$ is the maximum value of argument $\tau$, up to which the values of the correlation function of action are still of interest. Hence, the spectral density $S_g(\omega)$ must decrease faster, than $\exp(-\omega\tau_M)$ on high frequencies. This condition, in particular, is fulfilled for Gauss spectral density with a large margin even at $\tau_M\to\infty$, and also for all spectral densities, which are distinct from zero only in a limited band of frequencies. At a finite number of known derivative dispersions it is possible to recover only the initial part of curve $R_g(\tau)$.

**A relation between maximum values in various orders derivatives.** Let us find out, what type of conditions must satisfy the maximum absolute values of action *g*(*t*) and its *N* younger derivates overlaying restrictions (2.22) on action. For this purpose the correlation of variations in time *N*-th, (*N* - 1)-th and (*N* - 2)-th derivatives is considered.

Let in some instant $t_*$ the $(N-2)$-th derivative of action have a value $g^{(N-2)}(t_*)\in\left[0,\,g_M^{(N-2)}\right]$, (*N* - 1)-th derivative — a value $g^{(N-1)}(t_*)\in\left[0,g_M^{(N-1)}\right]$, and $N-1$ derivative — the maximum possible negative value $g^{(N)}(t_*)=-g_M^{(N)}$. Then for the $(N-1)$-th derivative to be cancelled down to zero not later than the function $g^{(N-2)}(t)$ exceeds the maximum permissible value $g_M^{(N-2)}$, the following condition should be satisfied:

$$g^{(N-2)}(t_*)+\tfrac{1}{2}g_M^{(N)}\left[g^{(N-1)}(t_*)/g_M^{(N)}\right]^2\le g_M^{(N-2)}$$

or, taking into account an arbitrary choice of moment $t_*$,

$$\left[g^{(N-1)}(t)\right]^2\le 2g_M^{(N)}\left[g_M^{(N-2)}-g^{(N-2)}(t)\right].$$

From here, accepting $g^{(N-1)}(t)=g_M^{(N-1)},g^{(N-2)}(t)=0$,

$$g_M^{(N-1)}/g_M^{(N-2)}\le 2\,g_M^{(N)}/g_M^{(N-1)} \tag{3.8}$$

can be obtained.

Inequality (3.8) can be written in a more expended, but equivalent form. That can be easily converted into:

$$g_M^{(N-1)} \big/ g_M^{(N-2)} \le \sqrt{2 g_M^{(N)} \big/ g_M^{(N-2)}} \le 2 g_M^{(N)} \big/ g_M^{(N-1)} . \qquad (3.9)$$

The equal signs in (3.8) and (3.9) are possible only at variation of $N$-th derivative of action under meander law. For this purpose the $(N + 1)$-th derivative must be unlimited. As a step, the modification of derivatives with indexes $i < N$ is impossible and the following strict inequalities are valid for them

$$g_M^{(i-1)} \big/ g_M^{(i-2)} < \sqrt{2 g_M^{(i)} \big/ g_M^{(i-2)}} < 2 g_M^{(i)} \big/ g_M^{(i-1)} , \; i = \overline{2, N-1} . \qquad (3.10)$$

If inequalities (3.8) — (3.10) are not fulfilled, then magnitudes $\left\{g_M^{(i)}\right\}_0^N$ cannot be the maximum of real action derivatives.

Note that the inequality (3.8) can be developed even from inequality of Landau-Hadamard. This paper which has reference to the following more common outcome obtained by A.N.Kolmogorov is also interesting:

$$\left(g_M^{(i)}\right)^N \le C_{Ni}^2 \left(g_M^{(N)}\right)^i \left(g_M\right)^{N-i}, 0 \le i < N, \qquad (3.11)$$

where

$$C_{Ni} = \alpha_{N-i} \big/ \alpha_N^{(N-i)/N},$$

$$\alpha_K = \frac{4}{\pi}\left[1 + \frac{1}{(-3)^{K+1}} + \frac{1}{5^{K+1}} + \frac{1}{(-7)^{K+1}} + \frac{1}{9^{K+1}} + \ldots\right].$$

Relations of maximum values of five derivatives of actions are investigated in his work. However, the bulkiness of obtained expressions makes their practical use complicated.

## §3.2. Experimental determination of spectral and correlation characteristics of actions

**Common properties of estimations in casual processes statistical characteristics**. The statistical characteristics of stationary casual processes and their derivatives are determined as an average by the indefinitely large number of realizations or, if the ergodicity condition is fulfilled, by realization of an indefinitely large duration. When characterizing the properties on whole sets of possible realizations, they are noncasual magnitudes or noncasual functions.

The amount of treated realizations and their duration is always restricted in practice. This is not only due to the complexity and worth of experiments, but also to finite duration of time intervals, of which it is possible to consider this or that casual process as a stationary one. Therefore, it is necessary to consider any statistical characteristic obtained by experimental data handling, only as an estimation of a true characteristic. The values of such an estimation calculated for different realizations, has a casual scatter.

Values of estimations should group more and more tightly around true characteristics at the increase of the duration for the treated realizations. Dispersion and mathematical expectation of its casual deviation from true characteristics are considered as measures of clustering or accuracy of estimation.

The estimation is called unbiased, if its expectation is equal to the evaluated characteristic. The estimate is called consistent, if it converges, by probability, to the evaluated characteristic at an unlimited increase of realization duration $T_O$. Expectation and error dispersion of consistent estimates tend to zero when $T_O \to \infty$.

The distribution law of estimation error is possible to consider, as a rule, as normal. Then, having set some maximum absolute values of error and knowing dispersion of error, it is easy to calculate the probability of error location within the limits of a preset confidence interval. This probability is considered confidential and is used as the measure for reliability of estimate.

The handling of action realizations necessary to derive the estimations can be carried out either on analogue, or on digital computing devices. Discrete (in time and on a level) values of realizations are treated in the latter case. Algorithms of calculation can be considered as an outcome of discretization of appropriate continuous algorithms. Note that the discrete algorithms sometimes ensure a smaller methodical error in estimation, than continuous algorithms, although casting-out a member contained in realization of information, which is connected to discretization hinders the rise of estimation accuracy.

Consider some estimates on the basis of continuous introduction of centered stationary casual action realizations.

**Estimation of correlation functions and dispersion**. Experimental determination of correlation functions and dispersions is one of the most developed and equipped with various kinds of instrumentation experimental analyses for casual processes. It is possible to evaluate the estimate $R_{gT}(\tau)$ for correlation function $R_g(\tau)$ of action $g(t)$, observed in interval $t \in [0, T_O]$, by the formula

$$R_{gT}(\tau) = \frac{1}{T_O - \tau} \int_0^{T_O - \tau} g(t) g(t + \tau) dt \,. \tag{3.12}$$

The correlation meter, used for this purpose, contains the delay line, multiplier and integrator among the basic units. A recorder of realization on magnetic tape, film or other tools of information storage is often included in its structure. It allows reproducing obtained realization of action for evaluation of estimation $R_{gT}(\tau)$ at different values of argument $t$, $\tau << T_O$.

The large number of indirect methods to experimentally determine the correlation function without immediate use of formula (3.12) is described in the literature.

It is necessary to analyze the accuracy of estimation (3.12). This estimation is unbiased, because

$$M\{R_{gT}(\tau)\} = \frac{1}{T_O - \tau} \int_0^{T_O - \tau} M\{g(t) g(t + \tau)\} dt = \frac{1}{T_O - \tau} \int_0^{T_O - \tau} R_g(\tau) d(t) = R_g(\tau).$$

For dispersion of the estimation error with reference to a case, such as when the magnitude $T_O$ essentially exceeds an interval of correlation of action, it is possible to obtain the formula [8]

$$\sigma_R^2(\tau) \cong \frac{2}{T_O - \tau} \int_0^\infty \left[ R_g^2(\Theta) + R_g(\Theta + \tau) R_g(\Theta - \tau) \right] d\Theta . \tag{3.13}$$

At $\tau = 0$ formula (3.13) gives

$$\sigma_R^2(0) = \sigma_D^2 \approx \frac{4}{T_O} \int_0^\infty R_g^2(\Theta) d\Theta , \tag{3.14}$$

and at large enough values $\tau$,
when $R_g^2(\Theta) >> R_g(\Theta + \tau) R_g(\Theta - \tau)$,

$$\sigma_R^2(0) \approx \frac{2}{T_O - \tau} \int_0^\infty R_g^2(\Theta) d\Theta = \frac{\sigma_D^2 T_O}{2(T_O - \tau)} . \tag{3.15}$$

It is obvious from (3.15), that at $\tau > 0.1 T_O$ the error of correlation function estimation increases sharply at increasing $\tau$ and becomes indefinitely large at $\tau \to T_O$. It determines the extended requirements of duration of realization necessary to derive values for correlation function with a large interval of argument variation.

Taking into consideration an error of estimation only at $\tau = 0$, then having set a small relative r.-m.-s. error of estimation of dispersion $\sigma_D / D_0 \leq 0.1$, from (3.14), the required duration of observable realization is developed as:

$$T_O \geq \frac{400}{D_0^2} \int_0^\infty R_g^2(\Theta) d\Theta . \tag{3.16}$$

For example, for an action with correlation function $R_g(\tau) = D_0 e^{-\alpha|\tau|} \cos(\beta\tau)$ the formula (3.16) gives

$$T_O \geq 100 \left( \frac{1}{\alpha} + \frac{\alpha}{\alpha^2 + \beta^2} \right) .$$

The practical use of expressions (3.13) — (3.16) is troubled by the fact that correlation function $R_g(\tau)$ is a priori unknown. Therefore, it is necessary to accept on certain accounts any rough priori model of this function obtained on the basis of theoretical analysis of action properties or available experience of experimental investigations. It is necessary to consider its acceptable concurrence to the experimentally obtained estimation as the validation criterion of a priori model choice.

For a rough estimation of required realization duration, the formula $T_O > 50/\omega_O$ is also used. Here $\omega_l$ is the lower frequency in spectrum of action [80].

It is possible to conclude from tatter, that in order to estimate the discrete values of the correlation function with relative errors in percent units, the duration of treated realizations should exceed an interval of action correlation of a thousand times. However, although the deriving and filing of such long realizations are technically feasible, it is often not justified due to a possibility of stationary condition violation of action.

**Estimation of spectral density**. It is possible to find the estimation for spectral density of action power by the known estimation of the correlation function $R_{gT}(\tau)$ in interval $\tau \in [-\tau_M, \tau_M]$ as

$$S_{gT}(\omega) = 2\int_0^{\tau_M} R_{gT}(\tau)\cos(\omega\tau)d\tau . \tag{3.17}$$

Estimation (3.17) is asymptotically unbiased, but inconsistent, because its dispersion at any $\tau_M$ has the order $S_g^2(\omega)$. It is possible to increase an accuracy of estimation by smoothing curve $S_{gT}(\omega)$.

The estimation of spectral density can be also obtained immediately at the time of realization of action, without a preliminary determination of estimation of the correlation function. A set of narrow-band filters is required for that. Each filter extracts the spectral components of actions in the defined frequencies interval. Consider a case for use as ideal band pass filters with rectangular gain plot.

The action is also applied to an input of filter with transparency band in interval from $\omega - \Delta\omega/2$ up to $\omega + \Delta\omega/2$ and the signal $x(t, \omega, \Delta\omega)$ is observed on the output of the filter. Then as an estimation of a spectral density of power, it is natural to take a casual variable

$$S_{gT}(\omega) = \frac{1}{T_O \Delta\omega}\int_0^{T_O} x^2(t, \omega, \Delta\omega)\, dt . \tag{3.18}$$

Thus magnitude

$$P_x = \frac{1}{\pi} S_{gT}(\omega)\Delta\omega \tag{3.19}$$

corresponds to an average power (dispersion) of action corresponding to an interval of frequencies $\Delta\omega$ in the neighborhood of frequency $\omega$.

Estimation (3.18) can be considered practically unbiased only at the very small interval $\Delta\omega$.

It is possible to analyze a dispersion of estimation in the supposition, so that the spectral density is almost constant in interval $[\omega - \Delta\omega/2, \omega + \Delta\omega/2]$. Then the correlation function of signal $x(t, \omega, \Delta\omega)$ is equal to

$$R_x(\tau) = R_x(0)\frac{\sin(\Delta\omega\tau/2)}{\Delta\omega\tau/2}\cos(\omega\tau), \tag{3.20}$$

where $R_x(0) = \pi^{-1} S_g(\omega)\Delta\omega$.

For dispersion of casual value $P_x$ according to (3.20) analogous to (3.14) the expression can be written as:

$$D_P \approx \frac{4}{T_O}\int_0^\infty R_x^2(\tau)d\tau = \\ = \frac{4}{\pi^2 T_O} S_g^2(\omega)(\Delta\omega)^2 \int_0^\infty \frac{\sin^2(\Delta\omega\tau/2)}{(\Delta\omega\tau/2)^2}\cos^2(\omega\tau)d\tau = \frac{S_g^2(\omega)\Delta\omega}{\pi^2 T_O}. \tag{3.21}$$

On the other hand, according to (3.19) the following relation is fulfilled

$$D_P = \frac{1}{\pi^2} D_S(\Delta\omega)^2, \tag{3.22}$$

where $D_S$ is the required dispersion of estimation (3.18).

From (3.21) and (3.22) it is accepted:

$$D_S = \frac{S_g^2(\omega)}{T_O\Delta\omega}, \tag{3.23}$$

i.e. the realization of condition $T_O\Delta\omega >> 1$ is necessary at enough small values $\Delta\omega$ for reaching the high accuracy of spectral density estimation, defining the resolution of spectrum analyzer on the frequency. The relative error $\sqrt{D_S}/S_g(\omega) \le 0.1$ is supplied at $T_O\Delta\omega \ge 100$. In practice this means, that the determination of good spectral density estimation needs to handle rather long realizations of action, as well as at an estimation of the correlation function.

**Approximation of correlation functions and spectral densities estimations by analytical expressions.** Constructing the mathematical models, suitable for analytical investigation of control systems should complete the handling of outcomes in experimental investigation of correlation functions or spectral densities of actions. Therefore after evaluating the large number of functions $R_{gT}(\tau)$ or $S_{gT}(\omega)$ ordinates, there is a necessity to approximate these functions by analytical expressions. With the investigation of stationary systems with the frequency methods, the very important problem is the determination of adequate analytical model of spectral density. It can be made immediately on the graph of function $S_{gT}(\omega)$ or, more accurate, by the Fourier transform of correlation function model.

A large number of methods for functions approximation diagrammed with analytical expressions have been developed. The approximation can be executed with high accuracy by interpolating methods, square-law or steady approximation, and by the expansion into a series, etc. Methods connected in order to passage to logarithmic spectral density differ by their greater simplicity.

Nevertheless, the selection of analytical expressions for correlation functions and spectral densities is one of the greatest "gorges" in solutions of systems synthesis problem by results of experimental analysis of actions. One matter is that the substitution of graphs constructed on individual points is connected to analytical expression into its extrapolation in the region of the large values of argument, for which the experimental data are absent. Besides by virtue of casual scatter of experimental curve points, their accurate reproduction can cause the essential complication of models, and attempts to smooth out an experimental curve may cause the loss of exact features in its shape.

For example, at approximation of correlation function its oscillations of small amplitude can be smoothed out. That is essentially reflected in a kind of spectral density, where the narrow peaks are cut off. Approximating the spectral density, the inevitable arbitrariness in the estimation of its damping rate on high frequencies causes the invalidation of the correlation function model at small values of argument and false reflecting in properties of action differentiability.

The enumerated facts force us to approach the choice of correlation and spectral models of input actions very crucially. It is necessary to pay special attention to those singularities, which render the strongest influence on precision factors for investigated control systems. According to it, the analysis of systems critical to inaccuracy in models of actions representation is rather useful.

An additional requirement to the model of spectral density is the necessity to choose it as a fractional rational function of a frequency quadrate. One matter is that the known analytical methods of systems analysis and synthesis to complete a priori information are developed only applied to such models of spectral densities. Justifying such a restriction of considered spectral densities class, it is acceptable to give a hypothesis that the spectral density of any stationary casual action can be treated as the outcome of white noise transiting through the filter with the fractional rational transfer function. Thus, the spectral density should look like

$$S_g(\omega)=\frac{P_{2m}(\omega)}{Q_{2n}(\omega)}, \tag{3.24}$$

where $P_{2m}(\omega)$ and $Q_{2n}(\omega)$ are polynomials of powers $2m$ and $2n$, containing only even powers of ω, and $m < n$. The correlation function

$$R_g(\tau)=\sum_{i=1}^{n} e^{-\mu_i|\tau|}\left(a_i e^{j\beta_i|\tau|}+a_i^* e^{-j\beta_i|\tau|}\right), \tag{3.25}$$

corresponds to such spectral density.

Here $\mu_i$ and $\beta_i$ are the absolute values of real and imaginary members of polynomial $Q_{2n}(\omega)$ roots, and $a_i$ and $a_i^*$ are the complex conjugate factors (in specific case — real and identical).

The action with spectral density (3.24) is differentiated $N$ times, where $N = n - m - 1$.

Expressions (3.24) and (3.25) are used in practice, as a rule, at $n \le 3$. Some most common models of spectral densities look like (indexes 1 — 6 at the left at $S_g(\omega)$ point the variants):

$$^1S_g(\omega) = \frac{2D_0T_1}{1+\omega^2T_1^2} = \frac{2\mu D_0}{\mu^2+\omega^2}, \text{ where } \mu = T_1^{-1};$$

$$^2S_g(\omega) = \mu D_0\left[\frac{1}{\mu^2+(\omega-\beta)^2} + \frac{1}{\mu^2+(\omega+\beta)^2}\right] =$$

$$= 2\mu D_0 \frac{\mu^2+\beta^2+\omega^2}{\left(\mu^2+\beta^2\right)^2 + 2\omega^2\left(\mu^2-\beta^2\right)+\omega^4};$$

$$^3S_g(\omega) = \frac{4 \qquad \mu^3 D_0}{\left(\mu^2+\omega^2\right)^2};$$

$$^4S_g(\omega) = \frac{\mu D_0}{\beta}\left[\frac{2\beta-\omega}{\mu^2+(\omega-\beta)^2} + \frac{2\beta+\omega}{\mu^2+(\omega+\beta)^2}\right] =$$

$$= 4\mu D_0 \frac{\mu^2+\beta^2}{\left(\mu^2+\beta^2\right)^2 + 2\omega^2\left(\mu^2-\beta^2\right)+\omega^4};$$

$$^5S_g(\omega) = \frac{A}{\left(1+T_1^2\omega^2\right)\left(1+T_2^2\omega^2\right)\left(1+T_3^2\omega^2\right)};$$

$$^6S_g(\omega) = \frac{A}{\left(1+\varepsilon_1^2T_0^2\omega^2\right)\left[\left(1-T_0^2\omega^2\right)^2+\varepsilon_2^2T_0^2\omega^2\right]};$$

$0 \le \varepsilon_1 \le \infty, 0 \le \varepsilon_2 \le 2.$

Some models of correlation functions corresponding to them are:

$$^1R_g(\tau) = D_0e^{-|\tau|/T_1} = D_0e^{-\mu|\tau|};$$

$$^2R_g(\tau) = D_0e^{-\mu|\tau|}\cos\beta\tau;$$

$$^3R_g(\tau) = D_0e^{-\mu|\tau|}\left(1+\alpha|\tau|\right);$$

$$^4R_g(\tau) = D_0e^{-\mu|\tau|}\left(\cos\beta\tau + \frac{\mu}{\beta}\sin\beta|\tau|\right);$$

$$^5R_g(\tau) = D_0K_1\left(\rho_1e^{-|\tau|/T_1} + \rho_2e^{-|\tau|/T_2} + \rho_3e^{-|\tau|/T_3}\right)$$

where

$$\rho_1 = \frac{T_3^2 - T_2^2}{(T_1 - T_2)(T_1 - T_3)},\ \rho_2 = \frac{T_2^3(T_1 + T_3)}{T_1^3(T_1 - T_3)},$$

$$\rho_3 = \frac{T_3^3(T_1 + T_2)}{T_1^3(T_1 - T_3)},\ K_1 = \frac{T_1^3}{(T_1T_2 + T_1T_3 + T_2T_3)(T_3 - T_2)};$$

$$^6R_g(\tau) = D_0 K_2 \left[\rho_4 e^{-|\tau|/\varepsilon_1 T_0} + \rho_5 e^{-\varepsilon_2|\tau|/2T_0} \sin\left(\frac{\sqrt{4-\varepsilon_2^2}}{2}\frac{|\tau|}{T_0} + \varphi\right)\right],$$

where $\rho_4 = \varepsilon_1^3\varepsilon_2,\ \rho_5 = 2\sqrt{\left[\left(1+\varepsilon_1^2\right)^2 - \varepsilon_1^2\varepsilon_2^2\right]/\left(4-\varepsilon_2^2\right)},$

$$\varphi = \mathrm{arctg}\frac{\left(1+\varepsilon_1^2 - \varepsilon_1^2 \quad \varepsilon_2^2\right)\sqrt{4-\varepsilon_2^2}}{\varepsilon_2\left(1+3\varepsilon_1^2 - \varepsilon_1^2\varepsilon_2^2\right)},\ K_2 = \frac{1}{\left(1+\varepsilon_1^2 - \varepsilon_1^2\varepsilon_2^2 + \varepsilon_1^3\varepsilon_2\right)}.$$

If the roots of polynomial $Q_{2n}(\omega)$ in expression (3.24) are real and not too much different than $Q_{2n}(0)-1$ and $n >> 1$, this expression looks like

$$S_g(\omega) \cong P_{2m}(\omega)\ e^{-\alpha^2\omega^2} \tag{3.26}$$

or, generally,

$$S_g(\omega) = P_{2k}(\omega)e^{-\alpha_1^2\omega^2} + P_{2l}(\omega)e^{-\alpha_2^2\omega^2} + \ldots \tag{3.27}$$

It is based on the limiting relation

$$\lim_{n\to\infty}\frac{1}{\left(1+\alpha^2\omega^2/n\right)^n} = e^{-\alpha^2\omega^2}.$$

An example of expressions (3.26) and (3.24) in comparison is given in [3].

The actions with spectral densities (3.26) and (3.27) are indefinitely differentiated. This makes these models rather convenient in the cases when the differentiability is the basic property of actions, and a fractional rational kind of spectral densities is not obligatory. However, it is necessary to relate their membership to a class of singular casual processes. For those casual processes the exact extrapolation on any time forwarded by indefinitely long observable realization is theoretically possible.

## §3.3. Experimental determination of numerical characteristics for action derivatives

**Estimation of dispersions and maximum values of derivatives.** Formula (3.28) is used to estimate the dispersion of $i$-th derivative of action by this derivative $g^{(i)}(t)$ realization, observable in finite time interval $t \in [0, T_O]$.

$$D_{iT} = \frac{1}{T_O} \int_0^{T_O} \left[g^{(i)}(t)\right]^2 dt . \tag{3.28}$$

The realization of the derivative action can be obtained by appropriate choice of metering equipment. For example, the radio or gyro vertical can be used to measure the roll and pitch of an airplane; for measurement of angular velocities — laser or rate gyroscopes, and for measurement of accelerations — accelerometers. The track irregularities measuring instrument can be implemented by the free gyroscope to measure the slope and by the two-power gyroscope to measure the rate of slope modification. A series of similar examples can be easily prolonged.

The most accessible way of derivative obtainment is by the successive differentiation of actions. The quality of making derivative on continuous realization essentially depends on whether the realization has interferences or not. With the lack of interference's, the estimation of derivative is fulfilled by means of any differentiating device. However, such estimation is incorrect. The presence of unaccounted high-frequency noise with any small dispersion will cause the noticeable errors in estimation of derivative. Noise can affect the accuracy of second and subsequent derivative elimination especially strong. Therefore, even if estimating the derivatives of action without noise, it is necessary to assume a presence of small noise in realization. In this case differentiator-smoothing devices are used. The elementary device has a transfer function

$$W_{ds}(s) = \frac{k_{ds} s}{1 + T_s s} .$$

If the first or second derivative of action, but not the action itself is directly observed, then when generating the action and series of its consecutive derivatives, the operations of differentiation and integration are used.

The device scheme for an estimation of action dispersions and its five derivatives by its realization of second-order action derivative is shown as an example in Fig. 3.1. The centered signal of the second derivative, for example, from an accelerometer, is here being integrated to estimate the first derivative and doubly integrated to recover the action itself. Consecutive triple derivation for evaluation of the third, fourth and fifth derivatives is realized at the same time. Then, each of the obtained signals is raised in quadrate and is integrated. This gives the required estimations $D_{iT}$ at $i = 0, 1, \ldots, 5$ (with particular scale). If it is necessary to evaluate the estimations for maximum values of derivatives, then squarers and integrators in output circuits should be replaced accordingly by devices for determination of a module and by peak detectors.

In a case, when the response time of the differentiator-smoothing device $T_S$ is much less than the correlation of the action interval, the smoothing practically does not cause the methodical errors into the determination of derivative characteristics. The estimate for characteristics of smoothed derivative of action will then be made otherwise.

Note that dispersion of a smoothed derivative, expressed by the formula (2.21); can be considered as the generalized moment of spectral density of action concerning the basis function $\left[k_{ds}^2\omega^2/\left(1+\omega^2T_c^2\right)\right]^i$. With the use of other dynamic filters instead of differentiator-smoothing devices, the other generalized moments will bc obtaincd. They can also contain essential information on properties of action.

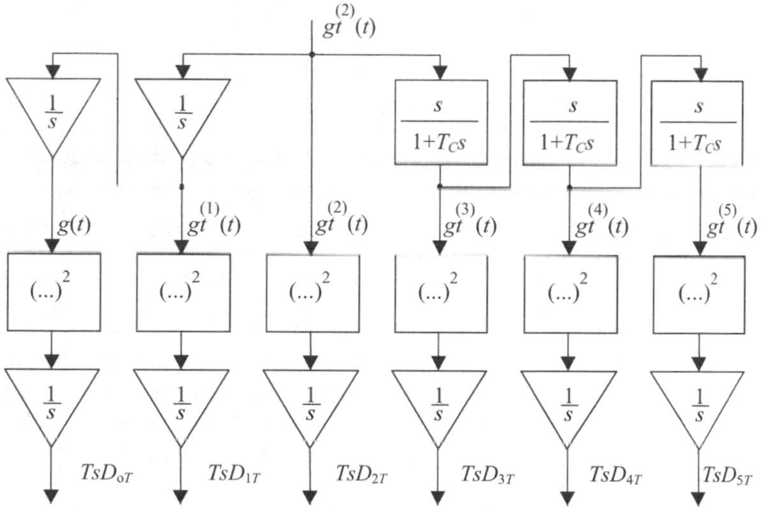


Fig. 3.1

**Analysis of estimations accuracy**. Find out the accuracy for estimation of derivatives dispersions by the formula (3.28). Similarly to §3.2 in order to estimate the dispersion of action it is easy to show that estimate (3.28) is unbiased. Its dispersion is expressed as

$$\sigma_{Di}^2 \approx \frac{4}{T_O}\int_0^\infty \left[R_{g(i)}(\tau)\right]^2 d\tau\,, \tag{3.29}$$

where $R_{g(i)}(\tau)$ is the correlation function of $i$-th derivative, for which the expression (3.1) is valid.

Consider, for example, the accuracy of estimation for derivatives dispersions of an indefinitely differentiable action with Gaussian spectral density

$$S_g(\omega) = A\exp\left(-\omega^2/\omega_*^2\right),$$

where $A$ and $\omega_*$ are some parameters. A correlation function of this action can be expressed by an inversion formula:

$$R_g(\tau) = \frac{1}{\pi}\int_0^\infty A\exp\frac{\omega^2}{\omega_*^2}\cdot\cos(\omega\tau)d\omega = D_0\exp\left(-\alpha^2\omega^2\right),$$

where $D_0 = \left(2\sqrt{\pi}\right)^{-1} A\omega_*, \alpha = \omega_*/2$.

The correlation functions for the first and the second derivatives result in the following expressions, according to (3.1):

$$R_{g(1)}(\tau) = 2\alpha^2 D_0\left(1-2\alpha^2\tau^2\right) e^{-\alpha^2\tau^2},$$

$$R_{g(2)}(\tau) = 4\alpha^4\left(3-12\alpha^2\tau^2+4\alpha^4\tau^4\right) e^{-\alpha^2\tau^2},$$

and it allows for the acceptance of the following expressions for estimations of relative errors dispersions after their substitution into (3.29), their integration, and rejection of the members of second order of smallness:

$$\frac{\sigma_{D0}^2}{D_0^2} \approx \frac{\sqrt{2\pi}}{\alpha T_*}, \frac{\sigma_{D1}^2}{D_1^2} \approx \frac{3}{4}\frac{\sqrt{2\pi}}{\alpha T_O}, \frac{\sigma_{D2}^2}{D_2^2} \approx \frac{35}{48}\frac{\sqrt{2\pi}}{\alpha T_O}.$$

Similar expressions can also be found with reference to estimations for dispersions of higher derivatives. This provides the foundation to conclude that the relative errors of estimates for dispersions of action and their several derivatives by the formula (3.29), are practically coincided among themselves and are close to a relative error of estimation for each point of action correlation function by the formula (3.12).

It is essential, that almost identical computing expenditures are required to estimate one point of correlation function and to disperse one of the action derivatives. Therefore, authentic estimation for dispersions of several action derivatives is represented by a simpler problem, than the construction of authentic analytical model of the correlation function, for which a few experimental points can be required.

**Use of singular points of action.** If the acceptance of the hypothesis on the Gaussian distribution law of action is valid, then the dispersions $\{D_i\}_0^N$ of action and its $N$ derivatives can be found easily by analyzing the experimentally obtained statistical characteristics of singular points.

An average number $n_{0c}$ of points where the curve $g(t)$ intersects the given level $C_0$, an average number $m_0$ of this curve extremums (maximums and minimums) and the average number $p_0$ of this curves inflections are referred to as statistical characteristics of singular points of action $g(t)$. An average number $n_{ic}$ of points where the curve $g^{(i)}(t)$ intersects the given level $C_i$, an average number

$m_i$ of extremums of this curve and an average number $p_i$ of this curves inflections are referred to as statistical characteristics for singular points of the $i$-th derivative of action (considering that the action is differentiated at least $N$ times, $N \geq i$). The validity of formulas $m_i = n_{i+1,0}$, $p_i = n_{i+2,0}$ at $C_i = 0$ and $n_{ic} = n_{i0}$ is obvious. Note that the average number of extremums, at separation into two equal parts, gives an average number of maximums and minimums.

The average number of singular points per time unit is evaluated as a result of the total number of accounts of singular points in the investigated realization, and its division by the duration of realization.

It is known [56, 81], that the average number of points where the curve $g(t)$ intersects the given level $C_0$, can be expressed by the formula:

$$n_{0c} = \frac{1}{\pi} \lim_{\tau \to 0} \left[ \frac{R'_g(\tau)}{\sqrt{D_0^2 - R_g^2(\tau)}} \right] \exp\left( -\frac{C_0^2}{2D_0} \right) = \frac{1}{\pi} \sqrt{\frac{-R''_g(0)}{D_0}} \exp\left( -\frac{C_0^2}{2D_0} \right),$$

taking (3.1) into account, it turns into:

$$n_{0c} = \frac{1}{\pi} \sqrt{\frac{D_1}{D_0}} \exp\left( -\frac{C_0^2}{2D_0} \right)$$

or, in relation to the $i$-th derivative of action,

$$n_{ic} = \frac{1}{\pi} \sqrt{\frac{D_{i+1}}{D_i}} \exp\left( -\frac{C_i^2}{2D_i} \right) \tag{3.30}$$

At $C_i = 0$ formula (3.30) gives an average number of intersections of zero level by the $i$-th derivative of action

$$n_{i0} = \frac{1}{\pi} \sqrt{\frac{D_{i+1}}{D_i}}. \tag{3.31}$$

The physical sense of expression (3.31) becomes clear when the considered magnitude $\sqrt{D_{i+1}/D_i} = \omega_{rmsi}$ is the r.-m.-s. frequency in spectrum of the $i$-th derivative of action.

From (3.30) and (3.31) it is easy to generate the following formulas as dispersions of derivative:

$$D_i = \frac{C_i^2}{2} \left( \ln \frac{n_{i0}}{n_{ic}} \right)^{-1}, \tag{3.32}$$

$$D_{i+1} = \pi^2 n_{i0}^2 D_i, \tag{3.33}$$

$$D_{i+2} = \pi^2 m_i^2 D_{i+1} = \pi^4 n_{i0}^2 m_i^2 D_i, \tag{3.34}$$

$$D_{i+3} = \pi^2 p_i^2 D_{i+2} = \pi^4 m_i^2 p_i^2 D_{i+1} = \pi^6 n_{i0}^2 m_i^2 p_i^2 D_i \,. \qquad (3.35)$$

Thus, the magnitudes $n_{ic}$, $n_{i0}$, $m_i$ and $p_i$ evaluated by a curve $g^{(i)}(t)$ allow calculation of the dispersions of $i$-th, $(i + 1)$-th, $(i + 2)$-th and $(i + 3)$-th derivatives of action. At $i = 0$ the formulas (3.32) — (3.35) provide the dispersions of action and three of its derivatives. The determination of dispersions of higher order derivatives implies the deriving and processing of curves $g^{(i+1)}(t)$, $g^{(i+2)}(t)$,..., $g^{(N-3)}(t)$.

Accuracy of dispersion $\{D_i\}_0^N$ estimations by singular points is determined by the accuracy of estimations for characteristics of singular points. The latter depends on correlation properties of action and on the duration of the treated realization $T_O$. The magnitude $T_O$ can be approximately chosen, according to the requirement, that the number of singular points within the limits of the considered realization must be about $10^2$.

The analysis on singular points concerns the simplest feasible kinds of experimental investigation of actions.

## §3.4. Determination of input action characteristics based on theoretical analysis

**General concepts.** The idea of the use of the theoretical analysis is to determine the characteristics of input action and to connect them to the characteristics of any other casual process, of which the mathematical model is known, or to reveal some restrictions imposed on action with the assumption that its physical nature, and constructive features and tactics of instrumentation also are used. Similar investigations allow obtainment of the necessary a priori information of system synthesis, without the experimental estimation of input action characteristics. It turns out rather useful when constructing the preliminary models of actions such as correctly permitting the choice of a procedure of their experimental investigation.

The strictness of theoretical investigation is especially conclusive in a case, when it is possible to reveal some initial casual process with known characteristics. The action can be obtained as a result of particular functional transformations. The passing of casual processes through the linear dynamic and nonlinear instantaneous elements is well analyzable [56]. This makes it possible to determine the characteristics of action with the same reliability, as the model of initial casual process.

Such a method for action research is widely used with synthesis of automatic systems in radiolocation and radio control, in marine instrumentation and in other areas, where the great amount of experimental data is generalized and accumulated.

With the analysis of interference, applied to an input of system, the essential circumstance is: if the spectral density $S_\upsilon(\omega)$ is practically uniform within the limits of supposed system bandpass, then the use of white noise is lawful as a model of interference. Accordingly, the analysis of interference is often reduced to a rough estimate of its spectrum width and determination of its unique parameter. That parameter is the level of spectral density near the zero frequency $S_\upsilon(0)$.

On the contrary, the theoretical analysis of reference action is usually completed by the construction of a more complicated model. Two or more numerical parameters are used to describe that model.

**Selection of spectrum for reference action with a deficiency of a priori information.** The complexity to construct the theoretical model for spectral density of reference action is connected to the necessity to decrease the rate analysis for spectral density $S_g(\omega)$ in middle and high frequencies. This rate essentially affects the control accuracy. The elementary fractional rational expression for spectral density, at which the problem of system synthesis is still of interest, i.e. the reference action basically can be recovered from the more broadband interference, looks like

$$S_g(\omega) = \frac{2D_0 T_1}{1+\omega^2 T_1^2}. \tag{3.36}$$

In conditions of full practicality of a priori indeterminacy, when the only restriction for action is the finite value of its dispersion $D_0$, the acceptance of model (3.36) minimizes the hazard of an error. The error may enable easier operation mode for synthesized system in comparison with processing of real action. The expression (3.36) is more preferable in this sense than the other models assigning a high decreasing rate of spectral density.

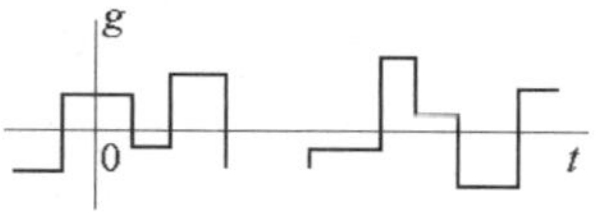


Fig. 3.2

With the selection of time constant $T_1$, defining the spectrum of action width, it is useful to take into consideration the following. It is easy to show [9], that, in particular, the process holding a constant value during some time interval, and then accepting another casual value by step (Fig.3.2), has a spectral density (3.36). These steps happen in casual instants. They are independent among themselves and distributed on a time axis with a constant average density $\lambda$ ($T_1 = 1/\lambda$ is an average time interval between steps). Supposing the possibility of action variation by such a law, the assumption of physical essence of action follows the necessity to assign the approximate number of steps in time unit $\lambda$, and find the value $T_1 = 1/\lambda$ from here.

The spectral density (3.36) can be used, for example, for description of processes, such as the action of atmosphere turbulence on the bearing surface of an airplane; the variation of acceleration at the maneuvering airplane, velocity of gyro drift, angular scintillation and fading of the radar target, velocity of unaccounted marine streams at vessels navigation, instability in thrust power of rocket engine.

If it is necessary to ensure the finite value of dispersion not only for the action itself but for its $i$-th derivative, then the expression (3.36) can be applied to a spectral density of this derivative. The spectral density of such an action looks like

$$S_g(\omega) = \frac{2D_i T_1}{\omega^{2i}\left(1+\omega^2 T_1^2\right)}. \tag{3.37}$$

At $i = 1$ the expression (3.37) corresponds to so-called typical input signal of servomechanism.

The estimation $T_1$ in (3.37) is often complicated, but dispersions or maximum values of two or more derivatives of action can be obtained on the basis of theoretical analysis. Then it is convenient to refuse the constructing of spectral model for an action all together and to accept the system synthesis procedure based directly on the dispersion or on maximum values of derivatives. Consider an example of such an investigation.

**Analysis of characteristics of coordinate derivatives for radar tracking target**. The target position (for example, an airplane) in polar scale is characterized by an azimuth angle, elevation angle and slope distance. Each of these casual time functions is the reference action for an appropriate servomechanism of radar.

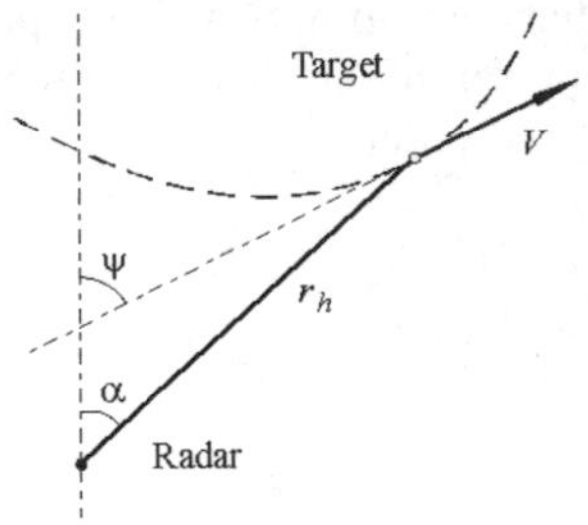


Fig. 3.3

Consider the azimuth derivatives and slope distance characteristics at the following principal restrictions.

1. The maneuvering capabilities of tracking targets are determined by the maximum possible values of linear velocity $\max|V(t)| = V_M$, longitudinal acceleration $\alpha_{i\min} \le \alpha_i \le \alpha_{i\max}$ and lateral acceleration $\max|a_1(t)| = a_{1M}$.

2. Radar is intended for interaction on targets with a slope distance $r(t)$, ranging from $r_{min}$ up to $r_{max}$.

Besides, in order to simplify the analysis it should be considered, that the point of radar installation is stationary, and the target maneuvers only in a horizontal plane practically without the variation of motion height $h(t)$ = const.

The relative position for velocity vector of target $V$ and its line-of-sight is shown in Fig. 3.3. The indications are α is the target azimuth, ψ is the course angle describing a direction of vector $V$ in a horizontal plane and $r_h = \sqrt{r^2 - h^2}$ is the horizontal distance to the target.

The following expressions are obtained for the first derivatives of slope distance and azimuth as an example of resolution of the target velocity vector into the radial and tangential component, accounting the relation between sloping and horizontal distances,:

$$\dot{r}(t) = V\cos(\psi - \alpha)\frac{r_h}{r} = V\sqrt{1 - \frac{h^2}{r^2}}\cos(\psi - \alpha), \tag{3.38}$$

$$\dot{\alpha}(t) = \frac{V\sin(\psi - \alpha)}{r_h} = \frac{V\sin(\psi - \alpha)}{\sqrt{r^2 - h^2}}, \tag{3.39}$$

where $V = V(t), r = r(t), \psi = \psi(t), \alpha = \alpha(t)$.

The derivation of the right and left hand sides in (3.38) and (3.39) gives

$$\ddot{r}(t) = \dot{V}\sqrt{1 - \frac{h^2}{r^2}}\cos(\psi - \alpha) + \frac{Vh^2\cos(\psi - \alpha)}{r^3\sqrt{1 - h^2/r^2}}\dot{r} - \\ - V\sqrt{1 - h^2/r^2}\sin(\psi - \alpha)(\dot{\psi} - \dot{\alpha}), \tag{3.40}$$

$$\ddot{\alpha}(t) = \frac{\dot{V}\sin(\psi - \alpha)}{\sqrt{r^2 - h^2}} - \frac{Vr\sin(\psi - \alpha)}{(r^2 - h^2)^{3/2}}\dot{r} + \frac{V\cos(\psi - \alpha)}{\sqrt{r^2 - h^2}}(\dot{\psi} - \dot{\alpha}). \tag{3.41}$$

The angular rate of target course variation $\psi(t)$ is determined by lateral acceleration $a_1(t)$ and is equal to

$$\psi(t) = \frac{a_1(t)}{V(t)}. \tag{3.42}$$

Taking into account, that $\dot{V}(t) = a(t)$, and substituting in (3.40) and (3.41) the values of $\dot{r}(t)$, $\dot{\alpha}(t)$ and $\psi(t)$ from (3.38), (3.39) and (3.42), for second derivatives of slope distance and azimuth, the following is valid:

$$\ddot{r}(t) = [a\cos(\psi - \alpha) - a_1\sin(\psi - \alpha)]\sqrt{1 - h^2/r^2} + \\ + \frac{V^2}{r}\left[\sin^2(\psi - \alpha) + \frac{h^2}{r^2}\cos^2(\psi - \alpha)\right], \tag{3.43}$$

$$\ddot{\alpha}(t) = \frac{a\sin(\psi - \alpha) + a_1\cos(\psi - \alpha)}{\sqrt{r^2 - h^2}} - \frac{V^2\sin 2(\psi - \alpha)}{r^2 - h^2}. \tag{3.44}$$

Note that the first addends in (3.43) and (3.44) are stipulated by true acceleration of target (longitudinal and lateral), and addends — by apparent "geometrical" acceleration, correlated with passage from Cartesian to a spherical frame. At a short slope distance, the "geometrical" acceleration can essentially exceed the true acceleration.

If the maximum values of derivatives in expressions (3.43) and (3.44) are of interest, then it is necessary to accept $r(t)=r_{\min}$, $V(t)=V_M, a(t)=a_M$, and $a_1(t)=\pm a_{1M}$. In common cases, the investigation of these expressions by maximum with the account of angular variation $\psi-\alpha$ is fulfilled by numerical methods. However, when accounting that the conditions $a_M \approx a_{lM}$, $h<r_{\min}$, $a_M < V_M^2/r_{\min}$ are usually valid in practice, the approximated analytical solution is possible as then:

$$\max|\ddot{r}(t)| \approx a_{1M}\sqrt{1-\frac{h^2}{r_{\min}^2}+\frac{V_M^2}{r_{\min}}}, \tag{3.45}$$

$$\max|\ddot{\alpha}(t)| \approx \frac{a_M + a_{1M}}{\sqrt{2\left(r_{\min}^2 - h^2\right)}} + \frac{V_M^2}{r_{\min} - h^2}. \tag{3.46}$$

Maximum values of radial and angular accelerations expressed by formulas (3.45) and (3.46), are reached with realization by a target, moving the course maneuver with a maximum path curvature with a maximum velocity in short-range boundary region of a radar operational area.

Maximums of first derivatives for slope distance and azimuth can be easily generated from expressions (3.38) and (3.39) as

$$\max|\dot{r}(t)| = V_M\sqrt{1-h^2/r_{\max}^2}, \tag{3.47}$$

$$\max|\dot{\alpha}(t)| = \frac{V_M}{\sqrt{r_{\min}^2 - h^2}}. \tag{3.48}$$

When determining derivative dispersions (average quadrates), the different solution options are possible. They depend on a set of parameters and on considered sections of target paths, selected for average-out. The greatest values of dispersions are obtained only if the heaviest section of target trajectory for processing at $r(t)\approx r_{\min}$, $V(t)\approx V_M$, $a(t)\approx a_M$, $a_1(t)\approx a_{lM}$ is considered. The average-out is carried out on a set of possible values of an angle $\psi-\alpha=\psi_1$. Let, for example, all values of this angle in an interval from zero up to $2\pi$ be equiprobable. The following equations can be obtained with the account of (3.38), (3.39), (3.43) and (3.44) for average quadrates of first and second derivatives of slope distance and azimuth:

$$\overline{\dot{r}^2} = \frac{V_M^2}{2}\left(1-\frac{h^2}{r_{\max}^2}\right), \tag{3.49}$$

$$\overline{\overline{\dot{\alpha}^2}} = \frac{V_M^2}{2\left(r_{\min}^2 - h^2\right)}, \tag{3.50}$$

$$\overline{\overline{\ddot{r}^2}} = \frac{a_M^2 + a_{1M}^2}{2}\left(1 - \frac{h^2}{r_{\min}^2}\right) + \frac{V_M^4}{8r_{\min}^2}\left(3 + \frac{2h^2}{r_{\min}^2} + \frac{3h^4}{r_{\min}^4}\right), \tag{3.51}$$

$$\overline{\overline{\ddot{\alpha}^2}} = \frac{\left(a_M^2 + a_{1M}^2\right)\left(r_{\min}^2 - h^2 + V_M^4\right)}{2\left(r_{\min}^2 - h^2\right)^2}. \tag{3.52}$$

The double line in expressions (3.49) — (3.52) means average-out by an angle $\psi_1 \in [0,2\pi)$. When deriving their equalities $\overline{\sin^2\psi_1} = \overline{\cos^2\psi} = \overline{\sin^2 2\psi_1} = 1/2$, $\overline{\sin^2\psi_1\cos^2\psi_1} = 1/8$,
$\overline{\sin\psi_1\cos\psi_1} = \overline{\sin^2\psi_1\cos\psi_1} = \overline{\sin\psi_1\cos^2\psi_1} = \overline{\sin^3\psi_1} = \overline{\cos^3\psi_1} = 0$,
$\overline{\sin^4\psi_1} = \overline{\cos^4\psi_1} = 3/8$ were used.

Note that the expectation of radial acceleration is nonzero and according to (3.43) results in

$$\overline{\overline{\ddot{r}}} = \frac{V_M^2}{2r_{\min}}\left(1 + \frac{h^2}{r_{\min}^2}\right). \tag{3.53}$$

Consider a numerical example. Let $V_M = 10^3$ m/sec, $a_M = a_{lM} = 30\ m/\text{sec}^2$, $h = 5\cdot 10^3\ m$, $r_{\min} = 10^4\ m$, $r_{\max} = 5\cdot 10^5\ m$. Then by formulas (3.45) — (3.53) the following is obtained:

$\max \dot{r}(t) = 10^3\ m/\text{sec}$,

$\left(\overline{\overline{\dot{r}^2}}\right)^{1/2} = 707\ m/\text{sec}$,

$\max \ddot{r}(t) = 126\ m/\text{sec}^2$,

$\left(\overline{\overline{\ddot{r}^2}}\right)^{1/2} = 72.6\ m/\text{sec}^2$,

$\overline{\overline{\ddot{r}}} = 62.5\ m/\text{sec}^2$,

$\max \dot{\alpha}(t) = 0.115\ \text{sec}^{-1} = 6.59$ grad/sec,

$\left(\overline{\overline{\dot{\alpha}^2}}\right)^{1/2} = 0.0816\ \text{sec}^{-1} = 4.68$ grad/sec,

$\max \ddot{\alpha}(t) = 0.0182\ \text{sec}^{-2} = 1.04$ grad/sec,

$\left(\overline{\overline{\ddot{\alpha}^2}}\right)^{1/2} = 0.0110\ \text{sec}^{-2} = 0.573$ grad/sec.

Similarly it is possible to analyze maximum values and the dispersion derivative of an elevation angle.

*Questions*

1. How are the spectral-correlation characteristics of stationary casual process derivatives connected to the process itself?
2. Impose on spectral density of casual process the necessary and sufficient conditions for its $N$-times differentiability.
3. Give the relations between dispersions of real-life action derivatives of different orders. What is the physical sense of these relations?
4. How can the spectrum of action width be estimated over dispersions of its derivatives?
5. Is it possible to uniquely recover the spectral-correlation characteristics of actions over dispersions of its derivatives?
6. How are the maximum values in action derivatives of different orders correlated?
7. Why is any experimentally obtained statistical characteristic of casual action only the estimation of a true characteristic?
8. What do the errors in experimental estimations of the correlation function and spectral densities of action depend on?
9. Why is the approximation of experimental estimates for correlation functions and for spectral densities of action by the analytical expressions rather complicated?
10. Give the simple analytical expressions for spectral-correlation characteristics of action.
11. How can the derivatives of action dispersions be experimentally estimated? What does the accuracy of those estimates depend on?
12. What is the singular point of action? How can the dispersions of action derivatives be estimated by the characteristics of its singular points?
13. What kinds of approaches can be used to determine the characteristics of action based on theoretical analysis?
14. What is the typical input signal of servomechanism? In what cases is it expedient to use on the input signal as models of action?
15. Explain the procedure for theoretical estimation of characteristics for derivative coordinates of a radar-tracking target.

# Chapter 4

## Robust Wiener filtering

### §4.1. Optimal filtering at full a priori information

**Problem definition.** Wiener filtering consists of the linear processing of the additive mixture of casual stationary signal and interference. The signal with the least possible r.-m.-s. error must be recovered in processing. In a more general case, it is necessary to estimate not only the signal itself, but also some of its linear functional transform, characterized by the ideal conversing operator $H_{id}$. The diagram of filtering error formation is shown in Fig. 4.1.

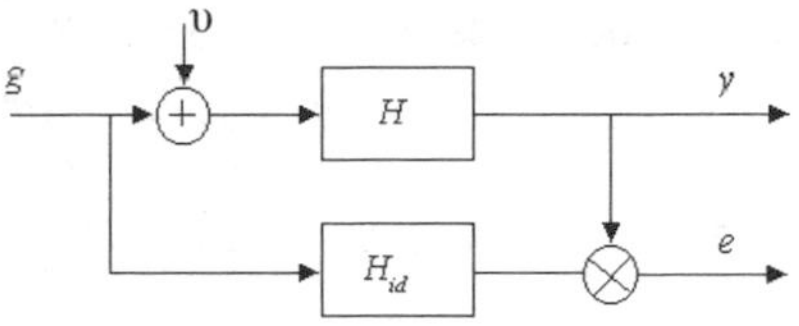


**Fig 4.1**

The requirements needed to minimize the measure of error

$$\overline{e^2(t)} = \overline{[H_{id} g(t) - y(t)]^2} \to \min_H \tag{4.1}$$

only in a stationary state at reference action stationarity determines the stationarity of the Wiener filter.

Taking into account the centrality of error in the linear filter at centered signal and interference, it is possible to consider the dispersion $D_e = \overline{e^2(t)}$, instead of the average square of error in (4.1), and to accept the criterion of optimality $D_e \to \min_H$.

The synthesis problem of the optimal filter with the transfer function $H_0 = \arg\min D_e$ can be set when giving information only about the spectral-correlation characteristics of signal and interference according to the requirements of Wiener filtering linearity. The probability density functions of input actions are not used to investigate the linear filter accuracy within the framework of the correlation theory and play no role in Wiener filtering problem. The knowledge of correlation functions for signal and interference, and also their mutual correlation functions, or knowledge of appropriate spectral densities allow calculation of the precise value of r.-m.-s. error for any linear filter and, therefore, include the full a priori information on input actions in the Wiener filtering problem.

Note with the refusal to accept the filter linearity in problem definition, the probability densities of input actions would be essential. The filter optimal by criterion (4.1), in this case, would really became linear only with Gaussian signal and interference, or with Gaussian interference and the non-Gaussian signal, but at a very high expected filtering accuracy, or with the lack of any information on probability distribution laws of signal and interference [34, 75, 79]. Therefore the Wiener filter can be considered as optimal by criterion (4.1) in the class of all stationary filters (linear and nonlinear) for Gaussian signal and interference.

**Physically unrealizable filter.** Consider the solution for the problem of determination of the frequency transfer function of Wiener filter $H_0(j\omega)=H_0(s)\big|_{s=j\omega}$. Thus with the purpose to calculate the simplification one should consider the fact that signal and interference are mutually uncorrelated. This is mostly typical for practical applications. The knowledge on spectral densities for a signal $S_g(\omega)$ and for interference $S_\upsilon(\omega)$ suffices for a full description of input action properties in frequency domain.

Write the expression for dispersion of the total centered filtering error as an integral of the sum of spectral densities for dynamic error and error from interference:

$$D_e=\frac{1}{2\pi}\int_{-\infty}^{\infty}\left[\left|H_{id}(j\omega)-H(j\omega)\right|^2 S_g(\omega)+\left|H(j\omega)\right|^2 S_\upsilon(\omega)\right]d\omega. \tag{4.2}$$

This expression should turn into the minimum at optimal frequency transfer function of filter $H(j\omega)=H_0(j\omega)$. At first, define it excluding the physical execution of the filter.

Write the frequency transfer functions $H(j\omega)$ and $H_{id}(j\omega)$ using the appropriate gain plot and phase plot:

$$H(j\omega)=A(\omega)e^{j\psi(\omega)}=A(\omega)\cos\psi(\omega)+jA(\omega)\sin\psi(\omega),$$

$$H_{id}(j\omega)=A_{id}(\omega)e^{j\psi_{id}(\omega)}=A_{id}(\omega)\cos\psi_{id}(\omega)+jA_{id}(\omega)\sin\psi_{id}(\omega).$$

Now it is easy to prove the validity of equality:

$$\left|H_{id}(j\omega)-H(j\omega)\right|^2=A_{id}^2(\omega)+A^2(\omega)-2A_{id}(\omega)A(\omega)\cos\left[\psi_{id}(\omega)-\psi(\omega)\right]$$

It also allows converting expression (4.2) as

$$D_e=\frac{1}{2\pi}\int_{-\infty}^{\infty}\Big\{\left[A_{id}^2(\omega)+A^2(\omega)-2A_{id}(\omega)A(\omega)\cos\left[\psi_{id}(\omega)-\psi(\omega)\right]\right]S_g(\omega)+ \\ +A^2(\omega)S_\upsilon(\omega)\Big\}d\omega. \tag{4.3}$$

As the functions $A_{id}(\omega)$, $A(\omega)$, $S_g(\omega)$ and $S_\upsilon(\omega)$, included in integrand, are not negative at any frequency ω, for minimization of the integral in (4.3) it is necessary for the unique negative addend in the integrand to be maximal by its absolute value. It occurs at $\cos[\psi_{id}(\omega)-\psi(\omega)]=1$, i.e. at

$$\psi(\omega)=\psi_{id}(\omega). \tag{4.4}$$

If the condition (4.4) is fulfilled, then expression (4.3) looks like:

$$D_e=\frac{1}{2\pi}\int_{-\infty}^{\infty}F(\omega)\,d\omega, \tag{4.5}$$

and where

$$F(\omega)=\left[A_{id}^2(\omega)+A^2(\omega)-2A_{id}(\omega)A(\omega)\right]S_g(\omega)+A^2(\omega)S_\upsilon(\omega). \tag{4.6}$$

In order to minimize value $e^2(t)$ it is necessary to find such a gain plot of filter $A(\omega)$, at which expression (4.6) accepts the minimal possible value for any frequency ω. It can be obtained, solving the equation $d\Im/dA(\omega)=0$. According to (4.6), the following expression is fulfilled

$$[2A(\omega)-2A_{id}(\omega)]\,S_g(\omega)+2A(\omega)S_\upsilon(\omega)=0,$$

whence the expression for optimal gain plot of the filter is obtained:

$$A_0(\omega)=\frac{S_g(\omega)}{S_g(\omega)+S_v(\omega)}A_{id}(\omega). \tag{4.7}$$

Write the required expression for an optimal frequency transfer function of the filter $H_0(j\omega)=A_0(\omega)e^{j\psi_0(\omega)}$ having combined expressions (4.4) and (4.7):

$$H_0(j\omega)=\frac{S_g(\omega)}{S_g(\omega)+S_\upsilon(\omega)}H_{id}(j\omega). \tag{4.8}$$

The physical sense of expression (4.8) is quite clear. In a frequency band, where $S_g(\omega)>>S_\upsilon(\omega)$, i.e. the power of signal essentially exceeds the power of interference, $H_0(j\omega)\approx H_{id}(j\omega)$ is fulfilled. It points out the necessity to process the signal according to an ideal conversing operator excluding the influence of interference. On the contrary, at $S_g(\omega)<<S_\upsilon(\omega)$ the expression $H_0(j\omega)<<H_{id}(j\omega)$ is obtained. This points out the necessity to suppress the interference anyway, paying no attention to the ideal conversing operator.

The substitution of expression (4.8) into (4.2) gives the minimum possible value of filtering error dispersion

$$D_{e\min}^{nr}=\frac{1}{2\pi}\int_{-\infty}^{\infty}\frac{S_g(\omega)S_\upsilon(\omega)}{S_g(\omega)+S_\upsilon(\omega)}\left|H_{id}(j\omega)\right|^2d\omega. \tag{4.9}$$

Generally a filter with frequency transfer function (4.8) is physically unrealizable. As the sum $S_g(\omega)+S_\upsilon(\omega)$ is a real non-negative frequency function, it can be decomposed in the complex conjugate multiplicands. One of the multiplicands has poles in the lower half-plane of the complex plane $\omega$ (or in the right half-plane of complex plane $s = j\omega$). The pulse response of the filter, found for such a frequency transfer function, exists at negative time intervals, i.e. up to the first moment of the action application. This means it violates the causality principle.

The synthesized filter is physically unrealizable because in actual dynamic systems there is a correlation between the gain plot $A(\omega)$ and phase plot $\psi(\omega)$, which was not taken into consideration in formula (4.8). Since equations (4.4) and (4.7) are generally incompatible, it is impossible to find one solution corresponding to both of these equations.

Note, if the particular case of filtering — the interpolation problem, is considered and recovery of the signal is required, which has been observed in a mixture with indefinitely long interference, and there is an entry for all values of this mixture in an interval $t \in (-\infty, \infty)$ on some hypothetical data carrier (for example, on indefinitely long magnetic tape), then all previous and following values of mixture are known in relationship to the arbitrary instant within the indicated interval. Therefore, the existence of nonzero values for pulse response of the filter-interpolator at $t<0$ formally is not an obstacle when realizing the filter with frequency transfer function (4.8). However, the survey of "future" values of mixture in the filter-interpolator is possible only in a finite time interval. In this case the dispersion of error is greater, than the dispersion calculated by formula (4.9), but still can be less, than in a physically realizable smoothing filter, where the following values of signal and interferences are generally inaccessible for observation.

**Physically realizable filter.** As the filter with frequency transfer function (4.8) in common cases cannot be realized, the problem in determining the realizable filter closest to it on dynamic properties is arising. Such filters should ensure the least mean square of filtering error among all physically realizable filters, i.e. must satisfy the criterion of (4.1). However, the definite value of error average square in the optimal realizable filter is greater than in the non-realizable one, because the additional requirements for zero values of pulse response at negative time (as well as any other additional requirement to dynamic properties of filter) are necessarily reflected in an accessible level of fundamental quality factors.

The synthesis problem of the physically optimum realizable filter can be solved in a time or frequency domain, and the results turn out to be identical. In the first case it is necessary to immediately research the Wiener-Hopf [9, 15] integral equation, written concerning the filter pulse response. In the second, more characteristic case of stationary filter synthesis, frequency transfer function (4.8) must be researched in order to recover the physically realizable part where the poles are located in an upper half-plane of complex plane $\omega$.

Usually, the methods suggested by G. Bode and K. Shannon are used. At first the denominator of expression (4.8) must be decomposed to the complex conjugate multiplicands, i.e. fulfill the factorization operation:

$$S_g(\omega)+S_\upsilon(\omega)=|\Psi(j\omega)|^2=\Psi(j\omega)\Psi(-j\omega). \tag{4.10}$$

Here all zeros and poles of function $\Psi(j\omega)$ lie in an upper half plane of complex plane $\omega$, and all zeros and poles of function $\Psi(-j\omega)$ lie in the lower one.

Then frequency transfer function (4.8) should be separated on realizable and non-realizable addends

$$H_0(j\omega)=\frac{1}{\Psi(j\omega)}\left[\frac{S_g(\omega)H_{id}(j\omega)}{\Psi(-j\omega)}\right]_+ +\frac{1}{\Psi(j\omega)}\left[\frac{S_g(\omega)H_{id}(j\omega)}{\Psi(-j\omega)}\right]_- . \tag{4.11}$$

Realizable part marked by sign plus, is appeared as the sum of elementary addends with poles lying in an upper half-plane. The non-realizable part is marked in (4.11) by sign minus and has poles in lower half-plane.

Truncating the non-realizable part out of (4.11) with the frequency transfer function of the physically realizable filter, optimal by criterion (4.1), the following expression is obtained

$$H_0(j\omega)=\frac{1}{\Psi(j\omega)}\left[\frac{S_g(\omega)H_{id}(j\omega)}{\Psi(-j\omega)}\right]_+ . \tag{4.12}$$

The dispersion of error in the optimal filter is calculated by formula (4.2) with the substitution $H(j\omega)=H_0(j\omega)$ from (4.12).

It is possible to show that at presence of cross correlation for signal and interference characterized by nonzero cross-correlation spectral densities $S_{g\upsilon}(\omega)$ and $S_{\upsilon g}(\omega)$,it is necessary to use the expression

$$H_0(j\omega)=\frac{1}{\Psi(j\omega)}\left[\frac{S_g(\omega)+S_{gv}(\omega)}{\Psi(-j\omega)}H_{id}(j\omega)\right]_+ , \tag{4.13}$$

instead of (4.12), and in comparison with (4.10),

$$\Psi(j\omega)\Psi(-j\omega)=S_g(\omega)+S_\upsilon(\omega)+S_{g\upsilon}(\omega)+S_{\upsilon g}(\omega).$$

It is necessary to note, that the frequency transfer function of the optimal realizable filter is determined by expression (4.12) or by more common expression (4.13) only at spectral densities of actions described by fractional rational functions of frequency square $\omega^2$ (in this case $\Psi(j\omega)$ is the fractional rational function of argument $j\omega$). If the other mathematical models of spectral density are preset, they should be approximated by fractional rational functions of frequency square. Otherwise, it is unable to find an analytical solution for problems of optimal filter synthesis.

The loss of the optimal realizable filter compared to the optimal nonexecutable one by the average square (dispersion) of error is of interest. The common analytical estimation of such loss is complicated. However, it is possible to make sure on definite examples that up to a hundred percent (more often 50-100%) of units depending on kinds of signal and interference spectral densities can be reached.

**Example 4.1.** Research realizable and non-realizable optimal smoothing ($H_{id}(j\omega)=1$) filters at spectral density of signal, as a kind (3.36)

$$S_g(\omega)=\frac{2D_0T_1}{1+\omega^2T_1^2},$$

evenly spectral density of interference is $S_\upsilon(\omega)=S_\upsilon$ and at absence of cross correlation between signal and interference.

It is possible to define the frequency transfer function for physically non-realizable optimal filter by the formula (4.8) as:

$$H_0(j\omega)=\frac{S_g(\omega)}{S_g(\omega)+S_\upsilon}=\frac{2D_0T_1}{2D_0T_1+S_\upsilon+\omega^2S_\upsilon T_1}=\frac{\rho_0}{\rho_0+1+\omega^2T_1^2}=\frac{k_0}{1+\omega^2T_*^2},$$

where $\rho_0=S_g(0)/S_\upsilon=2D_0T_1/S_\upsilon$, $k_0=\rho_0/(\rho_0+1)$, $T_*=T_1/\sqrt{\rho_0+1}$.

The theoretical minimum of error dispersion defined by formula (4.9) corresponds to $H_0(j\omega)$:

$$D_{e\min}^{nr}=\frac{1}{2\pi}\int_{-\infty}^{\infty}\frac{k_0S_\upsilon}{1+\omega^2T_*^2}d\omega=\frac{k_0S_\upsilon}{2T_*}=\frac{D_0}{\sqrt{\rho_0+1}}.$$

With the determination of the frequency transfer function of physically realizable filters formulas (4.10) — (4.12) are used.

The factorization of expression

$$S_g(\omega)+S_\upsilon=\frac{S_g(0)}{1+\omega^2T_1^2}+S_\upsilon=\left[S_g(0)+S_\upsilon\right]\frac{1+\omega^2T_*^2}{1+\omega^2T_1^2}$$

allows to accept the following:

$$\Psi(j\omega)=\sqrt{S_g(0)+S_\upsilon}\frac{1+j\omega T_*}{1+j\omega T_1},\quad \Psi(-j\omega)=\sqrt{S_g(0)+S_\upsilon}\frac{1-j\omega T_*}{1-j\omega T_1},$$

whence

$$\frac{S_g(\omega)}{\Psi(-j\omega)}=\frac{S_g(0)}{\sqrt{S_g(0)+S_\upsilon}(1+j\omega T_1)(1-j\omega T_*)}=$$
$$=\frac{S_g(0)}{\sqrt{S_g(0)+S_\upsilon}}\left(\frac{T_1}{T_1+T_*}\frac{1}{1+j\omega T_1}+\frac{T_*}{T_1+T_*}\frac{1}{1-j\omega T_*}\right).$$

With formula (4.12) the following expression is obtained truncating the second addend, which has a pole in the lower half-plane, out of the right hand side in the last expression:

$$H_0(j\omega)=\frac{S_g(0)}{S_g(0)+S_\upsilon}\frac{T_1}{T_1+T_*}\frac{1}{1+j\omega T_*}=\frac{k_0^*}{1+j\omega T_*},$$

where $k_0^*=\dfrac{\rho_0}{\sqrt{\rho_0+1}\left(\sqrt{\rho_0+1}+1\right)}=\dfrac{k_0\sqrt{\rho_0+1}}{\sqrt{\rho_0+1}+1}=1-\dfrac{1}{\sqrt{\rho_0+1}}$.

With such an optimal realizable frequency transfer function the filtering error dispersion makes

$$D_{e\min}^r=\frac{1}{2\pi}\int_{-\infty}^{\infty}\left\{\left|1-\frac{k_0^*}{(1+j\omega T_*)}\right|^2\frac{S_g(0)}{1+\omega^2T_1^2}+\left|\frac{k_0^*}{(1+j\omega T_*)}\right|^2 S_\upsilon\right\}d\omega=$$

$$=D_0\frac{T_1\left(1-k_0^*\right)+T_*}{T_1+T_*}+\frac{\left(k_0^*\right)^2N}{2T_*}=\frac{D_0}{\sqrt{\rho_0+1}}\left[1+\frac{\left(\sqrt{\rho_0+1}-1\right)^2}{\rho_0}\right].$$

For an estimation of loss of the realizable filter compared to a not realizable one on dispersion of error, the coefficient

$$\eta_r=\frac{D_{e\min}^r}{D_{e\min}^{nr}}=1+\frac{\left(\sqrt{\rho_0+1}-1\right)^2}{\rho_0}-\frac{2(\rho_0+1)}{\rho_0+1+\sqrt{\rho_0+1}}$$

is entered. This coefficient is the function of unique parameter $\rho_0=S_g(0)/S_\upsilon$ .

The graph of loss of the realizable filter compared to the non-realizable one by dispersion of error $\eta_r(\rho_0)$ is shown in Fig. 4.2 (curve 1).

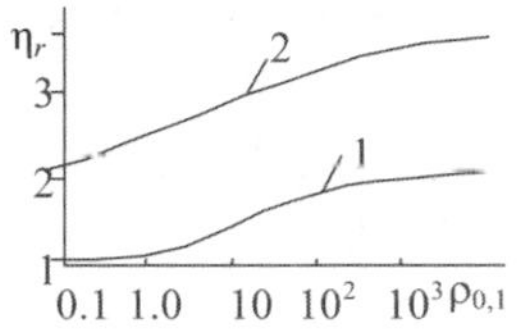


Fig. 4.2.

**Example 4.2.** It is necessary to repeat the investigations, which have been made in the previous example, with a signal spectral density of a kind (3.37)

$$S_g(\omega)=\frac{2D_1T_1}{\omega^2\left(1+\omega^2T_1^2\right)}.$$

The expression for the optimal physically non-realizable filter looks like:

$$H_0(j\omega)=\frac{2D_1T_1}{2D_1T_1+\omega^2S_\upsilon+\omega^4S_\upsilon T_1^2}=\frac{1}{1+\omega^2T_1^2/\rho_1+\omega^4T_1^4/\rho_1},$$

$$D_{e\min}^{nr}=\frac{1}{2\pi}\int_{-\infty}^{\infty}\frac{2D_1T_1S_\upsilon}{2D_1T_1+\omega^2S_\upsilon+\omega^4S_\upsilon T_1^2}d\omega=\frac{S_\upsilon}{T_1}\frac{\sqrt{\rho_1}}{2\sqrt{2\sqrt{\rho_1}+1}},$$

where $\rho_1 = S_{\dot{g}}(0)T_1^2/S_\upsilon = 2D_1T_1^3/S_\upsilon$, $S_{\dot{g}}(\omega) = \omega^2 S_g(\omega)$ is a signal $g(t)$ first derivative spectral density.

With the synthesis of the optimal physically realizable filter, the functions are defined as:

$$S_g(\omega) + S_\upsilon = \frac{2D_1T_1 + \omega^2 S_\upsilon + \omega^4 S_\upsilon T_1^2}{\omega^2(1+\omega^2T_1^2)} = 2D_1T_1\frac{(1+\omega^2\tau_1^2)(1+\omega^2\tau_2^2)}{\omega^2(1+\omega^2T_1^2)},$$

$$\Psi(j\omega) = \sqrt{2D_1T_1}\frac{(1+j\omega\tau_1)(1+j\omega\tau_2)}{j\omega\,(1+j\omega T_1)},$$

where $\tau_1\tau_2 = \frac{T_1^2}{\sqrt{\rho_1}}$, $\tau_1 + \tau_2 = \frac{T_1}{\sqrt{\rho_1}}\sqrt{2\sqrt{\rho_1}+1}$.

After the transformations according to expression (4.12)

$$H_0(j\omega) = \frac{1 + j\omega\tau_{eq}}{(1+j\omega\tau_1)(1+j\omega\tau_2)},$$

is acceptable,

where $\tau_{eq} = \tau_1 + \tau_2 - \frac{\tau_1\tau_2}{T_1} = \frac{T_1}{\sqrt{\rho_1}}\left(\sqrt{2\sqrt{\rho_1}+1} - 1\right)$.

The dispersion of the optimal physically realizable filter makes

$$D^r_{e\min} = \frac{1}{2\pi}\int_{-\infty}^{\infty}\left\{\left|1 - \frac{1+j\omega\tau_{eq}}{(1+j\omega\tau_1)(1+j\omega\tau_2)}\right|^2 \frac{2D_1T_1}{\omega^2(1+\omega^2T_1^2)} + \right.$$

$$\left. + \left|\frac{1+j\omega\tau_{eq}}{(1+j\omega\tau_1)(1+j\omega\tau_2)}\right|^2 S_\upsilon\right\}d\omega =$$

$$= \frac{D_1T_1^3}{\rho_1(\tau_1+\tau_2)} + \frac{S_\upsilon(\tau_{eq}^2 + \tau_1\tau_2)}{2\tau_1\tau_2(\tau_1+\tau_2)} = \frac{S_\upsilon}{T_1}\left(\sqrt{2\sqrt{\rho_1}+1} - 1\right).$$

The loss of the physically realizable filter compared to the non-realizable one by dispersion of error is characterized by the coefficient

$$\eta_r = \frac{D^r_{e\min}}{D^{nr}_{e\min}} = \frac{2}{\sqrt{\rho_1}}\left(\sqrt{2\sqrt{\rho_1}+1} - 1\right)\sqrt{2\sqrt{\rho_1}+1} = \frac{4\sqrt{2\sqrt{\rho_1}+1}}{\sqrt{2\sqrt{\rho_1}+1}+1},$$

depending on the unique parameter $\rho_1$.

The graph of $\eta_r(\rho_1)$ is shown in Fig. 4.2 (curve 2).

## §4.2. Minimax robust filtering with the band model of spectral indeterminacy

**The problem definition and preliminary notes.** The assigning of upper and lower boundaries for spectral densities of signal and (or) interference as some frequency function is the most simple and obvious form of giving the nonparametric classes of action spectral densities in a robust filtering problem. Concretizing the common mathematical expression for band model (1.13) with reference to spectral densities of signal $S_g(\omega)$ and interference $S_\upsilon(\omega)$, the admissible classes of these spectral densities are allocated by inequalities

$$S_{gl}(\omega) \le S_g(\omega) \le S_{gu}(\omega), \tag{4.14}$$

$$S_{\upsilon l}(\omega) \le S_\upsilon(\omega) \le S_{\upsilon u}(\omega). \tag{4.15}$$

Here $S_{gl}(\omega)$, $S_{gu}(\omega)$, $S_{\upsilon l}(\omega)$, $S_{\upsilon u}(\omega)$ are the known real non-negative functions. Consider, that they set the authentic boundaries for possible values of centered action spectral densities and are defined at the experimental or theoretical analysis of operation conditions in synthesized dynamic systems with all uncertain factors taken into account.

The dispersion of summarized centered errors will be considered as a basic characteristic of signal filtering accuracy from its additive mixture with interference. This error in a linear system at the fixed ideal converting operator (for example, at $H_{id} = 1$) according to expression (4.2), it is the functional of $S_g(\omega)$, $S_\upsilon(\omega)$ and $H(j\omega)$ at i.e. $D_e = D_e(S_g, S_\upsilon, H)$.

If the frequency transfer function $H(j\omega)$ is optimal for a pair of spectral densities $\{S_g(\omega), S_\upsilon(\omega)\}$, i.e. $H(j\omega) = \arg\min_H D_e(S_g, S_\upsilon, H) = H_0(j\omega)$, the dispersion of an error achieves its minimum $D_e(S_g, S_\upsilon, H_0) = D_{e\min}$.

As the spectral densities of signal and interference are not particularly defined, the value $D_{e\min}$ cannot be a precision factor of robust systems. However, the knowledge of its possible range variation and, especially, upper bounds or upper-bound estimates would allow consideration of the accuracy of systems quite objectively.

If admissible classes of action spectral densities are only restricted by inequalities (4.14) and (4.15), the problem to determine the range of values $D_e(S_g, S_\upsilon, H)$ variation can be solved simply. Actually, at any system transfer function the dispersion of dynamic error and dispersion of error from interference, according to (4.2), are increasing with the increment of values of signal and interference spectral densities, accordingly. Therefore, the upper bound of system total error dispersion is reached at $S_g(\omega) = S_{gu}(\omega)$ and at $S_\upsilon(\omega) = S_{\upsilon u}(\omega)$. This corresponds to the most unfavorable filtering condition. The lower bound in this case is reached at $S_g(\omega) = S_{gl}(\omega)$ and $S_\upsilon(\omega) = S_{\upsilon l}(\omega)$.

During the selection of a system transfer function, optimal for spectral densities $\{S_{gu}, S_{\upsilon u}\}$, i.e. at $H(j\omega) = H_0^u(j\omega) = \underset{H}{\operatorname{argmin}} D_e(S_{gu}, S_{\upsilon u}, H)$, the minimax system where the upper bound of error dispersion accepts the minimum value $D_e(S_{gu}, S_{\upsilon u}, H_0^u)$, is obtained.

If this value is acceptable for filtering accuracy requirements, then the system is a minimax robust one with the transfer function $H(j\omega) = H_0^u(j\omega)$. The range of possible values of filtering error dispersion in such systems at spectral densities of actions corresponding to conditions (4.14) and (4.15), is defined by inequalities:

$$D_e(S_{gu} S_{\upsilon l}, H_R) \le D_e(S_g, S_\upsilon, H_R) \le D_e(S_{gu}, S_{\upsilon u}, H_R).$$

Note, if the system were adaptive, then the lower boundary of error dispersion could be decreased down to value $D_e(S_{gl}, S_{\upsilon l}, H_0^l) = \min_H D_e(S_{gl}, S_{\upsilon l}, H)$. However, the upper bound would remain the same.

The problem with research of robust system accuracy at additional restrictions on spectral densities of action, except for inequalities (4.14) and (4.15), is considerably more complicated. For example, the restrictions on dispersions of signal and interference values can be preset as:

$$\frac{1}{\pi}\int_0^\infty S_g(\omega)\,d\omega \le D_0,\quad \frac{1}{\pi}\int_0^\infty S_\upsilon(\omega)\,d\omega \le D_\upsilon \tag{4.16}$$

or on the generalized moments of these spectral densities such as (2.21).

In order to avoid the contradiction between inequalities (4.14), (4.15) and (4.16) the magnitudes $D_0$ and $D_\upsilon$ must satisfy the conditions:

$$\begin{aligned} &\frac{1}{\pi}\int_0^\infty S_{gl}(\omega)\,d\omega < D_0 < \frac{1}{\pi}\int_0^\infty S_{gu}(\omega)\,d\omega \\ &\frac{1}{\pi}\int_0^\infty S_{\upsilon l}(\omega)\,d\omega < D_\upsilon < \frac{1}{\pi}\int_0^\infty S_{\upsilon u}(\omega)\,d\omega \end{aligned}, \tag{4.17}$$

The requirements of (4.17) do not allow the acceptance of the spectral densities of actions, which have the maximum possible values for all frequencies, and force us to search for the most unfavorable spectral densities of signal $S_{gmu}(\omega)$ and interference $S_{\upsilon mu}(\omega)$, distinct from functions $S_{gu}(\omega)$ and $S_{\upsilon u}(\omega)$. Note, in the case of strict solution, as a rule, it is necessary to deal with filters, which may not fulfill the condition of physical execution. The reason for this is that in the result of such a solution, the functions $S_{gmu}(\omega)$ and $S_{\upsilon mu}(\omega)$ usually have the first kind discontinuity. So, they cannot be described by fractional rational expressions.

The problem of determination is set with the most unfavorable spectral densities of actions and synthesis of the minimax robust filter. The spectral density classes are preset by inequalities (4.14) — (4.16). Consider an intuitive qualitative solution of this problem first. This allows for the understanding in a physical sense of research regularities. Then go on to strict mathematical solutions of the problem for the common case.

**Heuristic solutions for problems of filter synthesis.** Suppose the expression (4.9) for dispersion of total error in the optimal filter as

$$D_{e\min}^{nr}=\frac{1}{\pi}\int_0^\infty\frac{\left|\dot H_{id}(j\omega)\right|^2 d\omega}{\left[S_g(\omega)\right]^{-1}+\left[S_\upsilon(\omega)\right]^{-1}}\approx\frac{1}{\pi}\int_0^\infty\min\{S_g(\omega),S_\upsilon(\omega)\}\left|H_{id}(j\omega)\right|^2 d\omega. \qquad (4.18)$$

(4.18) is followed by the fact, that in optimal systems in each frequency range the dominant influence on error dispersion provides the action, in which the spectral density has the smaller values in this frequency range. Therefore, the fundamental tendency to form the most unfavorable spectral densities of signal and interference should consist of increasing the small values of spectral densities up to an upper bound in band model and of decreasing the large values down to the lower bound. The integrated restrictions (4.16) on actions dispersions must not be broken. Then the spectral densities of signal and interference will in some way be leveled and approached in the shape. To recover the signal by the spectral criterion will ensure the most unfavorable conditions.

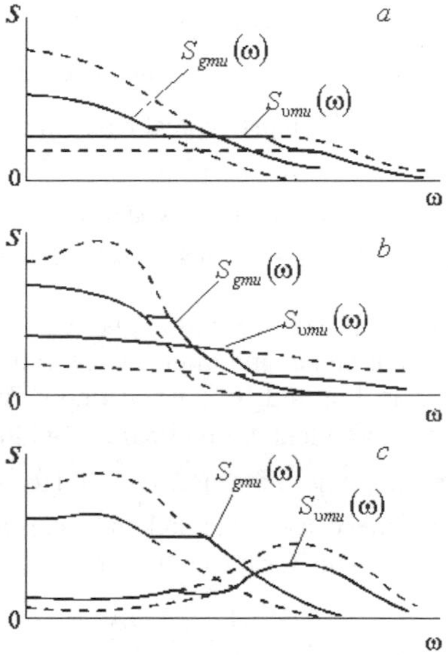


Fig.4.3

The most unfavorable spectral density of signals within the framework of appropriate band models must pass the lower boundary on low frequencies, and the upper bound on high frequencies. This occurs because in synthesis problems of automatic control systems, the standard situation is the problem where the signal has low frequency spectrum, and the interference is wide-ranged. On the contrary, the spectral density of interference must pass the upper bound on the low frequencies, and lower — on high frequencies.

The passage from when one of bounds to another must be sharp enough to be able to precisely catch the separation of dominant influence areas of functions $S_g(\omega)$ and $S_\upsilon(\omega)$ on value $D_{e\min}^{nr}$ in expression (4.18). However, it is also important to ensure the maximum similarity of signal and interference spectrums. In Fig. 4.3 three characteristic examples of the most unfavorable spectral densities of signal and interference are given due to the regularities mentioned above.

Practically, a quite acceptable law of passage from one boundary to another can be a vertical or horizontal step. Its elementary form simplifies the calculations. The location of the step on an abscissa axis can be defined from the condition for maximum admissible dispersion of action (4.16). Thus, spectral densities of signal and interference are researched without the correlation between them. The step can be vertical (discontinuity of the first kind), if it does not cause the additional extremums of spectral density that could cause the "mixing" of the areas of influence functions $S_g(\omega)$ and $S_\upsilon(\omega)$ in expression (4.18). Otherwise, the horizontal step is used, i.e. the fixing of spectral density level at the passage from one of its bounds to another.

If the midfrequency section, where graphs of bounds for spectral densities of signal and interference are intersected, has the decreasing bounds for function $S_g(\omega)$ and non-increasing ones for function $S_\upsilon(\omega)$ (Fig. 4.3*a*, *b*), then it is necessary to accept the step of function $S_{g\ mu}(\omega)$ horizontally, and a step of function $S_{\upsilon\ mu}(\omega)$ — vertically. This is the most typical case, but the others are also possible (Fig. 4.3*c*).

After determination the most unfavorable spectral densities of signal and interference, the frequency transfer function for the minimax robust filter can be found by formula (4.8).

**Example 4.3.** Using the heuristic method explained above, define the most unfavorable spectral densities of signal and interference with boundary functions of a triangular kind, as shown in Fig. 4.3 by dashed lines. The graphs for upper and lower boundary functions have identical inclination, width $\nu_u$, $\nu_l$ (for signal), $\lambda_u$, $\lambda_l$ (for interference), and maximum ordinates $A_u$, $A_l$ (for signal), $N_u$, $N_l$ (for interference). The preset dispersion values of signal and interference lie precisely in the middle intervals of possible values for these dispersions, i.e.

$$D_0 = \frac{A_u \nu_u + A_l \nu_l}{4\pi},$$

$$D_v = \frac{N_u \lambda_u + N_l \lambda_l}{4\pi}.$$

The type of boundary functions allows suggesting the standard law of most unfavorable spectral density variation through the step passage from one bound to another. This step is horizontal for signal and vertical for interference. The appropriate graphs of functions $S_{g\,mu}(\omega)$ and $S_{\upsilon\,mu}(\omega)$ are shown in Fig. 4.4 by solid lines.

For determination of the ordinate of horizontal step $A_{st}$ and abscissa of vertical step $\omega_{st}$, the following equations are obtained:

$$\frac{A_l \nu_l}{2} + \frac{2(\nu_u - \nu_l) - (A_u/\nu_u - A_l/\nu_l) A_{st}}{2} A_{st} = \frac{A_u \nu_u - A_l \nu_l}{4},$$

$$\frac{N_l \lambda_l}{2} + \frac{2(N_u - N_l) - (N_u/\lambda_u - N_l/\lambda_l) \omega_{st}}{2} \omega_{st} = \frac{N_u \lambda_u - N_l \lambda_l}{4}.$$

Here the left hand side maintains the squares of the figures, marked in Fig. 4.4 by solid lines, and the right hand side — the given values $\pi D_0$ and $\pi D_\upsilon$.

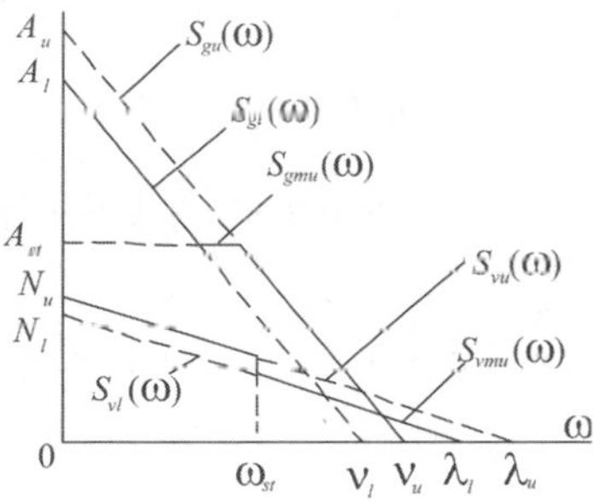


Fig. 4.4

As in $A_u/\nu_u = A_l/\nu_l$ and $N_u/\lambda_u = N_l/\lambda_l$, the solution of the written above equations equals:

$$A_{st} = \frac{A_u/\nu_u - A_l/\nu_l}{4(\nu_u - \nu_l)}, \quad \omega_{st} = \frac{N_u/\lambda_u - N_l/\lambda_l}{4(N_u - N_l)}.$$

Note, that the obtained expressions can be converted to

$$A_{st} \approx \frac{A_l}{2} \frac{1 + \nu_u/\nu_l}{2}, \quad \omega_{st} \approx \frac{\lambda_l}{2} \frac{1 + N_u/N_l}{2}.$$

From here at $\nu_u/\nu_l \approx 1$ and $N_u/N_l \approx 1$ the approximated formulas $A_{st} = A_l/2$ and $\omega_{st} = \lambda_l/2$ can be easily found.

## §4.3. Strict determination of the most unfavorable spectral densities of actions with band models of spectral indeterminacy

**Partition of frequencies axis on intervals, defined by the kind of boundary functions.** Consider the common analytical solution for the problem of determining the pair of the most unfavorable spectral densities $\{S_{gmu}(\omega), S_{\upsilon mu}(\omega)\}$, at which the dispersion of filtering error in a system with an optimal physically non-realizable transfer function of a kind (4.8) is maximized. The classes of admissible spectral densities of actions in robust systems correspond to the band model of spectral indeterminacy and are described by expressions (4.14) — (4.16).

The solution consists of a partition of frequency axis on defined intervals depending on the relation of values in boundary functions $S_{gl}(\omega)$, $S_{gu}(\omega)$, $S_{\upsilon l}(\omega)$, $S_{\upsilon u}(\omega)$ and determination of a kind of the most unfavorable spectral densities for each of indicated frequency intervals. Two following variants of such partitions can be required [43].

Four frequency areas $\alpha_1$, $\alpha_2$, $\beta_1$ and $\beta_2$ with an indication of inequalities fulfillment, written with the use of some positive finite coefficients $k_g$ and $k_\upsilon$ are determined in the first variant:

$$\omega \in \alpha_1 \text{, at } S_{gl}(\omega) < k_g S_{\upsilon l}(\omega) \le S_{gu}(\omega); \tag{4.19}$$

$$\omega \in \alpha_2 \text{, at } S_{gl}(\omega) \le S_{gu}(\omega) < k_g S_{\upsilon l}(\omega); \tag{4.20}$$

$$\omega \in \beta_1 \text{, at } k_\upsilon S_{\upsilon l}(\omega) \le S_{gl}(\omega) < k_\upsilon S_{\upsilon u}(\omega); \tag{4.21}$$

$$\omega \in \beta_2 \text{, at } k_\upsilon S_{\upsilon l}(\omega) \le k_\upsilon S_{\upsilon u}(\omega) \le S_{gl}(\omega). \tag{4.22}$$

Two combined frequency areas $\alpha = \alpha_1 \cup \alpha_2$ and $\beta = \beta_1 \cup \beta_2$ are also determined. The following inequalities are fulfilled for them

$$S_{gl}(\omega) < k_g S_{\upsilon l}(\omega) \text{ at } \omega \in \alpha, \tag{4.23}$$

$$k_\upsilon S_{\upsilon l}(\omega) \le S_{gl}(\omega) \text{ at } \omega \in \beta. \tag{4.24}$$

Note that at $k_g \le k_\upsilon$, the areas $\alpha$ and $\beta$ cannot be mutually intersected.

It is possible to graphically determine the areas of $\alpha_1$, $\alpha_2$, $\beta_1$ and $\beta_2$ by investigating the shape of graphs for boundary function relations $S_{gu}(\omega)/S_{\upsilon l}(\omega)$, $S_{gl}(\omega)/S_{\upsilon l}(\omega)$ and $S_{gl}(\omega)/S_{\upsilon u}(\omega)$. This is shown in Fig. 4.5.

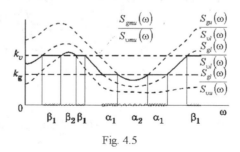

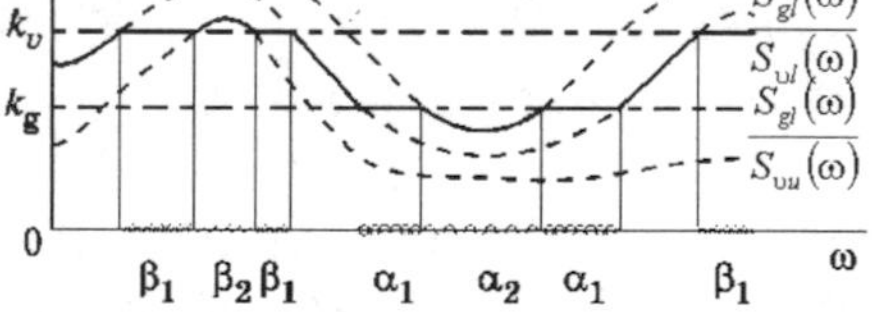


Fig. 4.5

The second variant for partition of the frequency axis on certain intervals differs from the first by the kind of considered inequalities. Four areas of frequency $\alpha_1$, $\alpha_2$, $\beta_1$ and $\beta_2$ are represented there. The position and sizes depend on values of some positive finite coefficients $l_g$ and $l_\upsilon$:

$$\omega \in a_1 \text{, at } l_\upsilon S_{\upsilon u}(\omega) \geq S_{gu}(\omega) > l_\upsilon S_{\upsilon l}(\omega), \tag{4.25}$$

$$\omega \in a_2 \text{, at } l_\upsilon S_{\upsilon l}(\omega) \geq l_\upsilon S_{gu}(\omega) \geq S_{gu}(\omega), \tag{4.26}$$

$$\omega \in b_1 \text{, at } S_{gu}(\omega) > l_g S_{\upsilon u}(\omega) \geq S_{gl}(\omega), \tag{4.27}$$

$$\omega \in b_2 \text{, at } S_{gu}(\omega) \geq S_{gl}(\omega) > l_g S_{\upsilon u}(\omega). \tag{4.28}$$

The position of areas $\alpha_1$, $\alpha_2$, $\beta_1$ and $\beta_2$ on the frequency axis is shown in Fig. 4.6. As well as in the first variant, it is possible to consider the integrated areas

$$a = a_1 \cup a_2,$$

$$b = b_1 \cup b_2,$$

The following inequalities are fulfilled according to them

$$l_\upsilon S_{\upsilon u}(\omega) > S_{gu}(\omega) \text{ at } \omega \in a, \tag{4.29}$$

$$S_{gu}(\omega) > l_g S_{\upsilon u}(\omega) \text{ at } \omega \in b. \tag{4.30}$$

It is natural that each of the inserted areas on the frequency axis depends on one of the coefficients $k_g$, $k_\upsilon$, $l_g$ or $l_\upsilon$, for example, $\alpha_1 = \alpha_1(k_g)$.

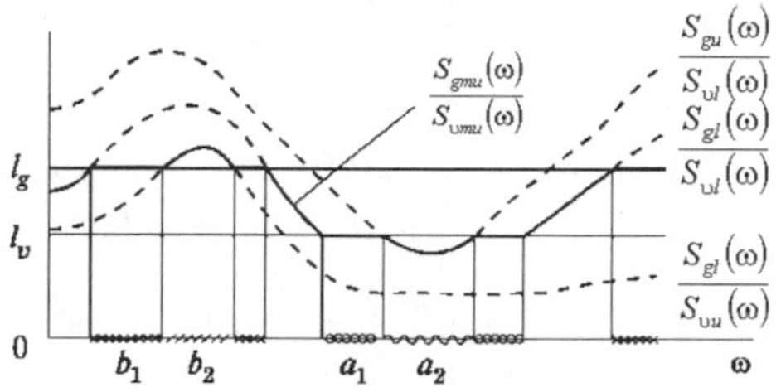


Fig. 4.6

The particular values of these coefficients are found from the solution of the equations constructed on account of the requirements for the restriction of signal and interference dispersions. At the first variant of partition, these equations look like

$$k_g \int_{\alpha_1(k_g)} S_{\upsilon l}(\omega)\,d\omega + \int_{\alpha_2(k_g)} S_{gu}(\omega)\,d\omega + \int_{\bar{\alpha}(k_g)} S_{gl}(\omega)\,d\omega = \pi D_0, \tag{4.31}$$

$$\frac{1}{k_\upsilon} \int_{\beta_1(k_\upsilon)} S_{gl}(\omega)\,d\omega + \int_{\beta_2(k_\upsilon)} S_{\upsilon u}(\omega)\,d\omega + \int_{\bar{\beta}(k_\upsilon)} S_{\upsilon l}(\omega)\,d\omega = \pi D_\upsilon, \tag{4.32}$$

$$l_g \int_{b_1(l_g)} S_{\upsilon u}(\omega)\,d\omega + \int_{b_2(l_g)} S_{gl}(\omega)\,d\omega + \int_{\bar{b}(l_g)} S_{gu}(\omega)\,d\omega = \pi D_0, \tag{4.33}$$

$$\frac{1}{l_\upsilon} \int_{a_1(l_\upsilon)} S_{gu}(\omega)\,d\omega + \int_{a_2(l_\upsilon)} S_{\upsilon l}(\omega)\,d\omega + \int_{\bar{a}(l_\upsilon)} S_{\upsilon u}(\omega)\,d\omega = \pi D_\upsilon. \tag{4.34}$$

The areas obtained by elimination of areas α, β, *a* and *b*, defined by expressions (4.25) — (4.28), from frequency axis $\omega \in [0,\infty)$ accordingly, are designated here through $\overline{\alpha}$, $\overline{\beta}$, $\overline{a}$ and $\overline{b}$.

**Theorem of the most unfavorable spectral densities of actions**. Now it is possible to consider a kind of the most unfavorable spectral densities of signal $S_{g\,mu}(\omega)$ and interference $S_{\upsilon\,mu}(\omega)$, pertinent to their admissible classes marked by relations (4.14) — (4.16), which deliver the maximum to function $D_e(S_g, S_\upsilon, H_0)$. This has been done on the basis of following theorem.

Theorem 4.1. The frequency transfer function for physically non-realizable minimax robust filters is determined by formula (4.8) as optimum for a pair of spectral densities of signal and interference $\{S_{gmu}(\omega), S_{\upsilon mu}(\omega)\}$, which is given at $H_{id}(j\omega)=1$ by the following expressions.

A. If the solutions of equations (4.31) and (4.32) correspond to the condition $\dot{V}(t)=d(t)$ (at $k_g>0$ and $k_\upsilon>0$), then

$$S_{gmu}(\omega)=\begin{cases} k_g S_{\upsilon l}(\omega) \text{ at } \omega\in\alpha_1(k_g), \\ S_{gu}(\omega) \text{ at } \omega\in\alpha_2(k_g), \\ S_{gl}(\omega) \text{ at } \omega\in\overline{\alpha}(k_g), \end{cases} \tag{4.35}$$

$$S_{\upsilon mu}(\omega)=\begin{cases} k_\upsilon^{-1} S_{gl}(\omega) \text{ at } \omega\in\beta_1(k_\upsilon), \\ S_{\upsilon u}(\omega) \text{ at } \omega\in\beta_2(k_\upsilon), \\ S_{\upsilon l}(\omega) \text{ at } \omega\in\overline{\beta}(k_\upsilon) \end{cases} \tag{4.36}$$

(The graph of ratio $S_{gmu}(\omega)/S_{\upsilon mu}(\omega)$ for this case is shown in Fig. 4.5 by a solid line).

B. If the condition $k_g \le k_\upsilon$ is not fulfilled, the given below equations (4.39) and (4.40) have a solution, which does not break the memberships of actions spectral densities to given classes, then

$$S_{gmu}(\omega)=\begin{cases} k_g S_{\upsilon l}(\omega)+S_{gx}(\omega) \text{ at } \omega\in\alpha_1(k), \\ S_{gu}(\omega) \text{ at } \omega\in\alpha_2(k), \\ S_{gl}(\omega)+S_{gx}(\omega) \text{ at } \omega\in\beta_1(k), \\ S_{gl}(\omega) \text{ at } \omega\in\beta_2(k), \end{cases} \tag{4.37}$$

$$S_{vmu}(\omega)=\begin{cases} k^{-1} S_{gl}(\omega)+S_{\upsilon x}(\omega) \text{ at } \omega\in\beta_1(k), \\ S_{\upsilon u}(\omega) \text{ at } \omega\in\beta_2(k), \\ S_{\upsilon l}(\omega)+S_{\upsilon x}(\omega) \text{ at } \omega\in\alpha_1(k), \\ S_{\upsilon l}(\omega) \text{ at } \omega\in\alpha_2(k), \end{cases} \tag{4.38}$$

where the positive coefficient *k* satisfies the equation

$$\int\limits_{\beta_2(k)}\left[kS_{\upsilon u}(\omega)-S_{gl}(\omega)\right]d\omega+\int\limits_{\alpha_2(k)}\left[kS_{\upsilon l}(\omega)-S_{gu}(\omega)\right]d\omega-\pi(kD_\upsilon-D_0)=0, \quad (4.39)$$

and $S_{gx}(\omega)$, $S_{\upsilon x}(\omega)$ are some non-negative functions fitting to a condition

$$S_{gx}(\omega)=kS_{\upsilon x}(\omega) \quad (4.40)$$

(the graph of the ratio $S_{gmu}(\omega)/S_{\upsilon mu}(\omega)$ in this case also looks like it is shown in Fig. 4.5, but at $k_g=k_\upsilon=k$ ).

C. If conditions $A$ and $B$ not are fulfilled, but the solutions of equations (4.33) and (4.34) obey to an inequality $l_g>l_\upsilon>0$, then

$$S_{gmu}(\omega)=\begin{cases} l_g S_{\upsilon u}(\omega) & \text{at } \omega\in b_1(l_g)\,, \\ S_{gl}(\omega) & \text{at } \omega\in b_2(l_g)\,, \\ S_{gu}(\omega) & \text{at } \omega\in \bar{b}(l_g)\,, \end{cases} \quad (4.41)$$

$$S_{\upsilon mu}(\omega)-\begin{cases} l_\upsilon^{-1} S_{gu}(\omega) & \text{at } \omega\in a_1(l_\upsilon), \\ S_{\upsilon l}(\omega) & \text{at } \omega\in a_2(l_\upsilon), \\ S_{\upsilon u}(\omega) & \text{at } \omega\in \bar{a}(l_\upsilon), \end{cases} \quad (4.42)$$

(the graph of ration $S_{gmu}(\omega)/S_{\upsilon mu}(\omega)$ for this case is shown in Fig.4.6 by a solid line).

D. At last, at nonfulfillment of conditions A, B and C of the expressions (4.41) and (4.42) are corrected, so if that equation (4.33) has no solution in relation to $l_g$, then $S_{gmu}(\omega)=S_{gu}(\omega)$ at $S_{\upsilon u}(\omega)>0$, and equation (4.34) has no solution in relation to $l_\upsilon$, $S_{\upsilon mu}(\omega)=S_{\upsilon u}(\omega)$ at $S_{gu}(\omega)>0$ .

Theorem proving. Following [81] the theorem proving these expressions can be presented as:

$$D_e(S_g,S_\upsilon,H_R)-D_e(S_{gmu},S_{\upsilon mu},H_R)\le 0 \quad (4.43)$$

for any $S_g(\omega)$ and $S_\upsilon(\omega)$, pertinent to their admissible classes. Here $H_R(j\omega)=H_0(j\omega)$ at $S_g(\omega)=S_{gmu}(\omega)$ and $S_\upsilon(\omega)=S_{\upsilon mu}(\omega)$, i.e. according to (4.8) at $H_{id}(j\omega)=1$

$$H_R(j\omega)=\frac{S_{gmu}(\omega)}{S_{gmu}(\omega)+S_{\upsilon mu}(\omega)}=\frac{S_{gmu}(\omega)/S_{\upsilon mu}(\omega)}{1+S_{gmu}(\omega)/S_{\upsilon mu}(\omega)}.$$

When with the absence of cross correlation of signal and interference, the equality $D_e(S_g,S_\upsilon,H_R)=D_{eg}(S_g,H_R)+D_{e\upsilon}(S_\upsilon,H_R)$ is true condition (4.43) will be obtained if

$$D_{eg}\left(S_g, H_R\right) - D_{eg}\left(S_{gmu}, H_R\right) \le 0 \tag{4.44}$$

and

$$D_{e\upsilon}\left(S_\upsilon, H_R\right) - D_{e\upsilon}\left(S_{\upsilon mu}, H_R\right) \le 0\,. \tag{4.45}$$

At first it is necessary to be convinced of the realization of inequality (4.44). In view of expression (4.8), its left side can be presented as:

$$\begin{aligned} &D_{eg}\left(S_g, H_R\right) - D_{eg}\left(S_{gmu}, H_R\right) = \frac{1}{\pi}\int_0^\infty \left[S_g(\omega) - S_{gmu}(\omega)\right] \times \\ &\times \left[\frac{1}{1 + S_{gmu}(\omega)/S_{\upsilon mu}(\omega)}\right]^2 d\omega = \frac{1}{\pi}\int_0^\infty \Delta S_g(\omega)\left|H_{eR}(j\omega)\right|^2 d\omega = \frac{1}{\pi}\int_0^\infty \Delta S_{eg}(\omega)\, d\omega, \end{aligned} \tag{4.46}$$

where

$$H_{eR}(j\omega) = 1 - H_R(j\omega) = \left[1 + S_{gmu}(\omega)/S_{\upsilon mu}(\omega)\right]^{-1},$$

$$\Delta S_g(\omega) = S_g(\omega) - S_{gmu}(\omega),\ \Delta S_{eg}(\omega) = \left|H_{eR}(j\omega)\right|^2 \Delta S_g(\omega).$$

According to expressions $A$ in the theorem, an integrand in (4.46) can be estimated for five areas of frequency $\alpha_1$, $\alpha_2$, $\beta_1$, $\beta_2$ and $\gamma = \overline{\alpha \cup \beta}$ not intersected, which cover the whole integration interval $\omega \in [0, \infty)$ in total. Please see the following examples:

At $\omega \in \alpha_1$ according to (4.35) and (4.36) (see Fig. 4.4 also), $S_{gmu}(\omega)/S_{\upsilon mu}(\omega) = k_g$, whence

$$\int_{\alpha_1} \Delta S_{eg}(\omega)\, d\omega = \left[\frac{1}{1 + k_g}\right]^2 \int_{\alpha_1} \Delta S_g(\omega)\, d\omega\,. \tag{4.47}$$

At $\omega \in \alpha_2$ according to (4.35) and (4.36)

$$S_{gmu}(\omega)/S_{\upsilon mu}(\omega) < k_g,\ \left|H_{eR}(j\omega)\right|^2 \ge \left[1 + k_g\right]^{-2},$$

but $\Delta S_g(\omega) \le 0$, in this case

$$\int_{\alpha_2} \Delta S_{eg}(\omega)\, d\omega \le \left[\frac{1}{1 + k_g}\right]^2 \int_{\alpha_2} \Delta S_g(\omega)\, d\omega\,. \tag{4.48}$$

At $\omega \in \beta_1$, according to (4.35) and (4.36)

$$S_{gmu}(\omega)/S_{\upsilon mu}(\omega) = k_\upsilon > k_g,\ \left|H_{eR}(j\omega)\right|^2 \le \left[1 + k_g\right]^{-2},$$

$$\Delta S_g(\omega) > 0,$$

$$\int_{\beta_1} \Delta S_{eg}(\omega)\,d\omega \le \left[\frac{1}{1+k_g}\right]^2 \int_{\beta_1} \Delta S_g(\omega)\,d\omega\,. \tag{4.49}$$

At $\omega \in \beta_2$ according to (4.35) and (4.36)

$$S_{gmu}(\omega)/S_{\upsilon mu}(\omega) \ge k_g\,,\; |H_{eR}(j\omega)|^2 \le [1+k_g]^{-2},\; \Delta S_g(\omega) \ge 0\,,$$

$$\int_{\beta_2} \Delta S_{eg}(\omega)\,d\omega \le \left[\frac{1}{1+k_g}\right]^2 \int_{\beta_2} \Delta S_g(\omega)\,d\omega\,. \tag{4.50}$$

At last, at $\omega \in \gamma$ according to (4.35) and (4.36) $S_{gmu}(\omega)/S_{\upsilon mu}(\omega) =$ $= S_{gl}(\omega)/S_{\upsilon l}(\omega) \ge k_g$, $|H_{eR}(j\omega)|^2 \le [1+k_g]^{-2}$,

$$\Delta S_g(\omega) \ge 0\,,$$

$$\int_{\gamma} \Delta S_{eg}(\omega)\,d\omega \le \left[\frac{1}{1+k_g}\right]^2 \int_{\gamma} \Delta S_g(\omega)\,d\omega\,. \tag{4.51}$$

Adding together the left and right hand sides of inequalities (4.47) — (4.51) accordingly, the following expression is fulfilled:

$$\int_0^\infty \Delta S_{eg}(\omega)\,d\omega \le \left[\frac{1}{1+k_g}\right]^2 \int_0^\infty \Delta S_g(\omega)\,d\omega\,. \tag{4.52}$$

As

$$\frac{1}{\pi}\int_0^\infty \Delta S_{eg}(\omega)\,d\omega = \frac{1}{\pi}\int_0^\infty S_g(\omega)\,d\omega - \frac{1}{\pi}\int_0^\infty S_{gmu}(\omega)\,d\omega = D_0 - D_0 = 0\,,$$

the expression (4.52) results in

$$\int_0^\infty \Delta S_{eg}(\omega)\,d\omega \le 0$$

or, according to (4.46), $D_{eg}(S_g, H_R) - D_{eg}(S_{gmu}, H_R) \le 0$, which was to be proven.

The validity of relation (4.45) can be similarly proved. This ensures the realization of inequality (4.43).

The procedure for proving B, C, D has no essential differences from the theorem proved for A.

**Example 4.4.** Synthesize the minimax robust filter with boundary functions assigning the admissible classes for spectral densities of signal and interference, with this kind:

$$S_{gu}(\omega)=\begin{cases}A_u(1-\omega/\nu_u)\text{ at } |\omega|\le\nu_u,\\ 0 \quad\text{ at } |\omega|>\nu_u,\end{cases}\qquad S_{\upsilon u}(\omega)=\begin{cases}N_u \text{ at } |\omega|\le\lambda_u,\\ 0 \text{ at } |\omega|>\lambda_u,\end{cases}$$

$$S_{gl}(\omega)=\begin{cases}A_l(1-\omega/\nu_l)\text{ at } |\omega|\le\nu_l,\\ 0 \quad\text{ at } |\omega|>\nu_l,\end{cases}\qquad S_{\upsilon l}(\omega)=\begin{cases}N_l \text{ at } |\omega|\le\lambda_l,\\ 0 \text{ at } |\omega|>\lambda_l,\end{cases}$$

where $A_u > A_l > N_u > N_l$, $\nu_l \le \nu_u < \lambda_l \le \lambda_u$.

Numerical values for parameters are: $A_u = 1.1$, $A_l = 0.8$, $\nu_u = 0.12\pi$, $\nu_l = 0.08\pi$, $N_u = 0.12$, $N_l = 0.095$, $\lambda_u = \lambda_l = \pi$.

Dispersions of signal and interference are: $D_0 = 0.005$, $D_\upsilon = 0.10$.

Here and in the following example it is accepted, that the signal and interference are metered in some identical dimensionless relative units (for example, signal and interference are the numerical code), and the dimension of frequency is conditionally assigned to the magnitude $\pi$, i.e. $\pi = 3.14\,\text{sec}^{-1}$.

Note, that

$$\frac{1}{\pi}\int_0^\infty S_{gu}(\omega)\,d\omega=\frac{A_u\nu_u}{2\pi}=\frac{1.1\cdot 0.12}{2}=0.066,$$

$$\frac{1}{\pi}\int_0^\infty S_{gl}(\omega)\,d\omega=\frac{A_l\nu_l}{2\pi}=\frac{0.8\cdot 0.08}{2}=0.032,$$

$$\frac{1}{\pi}\int_0^\infty S_{\upsilon u}(\omega)\,d\omega=\frac{N_u\lambda_u}{\pi}=0.12\cdot 1=0.12,$$

$$\frac{1}{\pi}\int_0^\infty S_{\upsilon l}(\omega)\,d\omega=\frac{N_l\lambda_l}{\pi}=0.095\cdot 1=0.095,$$

i.e. the conditions (4.17) are fulfilled.

Begin the determination of the most unfavorable spectral densities of actions with the construction of graphs for rations of boundary functions $S_{gu}/S_{\upsilon l}$, $S_{gl}/S_{\upsilon l}$, $S_{gl}/S_{\upsilon u}$ in order to define the areas of frequency $\alpha_1$, $\alpha_2$, $\beta_1$ and $\beta_2$ (by using Fig. 4.5). The indicated graphs are shown in Fig. 4.7. Conditionally, levels $k_g$ and $k_\upsilon$ are represented here.

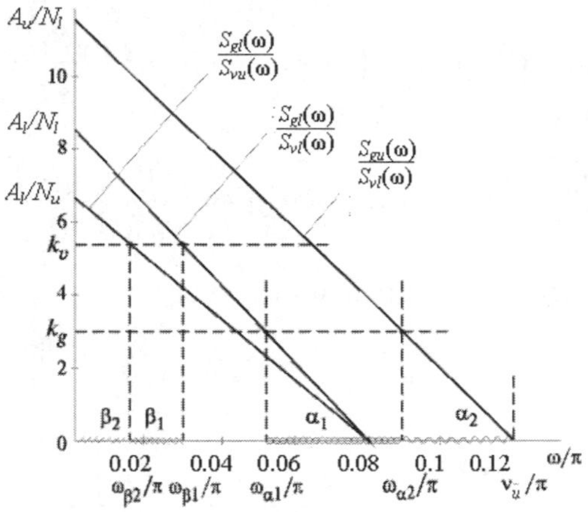


Fig. 4.7

It is visible, that the partition of the frequency axis on areas is carried out by points $\omega_{\alpha 1}$, $\omega_{\alpha 2}$, $\omega_{\beta 1}$ and $\omega_{\beta 2}$, by which abscissas satisfy equations

$$\frac{A_l}{N_l}\left(1-\frac{\omega_{\alpha 1}}{\nu_l}\right)=k_g ,$$

$$\frac{A_u}{N_l}\left(1-\frac{\omega_{\alpha 2}}{\nu_u}\right)=k_g ,$$

$$\frac{A_l}{N_l}\left(1-\frac{\omega_{\beta 1}}{\nu_l}\right)=k_\upsilon , \quad \frac{A_l}{N_u}\left(1-\frac{\omega_{\beta 2}}{\nu_l}\right)=k_\upsilon .$$

From here the following are accepted

$$\omega_{\alpha 1}=\nu_l\left(1-\frac{k_g N_l}{A_l}\right), \quad \omega_{\alpha 2}=\nu_u\left(1-\frac{k_g N_l}{A_u}\right),$$

$$\omega_{\beta 1}=\nu_l\left(1-\frac{k_\upsilon N_l}{A_l}\right), \quad \omega_{\beta 2}=\nu_l\left(1-\frac{k_\upsilon N_u}{A_l}\right).$$

Now determine the definite values of coefficients $k_g$ and $k_\upsilon$.

The expressions for the first, second and third addends in any part of (4.31) then can be written as:

$$k_g \int_{\alpha_1} S_{\upsilon l}(\omega)d\omega = k_g \int_{\omega_{\alpha 1}}^{\omega_{\alpha 2}} N_l \, d\omega = k_g N_l\left[\nu_u\left(1-\frac{k_g N_l}{A_u}\right)-\nu_l\left(1-\frac{k_g N_l}{A_l}\right)\right],$$

$$\int\limits_{\alpha_2} S_{gu}(\omega)\,d\omega = k_g \int\limits_{\omega_{\alpha 2}}^{\nu_u} A_u\left(1-\frac{\omega}{\nu_u}\right)d\omega = \frac{(k_g N_l)^2}{2}\frac{\nu_u}{A_u},$$

$$\int\limits_{\overline{\alpha}} S_{gl}(\omega)\,d\omega = \int\limits_{0}^{\omega_{\alpha 1}} A_l\left(1-\frac{\omega}{\nu_l}\right)d\omega = \frac{A_l^2-(k_g N_l)^2}{2A_l}\nu_l\,.$$

It allows specification of equation (4.31):

$$N_l^2\left(\frac{\nu_u}{2A_u}-\frac{\nu_l}{2A_l}-\frac{\nu_l}{A_u}+\frac{\nu_l}{A_l}\right)k_g^2+N_l(\nu_u-\nu_l)k_g+\frac{A_l\nu_l}{2}-\pi D_0=0$$

and determination of its roots

$$k_g=\frac{\nu_u-\nu_l\pm\sqrt{(\nu_u-\nu_l)^2-\left(\frac{\nu_l}{A_l}-\frac{\nu_u}{A_u}\right)(A_l\nu_l-2\pi D_0)}}{N_l(\nu_l/A_l-\nu_u/A_u)}\,.$$

The substitution of the parameter numerical values and the account of requirement $0<k_g\le A_l/N_l=8.42$ provide $k_g=5.01$.

Write the expressions for the first, second and third addend in the left hand side (4.32)

$$\frac{1}{k_\upsilon}\int\limits_{\beta_1} S_{gl}(\omega)\,d\omega=\frac{1}{k_\upsilon}\int\limits_{\omega_{\beta 2}}^{\omega_{\beta 1}} A_l\left(1-\frac{\omega}{\nu_l}\right)d\omega=\frac{k_\upsilon\nu_l}{2A_l}\left(N_u^2-N_l^2\right),$$

$$\int\limits_{\beta_2} S_{\upsilon u}(\omega)\,d\omega=\int\limits_{0}^{\omega_{\beta 2}} N_u\,d\omega=N_u\nu_l\left(1-\frac{k_\upsilon N_u}{A_l}\right),$$

$$\int\limits_{\overline{\beta}} S_{\upsilon l}(\omega)\,d\omega=\int\limits_{\omega_{\beta 1}}^{\lambda_l} N_l\,d\omega=N_l\left(\lambda_l-\nu_l+\frac{k_\upsilon N_u}{A_l}\right).$$

It allows for the specification of equation (4.32) as

$$\frac{\nu_l}{2A_l}\left(N_l^2-N_u^2\right)k_\upsilon+\nu_l(N_u-N_l)+N_l\lambda_l-\pi D_\upsilon=0$$

and to determine its root

$$k_\upsilon=2A_l\frac{N_u-N_l+(N_l\lambda_l-\pi D_\upsilon)/\nu_l}{N_u^2-N_l^2}=3.57\,.$$

As the condition $k_g\le k_\upsilon$ is not fulfilled, expression A cannot be used in the theorem and it is necessary to check the possibility of use of expression B.

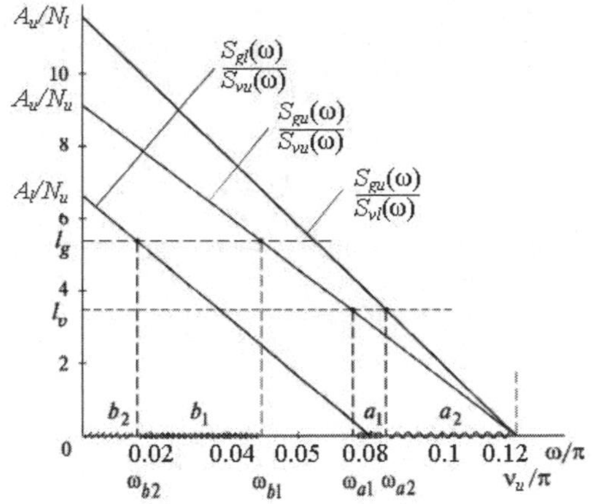


Fig. 4.8

Equation (4.39) is specified as

$$\int_0^{\omega_{\beta 2}}\left[kN_u - A_l\left(1-\frac{\omega}{\nu_l}\right)\right]d\omega + \int_{\omega_{\alpha 2}}^{\nu_u}\left[kN_l - A_u\left(1-\frac{\omega}{\nu_u}\right)\right]d\omega - \pi\left(kD_\upsilon - D_0\right) = 0$$

or, after transformations,

$$\left(\frac{N_u^2\nu_l}{A_l}+\frac{N_l^2\nu_u}{A_u}-\frac{N_uN_l\nu_u}{A_u}\right)k^2 - 2\left(N_u\nu_l - \pi D_\upsilon\right)k + A_l\nu_l - 2\pi D_0 = 0.$$

The solution of this equation with preset numerical values of parameters gives the positive root $k < A_l/N_u$, ensuring the defined finite sizes for areas $\alpha_1$, $\alpha_2$, $\beta_1$ and $\beta_2$ (see Fig. 4.7 at $k_g = k_\upsilon = k$). At the obtained value $k$ the functions $S_{gmu}(\omega)$ and $S_{\upsilon mu}(\omega)$ can be determined by the formulas (4.37) and (4.38) fitting to condition (4.40) functions $S_{gx}(\omega)$ and $S_{\upsilon x}(\omega)$. However, in this case functions $S_{gmu}(\omega)$ and $S_{\upsilon mu}(\omega)$ are not entered into the admissible boundaries, i.e. do not satisfy inequalities (4.14) and (4.15). This testifies the impossibility to use expression B in the theorem and forces us to check the possibility of using expression C.

In a view of expression C of theorem 4.1, construct the graphs for rations of boundary functions $S_{gu}/S_{\upsilon l}$, $S_{gu}/S_{\upsilon u}$, $S_{gl}/S_{\upsilon u}$ in order to define frequency areas $\alpha_1$, $\alpha_2$, $\beta_1$ and $\beta_2$ (by using Fig. 4.6). These graphs are shown in Fig. 4.8. Levels $l_g$ and $l_\upsilon$ are also conditionally indicated there.

It is visible, that the partition of the frequency axis on areas is carried out by points $\omega_{a1}$, $\omega_{a2}$, $\omega_{b1}$ and $\omega_{b2}$, which abscissas satisfy the equations

$$\frac{A_u}{N_u}\left(1-\frac{\omega_{a1}}{\nu_u}\right)=l_\upsilon, \quad \frac{A_u}{N_l}\left(1-\frac{\omega_{a2}}{\nu_u}\right)=l_\upsilon,$$

$$\frac{A_u}{N_u}\left(1-\frac{\omega_{b1}}{\nu_u}\right)=l_g, \quad \frac{A_l}{N_u}\left(1-\frac{\omega_{b2}}{\nu_l}\right)=l_g.$$

From here it is possible to collect expressions for $\omega_{a1}, \omega_{a2}, \omega_{b1}, \omega_{b2}$:

$$\omega_{a1}=\nu_u\left(1-\frac{l_\upsilon N_u}{A_u}\right), \quad \omega_{a2}=\nu_u\left(1-\frac{l_\upsilon N_l}{A_u}\right),$$

$$\omega_{b1}=\nu_u\left(1-\frac{l_g N_u}{A_u}\right), \quad \omega_{b2}=\nu_l\left(1-\frac{l_g N_u}{A_l}\right).$$

This allows for the specification of equations (4.33) and (4.34) after their transformations as

$$N_B^2\left(\frac{\nu_l}{A_l}-\frac{\nu_u}{A_u}\right)l_g^2+2N_u(\nu_u-\nu_l)l_g+A_l\nu_\upsilon-2\pi D_0=0,$$

$$\frac{\nu_u}{2A_u}\left(N_u^2-N_l^2\right)l_\upsilon-N_u\lambda_u+\pi D_\upsilon=0$$

and to determine its roots

$$l_{g1,2}=\frac{-(\nu_u-\nu_l)\pm\sqrt{(\nu_u-\nu_l)^2-\left(\frac{\nu_l}{A_l}-\frac{\nu_u}{A_u}\right)(A_l\nu_l-2\pi D_0)}}{N_u\left(\nu_l/A_l-\nu_u/A_u\right)},$$

$$l_\upsilon=2A_u\frac{N_u\lambda_u-\pi D_\upsilon}{\nu_u\left(N_u^2-N_l^2\right)}.$$

The substitution of numerical values for parameters provides $l_g=4.0$ (the second root exceeding the value $A_u/N_l$, is truncated) and $l_\upsilon=68.2$. As $l_g<l_\upsilon$ and, moreover, $l_\upsilon>A_u/N_l$, i.e. the equation (4.34) has no acceptable solution, and expression C of the theorem cannot be used either.

According to expression D of theorem 4.1, define the most unfavorable spectral density of signal by formula (4.41):

$$S_{gmu}(\omega)=\begin{cases}A_l(1-\omega/\nu_l) & \text{at } 0<|\omega|<\omega_{b2},\\ l_g N_u & \text{at } \omega_{b2}<|\omega|<\omega_{b1},\\ A_u(1-\omega/\nu_u) & \text{at } \omega_{b1}<|\omega|<\nu_u,\end{cases}$$

where $\omega_{b2}=\nu_l\left(1-l_g N_u/A_l\right)=0.032\pi$, $\omega_{b1}=\nu_u\left(1-l_g N_u/A_u\right)=0.068\pi$ (at $l_g=4.0$). Accept the most unfavorable spectral density of interference in the frequency area $|\omega|\le \nu_u$, where $S_{gu}(\omega)<0$, as $S_{\upsilon mu}(\omega)=S_{\upsilon u}=N_u$. At $|\omega|>\nu_u$, when $S_{gu}(\omega)=0$, the values of spectral density of interference do not effect the properties of physically non-realizable minimax robust filters and therefore can be found, for example, on the basis of deriving the preset dispersion of interference $D_\upsilon$, by formula

$$S_{\upsilon mu}(\omega)=\left(\pi D_\upsilon - N_u\nu_u\right)/\lambda_u=\text{const}\ .$$

Therefore the result the expression for $S_{\upsilon mu}(\omega)$ is:

$$S_{\upsilon mu}(\omega)=\begin{cases} N_u & \text{at} \quad |\omega|<\nu_u, \\ \left(\pi D_\upsilon - N_u\nu_u\right)/\lambda_u & \text{at} \quad \nu_u<|\omega|\le\lambda_u. \end{cases}$$

The graphs for functions $S_{gmu}(\omega)$, $S_{\upsilon mu}(\omega)$ are shown in Fig. 4.9.

Now it is possible to determine the frequency transfer function of the physically non-realizable minimax robust filter $H_R(j\omega)$ by formula (4.8). The graph for this real function of frequency is shown in Fig. 4.10.

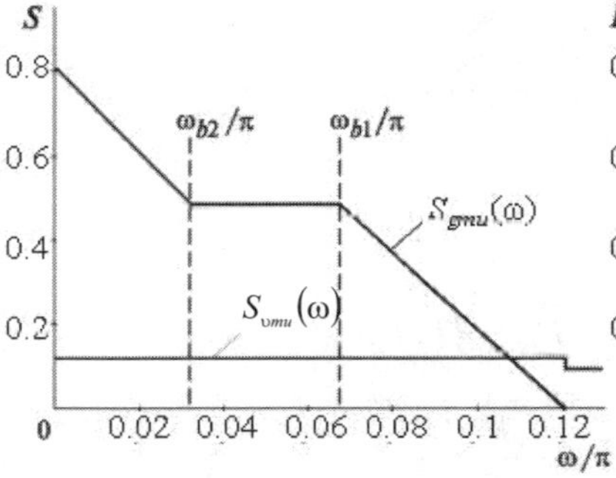


Fig. 4.9

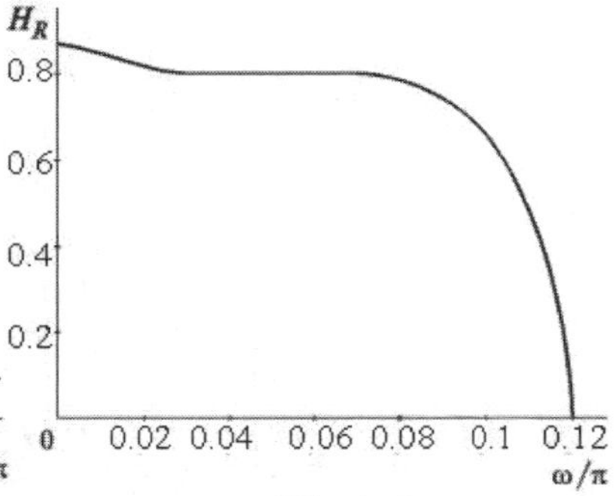


Fig. 4.10

The synthesized filter ensures the minimum possible dispersion of error, for which according to formula (4.9), the following expression is obeyed

$$D_{e\min}^{nr}=\frac{1}{\pi}\int_0^{\nu_u}\frac{S_{gmu}(\omega)N_u}{S_{gmu}(\omega)+N_u}d\omega=$$

$$=N_u\left[\frac{N_u\nu_l}{A_l}\ln\frac{A_l+N_u-A_l\omega_{b2}/\nu_l}{A_l+N_u}+\frac{N_u\nu_u}{A_u}\ln\frac{N_u}{A_u+N_u-A_u\omega_{b1}/\nu_u}+\right.$$

$$\left.+\frac{A_l\left(\nu_l-\omega_{b2}\right)\left(\omega_{b1}-\omega_{b2}\right)}{\left(A_u+N_u\right)\nu_1-A_l\omega_{b2}}-\omega_{b1}+\omega_{b2}+\nu_u\right].$$

The substitution of numerical values for parameters provides the value $D_{e\min}^{nr} = 0.0104$.

**Example 4.5.** Synthesize the minimax robust filter at boundary functions assigning the admissible classes of spectral densities for signal and interference, as a kind:

$$S_{gu}(\omega) = \frac{2D_{0u}T_{gu}}{1+\omega^2 T_{gu}^2}, \quad S_{\upsilon u}(\omega) = \frac{2D_{\upsilon u}T_{\upsilon u}}{1+\omega^2 T_{\upsilon u}^2},$$

$$S_{gl}(\omega) = \frac{2D_{01}T_{gl}}{1+\omega^2 T_{gl}^2}, \quad S_{\upsilon l}(\omega) = \frac{2D_{\upsilon l}T_{\upsilon l}}{1+\omega^2 T_{\upsilon l}^2},$$

where $D_{0u} = 1.1$, $D_{0l} = 0.8$, $D_{\upsilon u} = 1.0$, $D_{\upsilon l} = 0.6$, $T_{gu} = 8.0/\pi$, $T_{gl} = 10/\pi$, $T_{\upsilon u} = 0.8/\pi$, $T_{\upsilon l} = 1.0/\pi$. Signal and interference dispersions are: $D_0 = 0.85$, $D_\upsilon = 0.625$.

The graphs for the rations of boundary functions $S_{gu}/S_{\upsilon l}$, $S_{gl}/S_{\upsilon l}$, $S_{gl}/S_{\upsilon u}$ are shown in Fig. 4.11, the levels $k_g$ and $k_\upsilon$ are conditionally indicated there.

The abscissas of points $\omega_{\alpha 1}$, $\omega_{\alpha 2}$, $\omega_{\beta 1}$ and $\omega_{\beta 2}$, the frequency axis on areas $\alpha_2$, $\alpha_2$, $\beta_1$ and $\beta_2$, should satisfy the equations:

$$\frac{S_{gl}(\omega_{\alpha 1})}{S_{\upsilon l}(\omega_{\alpha 1})} = k_g, \quad \frac{S_{gu}(\omega_{\alpha 2})}{S_{\upsilon l}(\omega_{\alpha 2})} = k_g, \quad \frac{S_{gl}(\omega_{\beta 1})}{S_{\upsilon l}(\omega_{\beta 1})} = k_\upsilon, \quad \frac{S_{gl}(\omega_{\beta 2})}{S_{\upsilon u}(\omega_{\beta 2})} = k_\upsilon,$$

whence

$$\omega_{\alpha 1} = \sqrt{\frac{D_{0l}T_{gl} - k_g D_{\upsilon l}T_{\upsilon l}}{(k_g D_{\upsilon l}T_{gl} - D_{0l}T_{\upsilon l})T_{\upsilon l}T_{gl}}},$$

$$\omega_{\alpha 2} = \sqrt{\frac{D_{0u}T_{gu} - k_g D_{\upsilon l}T_{\upsilon l}}{(k_g D_{\upsilon l}T_{gu} - D_{0u}T_{\upsilon l})T_{\upsilon l}T_{gu}}},$$

$$\omega_{\beta 1} = \sqrt{\frac{D_{0l}T_{gl} - k_\upsilon D_{\upsilon l}T_{\upsilon l}}{(k_\upsilon D_{\upsilon l}T_{gl} - D_{0l}T_{\upsilon l})\, T_{\upsilon l}T_{gl}}},$$

$$\omega_{\beta 2} = \sqrt{\frac{D_{0l}T_{gl} - k_\upsilon D_{\upsilon u}T_{\upsilon u}}{(k_\upsilon D_{\upsilon u}T_{gl} - D_{0l}T_{\upsilon u})\, T_{\upsilon u}T_{gl}}}.$$

Write equations (4.31) and (4.32), after the transformations, as follows:

$$k_g D_{\upsilon l}\left[\operatorname{arctg}(\omega_{\alpha 2}T_{\upsilon l}) - \operatorname{arctg}(\omega_{\alpha 1}T_{\upsilon l})\right] -$$

$$- D_{0u}\operatorname{arctg}(\omega_{\alpha 2}T_{gu}) + D_{0l}\operatorname{arctg}(\omega_{\alpha 1}T_{gl}) + \frac{\pi}{2}(D_{0u} - D_0) = 0,$$

$$\frac{1}{k_\upsilon} D_{01}\left[\operatorname{arctg}\left(\omega_{\beta 1} T_{gl}\right) - \operatorname{arctg}\left(\omega_{\beta 2} T_{gl}\right)\right] +$$

$$+ D_{\upsilon u} \operatorname{arctg}\left(\omega_{\beta 2} T_{\upsilon u}\right) - D_{\upsilon l} \operatorname{arctg}\left(\omega_{\beta 1} T_{gl}\right) - \frac{\pi}{2}\left(D_\upsilon - D_{\upsilon l}\right) = 0$$

Their numerical solution at preset parameters equals $k_g = 7.37$ and $k_\upsilon = 8.21$. The appropriate values for singular points on the frequency axis make

$$\omega_{\alpha 1} = 0.089\pi,\ \omega_{\alpha 2} = 0.124\pi,\ \omega_{\beta 1} = 0.080\pi,\ \omega_{\beta 2} = 0.047\pi.$$

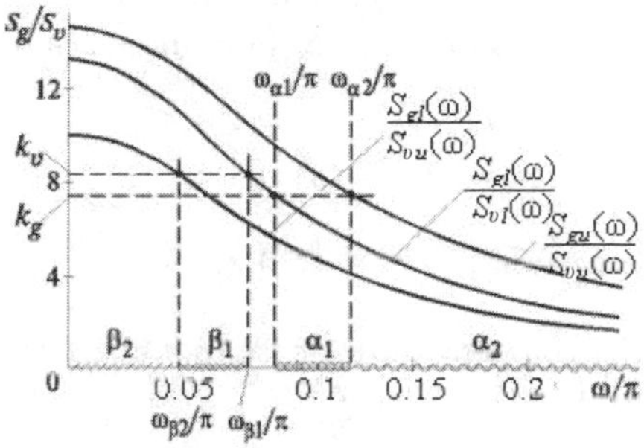


Fig. 4.11

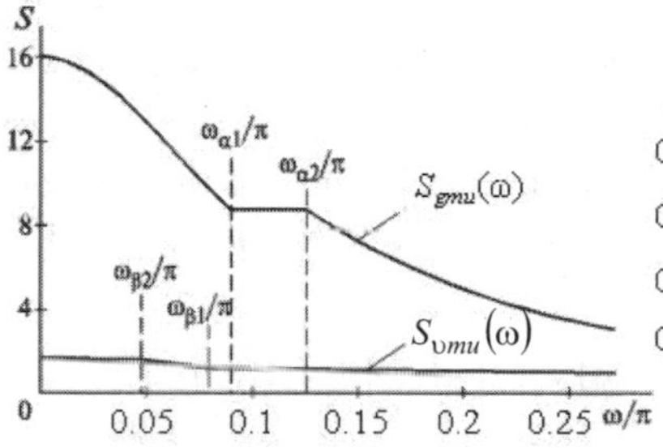


Fig. 4.12

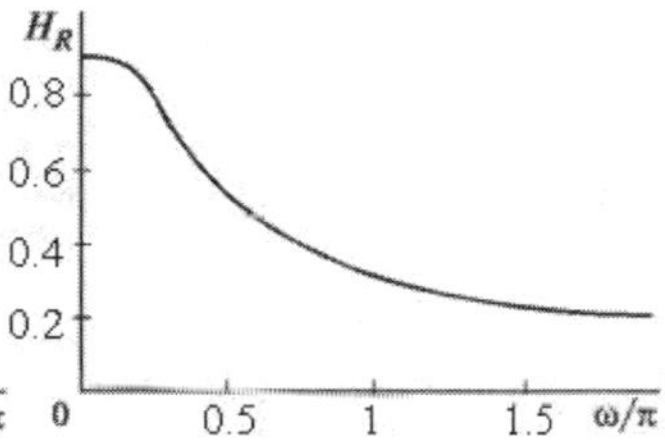


Fig. 4.13

As $k_g < k_\upsilon$, it is possible to use expression A of theorem 4.1 to determine the most unfavorable spectral densities of signal and interference by formulas (4.35) and (4.36), such as

$$S_{gmu}(\omega) = \begin{cases} S_{gl}(\omega) & \text{at } |\omega| \le \omega_{\alpha 1}, \\ k_g S_{\upsilon l}(\omega) & \text{at } \omega_{\alpha 1} \le |\omega| \le \omega_{\alpha 2}, \\ S_{gu}(\omega) & \text{at } |\omega| \ge \omega_{\alpha 2}, \end{cases}$$

$$S_{\upsilon mu}(\omega) = \begin{cases} S_{\upsilon u}(\omega) & \text{at } |\omega| \le \omega_{\beta 2}, \\ k_\upsilon^{-1} S_{gl}(\omega) & \text{at } \omega_{\beta 2} \le |\omega| \le \omega_{\beta 1}, \\ S_{\upsilon l}(\omega) & \text{at } |\omega| \ge \omega_{\beta 1}. \end{cases}$$

The graphs for functions $S_{gmu}(\omega)$ and $S_{\upsilon mu}(\omega)$ are shown in Fig. 4.12. The graph for real transfer function $H_R(\omega)$ in the minimax robust filter defined by the formula (4.8), is given in Fig. 4.13.

## §4.4 Minimax robust filtering with models of spectral density with ε-pollution

**Problem definition and preliminary notes.** The model of indeterminacy as a kind of ε-pollution was widely used in the fundamental work by P. Huber [35] to describe the probability density laws and therefore has become very popular. With reference to spectral densities of signal $S_g(\omega)$ and interference $S_\upsilon(\omega)$ it can be defined as

$$S_g(\omega) = (1-\varepsilon_g)S_{gb}(\omega) + \varepsilon_g S_{gp}(\omega), \tag{4.53}$$

$$S_\upsilon(\omega) = (1-\varepsilon_\upsilon)S_{\upsilon b}(\omega) + \varepsilon_\upsilon S_{\upsilon p}(\omega). \tag{4.54}$$

Here $S_{gb}(\omega)$ and $S_{\upsilon b}(\omega)$ are known base (nominal) spectral densities, $S_{gp}(\omega)$ and $S_{\upsilon p}(\omega)$ are unknown "polluting" spectral densities, $\varepsilon_g$ and $\varepsilon_\upsilon$ are preset factors describing the considered measure of indeterminacy in a model, and $\varepsilon_g, \varepsilon_\upsilon \in [0, 1)$. All spectral densities in (4.53) and (4.54) satisfy integral restrictions such as (4.16), i.e. the dispersions of signal and interference are finite and do not depend on values $\varepsilon_g$ and $\varepsilon_\upsilon$.

The ε-pollution model can be treated as a special case of the considered above band model of indeterminacy if the values $S_{gh}(\omega)$ and $S_{\upsilon h}(\omega)$ in (4.14) and (4.15) are equal to

$$S_{gl}(\omega) = (1-\varepsilon_g)S_{gb}(\omega),\ S_{\upsilon l}(\omega) = (1-\varepsilon_\upsilon)S_{\upsilon b}(\omega), \tag{4.55}$$

The upper bounds $S_{gu}(\omega)$ and $S_{\upsilon u}(\omega)$ are not given in this case, i.e. their values are formally considered indefinitely large. Therefore, it is possible to apply the results of §4.2 and 4.3 to determine the most unfavorable spectral densities $S_{gmu}(\omega)$ and $S_{\upsilon mu}(\omega)$ at the synthesis of the minimax robust filter. The problem statement for synthesis remains the same, and its heuristic solution is connected to the available freedom in the choice of the "polluting" component of spectral densities for the best approach of curves $S_g(\omega)$ and $S_\upsilon(\omega)$ by their shapes within the preset restrictions on power of signal and interferences.

**The strict solution to the filter synthesis problem.** The strict solution to the given problem requires, similar to the case of the band model of indeterminacy, the partition of the frequency axis on the intervals at an indication of realization inequalities reduced in §4.3 written with coefficients $k_g$ and $k_\upsilon$. Such positive values of coefficients are determined from the solution of the following equations, to which equations (4.31) and (4.32) are transformed:

$$k_g \int_{\alpha(k_g)} (1-\varepsilon_\upsilon) S_{\upsilon b}(\omega)\, d\omega + \int_{\overline{\alpha}(k_g)} (1-\varepsilon_g) S_{gb}(\omega)\, d\omega = \pi D_0 , \tag{4.56}$$

$$\frac{1}{k_\upsilon} \int_{\beta(k_\upsilon)} (1-\varepsilon_g) S_{gb}(\omega)\, d\omega + \int_{\overline{\beta}(k_\upsilon)} (1-\varepsilon_\upsilon) S_{\upsilon b}(\omega)\, d\omega = \pi D_\upsilon . \tag{4.57}$$

In order to determine the strict solution it is possible to use theorem 4.1 (§4.3), from which with the account of (4.55) follows (special proof is not required) that if $k_g \geq k_\upsilon$, then

$$S_{gmu}(\omega) = \begin{cases} k_g (1-\varepsilon_\upsilon) S_{\upsilon b}(\omega) & \text{at } \omega \in \alpha(k_g) , \\ (1-\varepsilon_g) S_{gb}(\omega) & \text{at } \omega \in \overline{\alpha}(k_g) , \end{cases} \tag{4.58}$$

$$S_{\upsilon mu}(\omega) = \begin{cases} k_\upsilon^{-1} (1-\varepsilon_g) S_{gb}(\omega) & \text{at } \omega \in \beta(k_\upsilon), \\ (1-\varepsilon_\upsilon) S_{\upsilon b}(\omega) & \text{at } \omega \in \overline{\beta}(k_\upsilon). \end{cases} \tag{4.59}$$

Otherwise

$$S_{gmu}(\omega) = \begin{cases} k(1-\varepsilon_\upsilon) S_{\upsilon b}(\omega) + S_{gx}(\omega) & \text{at } \omega \in \alpha(k), \\ (1-\varepsilon_g) S_{gb}(\omega) + S_{gx}(\omega) & \text{at } \omega \in \overline{\alpha}(k), \end{cases} \tag{4.60}$$

$$S_{\upsilon mu}(\omega) = \begin{cases} k^{-1}(1-\varepsilon_g) S_{gb}(\omega) + S_{\upsilon x}(\omega) & \text{at } \omega \in \beta(k), \\ (1-\varepsilon_\upsilon) S_{\upsilon b}(\omega) + S_{\upsilon x}(\omega) & \text{at } \omega \in \overline{\beta}(k), \end{cases} \tag{4.61}$$

where $k = D_0 / D_\upsilon$, and $S_{gx}(\omega)$ and $S_{\upsilon x}(\omega)$ are some non-negative functions, and $S_{gx}(\omega) = k S_{\upsilon x}(\omega)$. Naturally, the spectral densities $S_{gx}(\omega)$ and $S_{\upsilon x}(\omega)$ must satisfy the preset restrictions by the power of signal and interference.

After determining the pair of most unfavorable spectral densities $\{S_{gmu}(\omega), S_{\upsilon mu}(\omega)\}$, the frequency transfer function for physically realizable minimax filter can be determined by formula (4.8).

*Questions*

1. How is the Wiener filtering problem defined at the complete a priori information?
2. How can the robust Wiener filtering problem providing the nonparametric classes of spectral densities for signals be defined?
3. How is the accuracy of physically non-realizable Wiener filter estimated?
4. What does the physically realizable filter accuracy depend on?
5. Describe the procedure for the synthesis of the robust Wiener filter with the band model of spectral indeterminacy.

6. How can the most unfavorable spectrum of action for the robust Wiener filter with the band model of spectral indeterminacy be found?
7. Describe the procedure for synthesis of the robust Wiener filter with the spectral indeterminacy model with $\varepsilon$-pollution.
8. What inferences can be made when comparing the types of synthesis of robust Wiener filters considered above?

Chapter 5

# Analysis of control error dispersion by numerical characteristics of actions

## §5.1. Introduction

**Problem statement.** Before applying the synthesis problem of the automatic control system corresponding to the defined accuracy requirements, it is necessary to be convinced that the available a priori information about the properties of external actions is enough to estimate the accuracy of any considered variant of the system. It is also desirable, that deriving the numerical value for accepted precision factor be produced simply enough or, at least, be easily calculated algorithmically. Therefore, the present and next chapters are devoted to the methods of accuracy analysis.

Here the dispersion of stationary centered error is accepted as the fundamental factor of control accuracy. Reference action $g(t)$ and interference $\upsilon(t)$, reduced to a system input, are implicated as their casual centered components. These components can be either stationary, or have stationary casual increments. Linear systems with frequency transfer functions of the open loop are considered as:

$$W(j\omega)=\frac{1+b_1 j\omega+...+b_{n-1}(j\omega)^{n-1}}{a_k(j\omega)^k+a_{k+1}(j\omega)^{k+1}+...+a_n(j\omega)^n}. \quad (5.1)$$

The parameters $\{a_i\}_k^n,\{b_j\}_1^{n-1}\in[0,\infty)$ are considered constant and known. At the unit principal feedback, the frequency transfer functions for the closed loop system by controlled magnitude and by error are determined by the formulas

$$H(j\omega)=W(j\omega)[1+W(j\omega)]^{-1}, H_e(j\omega)=[1+W(j\omega)]^{-1}.$$

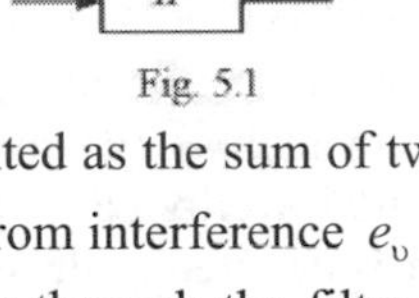


Fig. 5.1

The resultant error is represented as the sum of two its components $e=e_g+e_\upsilon$. The dynamic error $e_g$ and error from interference $e_\upsilon$ are formed accordingly when transiting of the reference action through the filter with transfer function $H_e(s)$

and disturbing action through the filter with transfer function $H(s)$ (Fig. 5.1). If the cross correlation between the reference action and interference is absent, then the dispersion of resultant error is equal to the dispersion of dynamic error added to the dispersion of error from perturbation

$$D_e = D_{eg} + D_{e\upsilon} . \tag{5.2}$$

The number of addends in the right hand side of (5.2) can be more than two with the presence of several external actions or with their separately considered components.

The dispersions $D_{eg}$ and $D_{e\upsilon}$ are connected to spectral densities of reference and disturbing actions $S_g(\omega)$ and $S_\upsilon(\omega)$ by integral relations:

$$D_{eg} = \frac{1}{\pi}\int_0^\infty |H_e(j\omega)|^2 S_g(\omega)\, d\omega , \tag{5.3}$$

$$D_{ev} = \frac{1}{\pi}\int_0^\infty |H_e(j\omega)|^2 S_\upsilon(\omega)\, d\omega . \tag{5.4}$$

The spectral densities $S_g(\omega)$ and $S_\upsilon(\omega)$ give the complete a priori information for an accurate and univalent calculation of error dispersion in the linear system. If both or one of them is considered unknown, then it is expedient to put the problem for determining the error dispersion $[D_{e\min}, D_{e\max}]$ with the presence of some deficient, but with authentic information about the spectral content of actions. The methods to derive such authentic a priori information are described in chapter 4. The power or generalized moments of spectral density, upper bound frequency of the spectrum, and information about the shape of the spectral density curve (unimodal, nonincreasing, concave) etc. are meant for the reference action.

Similar information can exist for the properties of interference. However, unlike for reference action, the case with the absence of complete a priori information for interference is not characteristic. Within the narrow bandpass limits of a system it is often possible to consider the spectral density of broadband interference as uniform.

Determination of the upper and lower boundaries of error dispersion $D_e \in [D_{e\min}, D_{e\max}]$ enables a quite definite description of the system accuracy. The comparison of an upper bound $D_{e\max}$ with preset admissible error dispersion $D_e^0$ answers the question, whether the system satisfies to the accuracy requirements applied to it. The knowledge on lower boundary $D_{\min}$ allows for the estimation of the maximum possible loss in accuracy at the expense of a priori information incompleteness. With an increasing amount of a priori information, the difference between the upper and lower boundaries should decrease.

The majority of this chapter is devoted to the estimation of boundaries for error dispersion by dispersions of action ideal derivatives (further word "ideal" will be omitted), expressed by formula (3.20). There are three reasons for this. First, the procedure for estimation of error dispersion boundaries usable for more common case, when the generalized moments of spectral density are preset can be explained, with an example of power moments for action spectral density with greatest ease and obviousness. Secondly, the estimation of dynamic error dispersion by dispersions of reference action derivatives is widely used in engineering practice. However, it is not always done correctly enough. Thirdly, it is possible to use the series of outcomes obtained with reference to the accuracy analysis by derivatives of actions dispersions by analyzing real smoothed derivatives of actions by dispersions expressed by formula (3.21).

Notice that the dispersions of action derivatives, determined by formula (3.20), are uniquely connected to the expansion coefficients of its correlation function in the McLoren series (3.7). Therefore, if the infinite sequence of dispersions $\{D_i\}_0^\infty$ is preset, then the correlation function and the spectral density of action, basically can be uniquely recovered at certain conditions (see §3.1). This provides the complete a priori information to calculate the error dispersion by formula (5.3). At the finite number $N$ of known action dispersions this number can measure the amount of a priori information on the dynamic properties of action.

**Characteristic for solving methods.** Determining the boundaries of dispersion $D_{eg}$ or $D_{ev}$ can be considered as the special cases of the common research problem for extremums of integral

$$I = \frac{1}{\pi}\int_0^{\omega*} A^2(\omega)\,S(\omega)\,d\omega,\ \omega_* \in (0,\infty) \tag{5.5}$$

of the product of some known function $A^2(\omega)\in R_1^+$ by function $S(\omega)\in R_1^+$, preset by the generalized moments such as;

$$M_i = \frac{1}{\pi}\int_0^{\beta} u_i(\omega)\,S(\omega)\,d\omega,\ i=\overline{0,N}, \beta\in[\omega_*,\infty). \tag{5.6}$$

The choice of finite upper limits of integration $\omega_*$ in (5.5) and $\beta$ in (5.6) can be of interest when the hypothesis about the action spectrum, restricted by its width, or when researching the structure of output signal (error) frequency is accepted. In special cases $\omega_*\to\infty$ and $\beta\to\infty$, but $\beta\ge\omega_*$.

The estimation of integral (5.5) exact boundaries at the known generalized (in the special case — power) moments of function $S(\omega)$ concerns the class of problems which are called in mathematics the generalized (classical) problems of moments. In chapter two approaches for such problems solution are considered.

The first approach is connected to the choice of definite functions $S(\omega)$, which is a traditional problem of moments formulated and investigated by P.L. Chebyshev and A.A. Marcov for the first time. The outcomes obtained by them permit us to claim, that the boundaries of integral (5.5) can be reached at functions $S(\omega)$

similar to $\delta$-impulses having nonzero values in finite number of points on a semi-axis $\omega \in [0,\infty)$. Such functions of the elementary kind are called the canonical presentation of generalized moments. However, function $A^2(\omega)$ for this purpose should satisfy certain conditions, which, as it is shown in §5.2, are usually not fulfilled at investigation of control systems.

The other approach suggested in §5.3 is more convenient for solving the problems of investigating the accuracy of control systems and is free from any restrictions on function $A^2(\omega)$. It is grounded on the approximation of this function by the linear combinations of basis functions, majorizing it. Therefore, this approach is called the approximative. Such methods allow the finding generally of strict upper and lower estimations $\bar{I}$ and $\underline{I}$, and $[I_{\min}, I_{\max}] \in [\underline{I}, \bar{I}]$ but not the exact boundaries $I_{\max}$ and $I_{\min}$. These estimations are the strongest at optimal, in the defined sense, choice of majorizing functions.

The singularities and legitimacies of using an approximate method to determine the estimations for dispersions of the control resultant error component, are considered in §5.4 — 5.7. As it will be shown in chapter 6, this material is useful also to determine the maximum control error. The synthesis methods of robust systems stated in chapter 7 are based upon it.

§5.8 is devoted to the determination of the correlation between the approximate method and method of canonical presentations, or derivation of conditions at which the strongest estimations, found by the approximate method, coincide with the appropriate exact boundaries. The answer to the important problem about the cases where the extreme value of error dispersion is reached lies in the processing of the "equivalent" harmonic action, i.e. the legitimacy of known "equivalent" harmonic action method, is evaluated.

It is necessary to note the fruitfulness of use in the idea stated here and in the book [8], which includes research on the problem of linear system accuracy by dispersions of action derivatives with the classical problem of moments, which allows for the adding of sufficient strictness into the definition and solution of this problem. Earlier in the control theory only some outcomes of the abstract *L*-problem of moments were used, to which it was possible to reduce the problems of optimal control with restrictions, shown by N. N. Krassovski, [21, 49].

## §5.2. Use of canonical presentations for sequence of spectral density moments

**Canonical presentation concept.** The problem with recovering the spectral density $S(\omega)$ by the finite number of its generalized moments like (5.6) supposes the solution set. Select the solutions that have a kind of discrete set of spectral lines on frequencies $\{\omega_j\}_1^{\nu}$. Dispersion $\rho_j \geq 0$ must correspond to each spectral line there. Thus the relations (5.6) accept the kind:

$$M_i = \sum_{j=1}^{\nu} \rho_i u_i(\omega_j),\ i = \overline{0, N}. \tag{5.7}$$

The choice of unknown values $\{\rho_j\}_1^\nu$ and $\{\omega_j\}_1^\nu$ can be produced in a unique way, if the number of magnitudes is equal to the number of equations in (5.7), i.e. it equals $N+1$. This is fulfilled for even $N$ at $\nu = \frac{N}{2}+1$ and when assigning one of the frequencies $\omega_j$. For odd $N$ it is fulfilled at $\nu = \frac{N+1}{2}$, or at $\nu = \frac{N+3}{2}$ and when assigning two extra conditions, for example, makes $\omega_1 = 0, \omega_\nu \to \infty$. The appropriate presentations of sequence $\{M_i\}_0^N$ such as (5.7) are canonical (the strict determination is connected to the concept of the presentation index [1, 12, 53]).

The canonical presentations theory, based on the works of outstanding Russian mathematicians P.L. Chebyshev (1821 — 1894) and A.A. Markov (1856 — 1922), represent the problems of moments [52] and some of their results can be used to solve the problems stated here. In particular, it is shown [53], that the entry (5.6) is valid only at values of generalized moments $\{M_i\}_0^N$, corresponding to the following (positivity) condition. The inequality $\sum_i \alpha_i M_i \geq 0$ should be fulfilled for any real coefficients $\{\alpha_i\}_0^N$, at which $\sum_i \alpha_i u_i(\omega) \geq 0, \omega \in [0, \beta]$. The power moments satisfy this condition, if inequality (3.3) is valid.

**Main canonical presentations.** Consider that the system of basis functions $\{u_i(\omega)\}_0^N$ is the Chebyshev system, i.e. any polynomial $\sum_i \alpha_i u_i(\omega) \left(\sum_i \alpha_i^2 > 0\right)$ has not more then $N$ roots. The following condition must be obeyed at any $\omega_0 < \omega_1 < \ldots < \omega_N \in [0, \beta]$

$$\begin{vmatrix} u_0(\omega_0) & u_1(\omega_0) & \ldots & u_N(\omega_0) \\ u_0(\omega_1) & u_1(\omega_1) & \ldots & u_N(\omega_1) \\ \ldots & \ldots & \ldots & \ldots \\ u_0(\omega_N) & u_1(\omega_N) & \ldots & u_N(\omega_N) \end{vmatrix} > 0. \tag{5.8}$$

It is easy to make sure that the sequence of power functions $u_1(\omega) = \omega^{2i}, i = \overline{0, N}$ satisfies (5.8) when $\omega \leq [0, \infty)$. The other example for Chebyshev function system in a semi-infinite interval of frequencies looks like

$$u_i(\omega) = 1/\left(1 + \omega^2 T_i^2\right), i = \overline{0, N}, T_N < T_{N-1} < \ldots < T_0 < \infty. \tag{5.9}$$

Let the adjunction of one more function $u_{N+1}(\omega)=A^2(\omega)$ to the Chebyshev function system $\{u_i(\omega)\}_0^N$ also provide the Chebyshev function system $\{u_i(\omega)\}_0^{N+1}$. Name the function $A^2(\omega)$ as Chebyshev prolongation of function system $\{u_i(\omega)\}_0^N$. Then, there are the strict upper and lower boundaries for integral (5.5), which can be found by the canonical presentations of the moments $\{M_i\}_0^N$.

In particular, if $\omega_* = \beta$ then it is necessary to find the so-called upper and lower main canonical presentations. The upper main presentation $\{\overline{\omega}_j\}_1^{\overline{\nu}^*}, \{\overline{\rho}_j\}_1^{\overline{\nu}^*}$ assumes the presence of spectral components at even $N$ on frequency $\overline{\omega}_{\overline{\nu}^*} = \beta$, and at odd $N$ — on frequencies $\overline{\omega}_1 = 0$ and $\overline{\omega}_{\overline{\nu}^*} = \beta$. The lower main presentation $\{\underline{\omega}_j\}_1^{\underline{\nu}^*}, \{\underline{\rho}_j\}_1^{\underline{\nu}^*}$ at even $N$ includes the point $\underline{\omega}_1 = 0$, and at odd $N$ the spectral components on interval $\omega \in [0,\beta]$ boundaries must be absent. It is proved [53] that the extreme values of integral (5.5) are reached at spectral densities $S(\omega)$ corresponding to main presentations, i.e. the relations

$$\sum_{j=1}^{\underline{\nu}*} \underline{\rho}_j A^2(\underline{\omega}_j) \le \frac{1}{\pi}\int_0^{\beta} A^2(\omega)\, S(\omega)\, d\omega \le \sum_{j=1}^{\overline{\nu}^*} \overline{\rho}_j A^2(\overline{\omega}_j). \tag{5.10}$$

are strictly fulfilled.

If $\omega_* < \beta$ then it is necessary to find the canonical presentation that includes the point $\omega = \omega_*$, i.e. to find the frequencies $\omega_1 < \omega_2 < .. < \omega_{l-1} < \omega_l = \omega_* < \omega_{l+1} < .. < \omega_\nu$ and the appropriate dispersions $\{\rho_j\}_1^\nu$ solving the system of equations (5.7). Further the relations:

$$\frac{1}{\pi}\int_0^{\omega_*+0} A^2(\omega)\, S(\omega)\, d\omega \le \sum_{j=1}^{l} \rho_j A^2(\omega_j), \tag{5.11}$$

$$\frac{1}{\pi}\int_0^{\omega_*-0} A^2(\omega)\, S(\omega)\, d\omega \ge \sum_{j=1}^{l-1} \rho_j A^2(\omega_j) \tag{5.12}$$

are used.

**Examples of canonical presentations.** In order to be able to consider function $A^2(\omega)$ as Chebyshev prolongation of even power basis function system $u_i(\omega) = \omega^{2i}$, $i = \overline{0,N}$, it should have the non-negative $(N+1)$-th derivative by $\omega^2$

$$d^{N+1} A^2(\omega) / d\omega^{2(N+1)} \ge 0\,. \tag{5.13}$$

For practical use of formulas (5.10) — (5.12) this derivative of fixed sign is enough.

Select the power functions and fractional rational functions of $\omega^2$ of the first order among the functions corresponding to condition (5.13) at $\omega \in [0,\infty)$. Consider, for example, the outcomes that can be obtained with reference to such functions in a characteristic case, when $N = 2$ and the dispersions of action and two of its derivatives $D_0$, $D_1$ and $D_2$, expressed by the common formula (5.6) at $u_0(\omega) = 1, u_1(\omega) = \omega^2$, $u_2(\omega) = \omega^4$ are known, and at spectrum of action concentrated in frequencies interval $\omega \in [0, \beta]$, $\sqrt{D_2/D_1} \le \beta < \infty$.

It is necessary to write the system of equations in (5.7) to determine the canonical presentations. This system at $\nu = N/2 + 1 = 2$ looks like:

$$\rho_1 + \rho_2 = D_0, \rho_1\omega_1^2 + \rho_2\omega_2^2 = D_1, \rho_1\omega^4 + \rho_2\omega_2^4 = D_2 . \tag{5.14}$$

The canonical presentation, which includes some point $\omega = \omega_* \in \left[\sqrt{D_2/D_1}, \beta\right]$, can be found by solving (5.14) at $\omega_2 = \omega_*$:

$$\begin{gathered} \omega_1 = \sqrt{\left(D_1\omega_*^2 - D_2\right)/\left(D_0\omega_*^2 - D_1\right)}, \\ \rho_1 = \left(D_0\omega_*^2 - D_1\right)^2/\left(D_0\omega_*^4 - 2D_1\omega_*^2 + D_2\right), \\ \omega_2 = \omega_*, \rho_2 = \left(D_0 D_2 - D_1^2\right)/\left(D_0\omega_*^4 - 2D_1\omega_*^2 + D_2\right) \end{gathered} \tag{5.15}$$

It enables the use of formulas (5.11) and (5.12) at $l = 2$ and determination of the boundaries of integral (5.5) with the arbitrary upper limit $\omega_*$.

If the upper limits of integration in (5.5) and (5.6) are coincided, i.e. $\omega_* = \beta$, then the main canonical presentations are necessary for further investigation. Determine the lower main presentation from (5.14) at $\omega_1 = 0$:

$$\underline{\omega}_1 = 0, \rho_1 = D_0 - D_1^2/D_2, \underline{\omega}_2 = \sqrt{D_2/D_1}, \rho_2 = D_1^2/D_2 . \tag{5.16}$$

The upper main presentation can be obtained by the formula (5.15) at $\omega_* = \beta$.

The position of spectral components, appropriate to the upper and lower main presentations at $k = 0, N = 2, \beta \to \infty$, on frequencies axis is shown in Fig. 5.2.

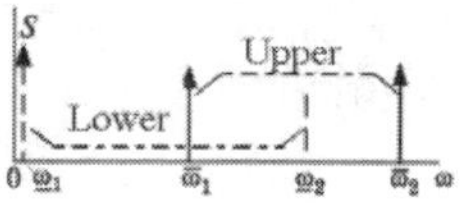


Fig. 5.2

Now it is possible to find the accurate boundaries for dispersions of casual order action derivatives $i > 2$ as $\underline{D}_i \le D_i \le \overline{D}_i$ by the dispersions $D_0, D_1$ and $D_2$ using formula (5.10) at $A^2(\omega) = \omega^{2i}$:

$$D_{i\max}=\left[\left(D_1\beta^2-D_2\right)^i\left(D_0\beta^2-D_1\right)^{2-i}+\left(D_0D_2-D_1^2\right)\beta^{2i}\right]\times$$
$$\times\left(D_0\beta^4-2D_1\beta^2+D_2\right)^{-1}, \tag{5.17}$$
$$D_{i\min}=D_2^{i-1}D_1^{2-i}, i=3,4,5,\ldots$$

It is visible, that the upper bounds of derivative dispersions infinitely increasing at $\beta\to\infty$ depend only on the width of action spectrum $\beta$.

**Boundaries of error dispersion.** Determine the boundaries for dispersion of dynamic error in the first order system, accepted in (5.5) $S(\omega)=S_g(\omega),\omega_*=\beta$ and

$$A^2(\omega)=\left|H_e(j\omega)\right|^2=\left|\frac{1}{1+W(j\omega)}\right|=\frac{a_0+a_1^2\omega^2}{\left(1+a_0\right)^2+a_1^2\omega^2}, \tag{5.18}$$

where $W(j\omega)=1/(a_0+a_1j\omega)$. As (5.18) satisfies to the condition (5.13) at $\omega\in[0,\infty)$, it is possible to use formula (5.10). With the account of (5.15) and (5.16) they present the following expressions for boundaries of error dispersion $\underline{D}_{eg}\le D_{eg}\le\overline{D}_{eg}$

$$D_{eg\max}=\frac{D_0\beta^2-D_1^2}{D_0\beta^4-2D_1\beta^2+D_2}\frac{a_0^2\left(D_0\beta^2-D_1\right)+a_1^2\left(D_1\beta^2-D_2\right)}{\left(1+a_0\right)^2\left(D_0\beta^2-D_1\right)+a_1^2\left(D_1\beta^2-D_2\right)}+$$
$$+\frac{D_0D_2-D_1^2}{D_0\beta^4-2D_1\beta^2+D_2}\frac{a_0^2+a_1^2\beta^2}{\left(1+a_0\right)^2+a_1^2\beta^2}, \tag{5.19}$$

$$D_{eg\min}=\left(D_0-\frac{D_1^2}{D_2}\right)\frac{a_0^2}{1+a_0^2}+\frac{D_1^2}{D_2}\frac{a_0^2D_1+a_1^2D_2}{\left(1+a_0\right)^2D_1+a_1^2D_2}. \tag{5.20}$$

At the unlimited width of the action spectrum, when $\beta\to\infty$, (5.19) is the expression independent of the magnitude $D_2$:

$$D_{eg\max}=D_0\left(a_0^2D_0+a_1^2D_1\right)/\left[\left(1+a_0\right)^2D_0+a_1^2D_1\right]. \tag{5.21}$$

For a system with the astatism, which has the transfer function $W(s)=K_1/s=\left(a_1s\right)^{-1}$ in open state, where $K_1=a_1^{-1}$ is the gain-factor by the velocity, and from (5.21) and (5.20) at $a_0=0$ the elementary, however nontrivial formulas can be accepted as:

$$D_{eg\max}=\frac{D_1}{K_1^2+D_1/D_0}, D_{eg\min}=\frac{D_1}{K_1^2+D_2/D_1}. \tag{5.22}$$

It is interesting, that the upper and lower boundaries of error dispersion in (5.22) are reached when processing the harmonic actions found accordingly by dispersions $D_0, D_1$ and $D_1, D_2$. If the dispersion of only the first derivative of action is known, i.e. $D_0 \to \infty$ and $D_2 \to \infty$, then the formula (5.22) equals the trivial outcome $D_{eg\max} = D_1 / K_1^2$, $D_{eg\min} = 0$.

For the gain plot of $n > 1$ order systems the condition (5.13) is fulfilled only on an individual interval of the frequency axis. That restricts the abilities of the canonical presentation method at the analysis for dispersion of control error. It does not allow for finding out the exact boundaries of dispersion in the common case, and does not eliminate the possibility to derive some approximated solutions. They are described in detail in [8].

Nevertheless, it is necessary to claim that the canonical presentation method can not be the universal method to investigate the accuracy of dynamic system by the generalized or power moments of action spectral density.

## §5.3. Approximative method of estimation

**Common expressions.** Consider the following universal method to determine the upper and lower estimations of an integral (5.5) by known generalized moments $\{M_i\}_K^N$ like in (5.6), $0 \le K \le N$.

Assume, that some linear combinations of basis functions are found

$$C(\omega) = \sum_{i=K}^{N} c_i u_i(\omega),\ Q(\omega) = \sum_{i=K}^{N} q_i u_i(\omega). \qquad (5.23)$$

$\{c_i\}_K^N, \{q_i\}_K^N \in (-\infty, \infty)$ satisfies the condition

$$Q(\omega) \le A^2(\omega) \le C(\omega),\ \omega \in [0, \omega_*], \qquad (5.24)$$

i.e. majorizing the function $A^2(\omega)$ in the frequency area in which (5.5) is being integrated. Suppose also

$$Q(\omega) \le 0 \le C(\omega),\ \omega \in (\omega_*, \beta], \qquad (5.25)$$

where $\beta \in (\omega_*, \infty)$ is the upper bound frequency in spectrum of action, i.e. $S(\omega) \equiv 0$ at $\omega \in (\beta, \infty)$. The possible kind of functions $C(\omega)$ and $Q(\omega)$ is shown in Fig. 5.3 ($a$ at $\omega_* < \beta < \infty$, $b$ at $\omega_* = \beta < \infty$, $c$ at $\omega_* \to \infty, \beta \to \infty$, $A^2(\omega) = |H_e(j\omega)|^2$, $d$ at $A^2(\omega) = |H(j\omega)|^2$).

Having multiplied all members of complicated inequality (5.24) by the spectral density $S(\omega)$, having integrated them by the frequency $\omega \in [0, \omega_*]$ and by using (5.5) and (5.23), it is obtained as the result:

$$\frac{1}{\pi} \int_0^{\omega_*} \left( \sum_{i=K}^{N} q_i u_i(\omega) \right) S(\omega)\, d\omega \le I \le \frac{1}{\pi} \int_0^{\omega_*} \left( \sum_{i=K}^{N} c_i u_i(\omega) \right) S(\omega)\, d\omega,$$

also according to (5.6) and (5.25) gives the required estimations

$$\overline{I}=\sum_{i=K}^{N}c_iM_i,\ \underline{I}=\sum_{i=K}^{N}q_iM_i. \tag{5.26}$$

Estimations in (5.26) are the strongest, if the choice of coefficients $\{c_i\}_K^N, \{q_i\}_K^N$ was made optimally by criteria

$$\overline{I}\to\min_C \overline{I},\quad \underline{I}\to\max_Q \underline{I} \tag{5.27}$$

with restrictions similar to (5.24) and (5.25). These are the linear programming problems with restrictions as the function which, basically, can be treated as dual in relationship to initial infinite-dimensional problems for determination the spectral densities $S(\omega)$, extremizing the integral (5.5). However, such treatment does not cause the obtainment of any additional constructive results because the common duality theory for infinite-dimensional problems does not exist.

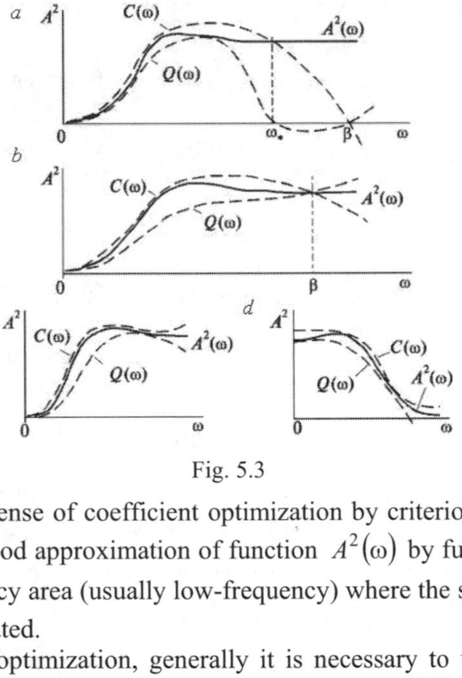


Fig. 5.3

The geometrical sense of coefficient optimization by criterion (5.27) is, that it ensures especially good approximation of function $A^2(\omega)$ by functions $C(\omega)$ and $Q(\omega)$ in that frequency area (usually low-frequency) where the spectrum of action is basically concentrated.

At realization of optimization, generally it is necessary to use the numerical methods. However, sometimes it can be made analytically. Possibilities to produce the analytical result increase strongly at the assumption of an approximate solution, which accords not to the strongest, but to practically acceptable estimations $\overline{I}$ and $\underline{I}$. The choice of method for such research depends on the type of function $A^2(\omega)$ and on other circumstances described in §5.4, 5.5.

**Case of power moments.** For the reasons explained in §5.1, but not due to the limited possibilities of the method, further consider the specific case which is the most interesting for applied problems, when $u_i(\omega)=\omega^{2i}$, i.e. at known dispersions of action derivatives $\{D_i\}_K^N$. The upper bound of integration $\omega_*$ in (5.5) in this case coincides with the spectrum of action width $\beta$. The condition $\omega_*=\beta$ is not principal, because the choice of finite function $A^2(\omega)$ can provide any limits of integration in (5.5) anyway. Naturally, the case with $\beta\to\infty$ is possible and rather typical.

At power basis functions and $\omega_*=\beta$ relations (5.23) — (5.26) become:

$$C(\omega)=C_{2N}(\omega)=\sum_{i=K}^{N}c_i\omega^{2i},\ Q(\omega)=Q_{2N}(\omega)=\sum_{i=K}^{N}q_i\omega^{2i}, \tag{5.28}$$

$$C_{2N}(\omega)\geq A^2(\omega),\ Q_{2N}(\omega)\leq A^2(\omega),\ \omega\in[0,\beta], \tag{5.29}$$

$$\overline{I}=\sum_{i=K}^{N}c_iD_i,\ \underline{I}=\sum_{i=K}^{N}q_iD_i\ . \tag{5.30}$$

Clearly, that for each definite function $A^2(\omega)$ there is some minimal set of derivative dispersions with orders from $K$ up to $N$ at which the inequality (5.29) may be basically fulfilled. In particular, when realizing the first of them, the order $K$ should not exceed the number of younger zero coefficients of expansion for function $A^2(\omega)$ in the McLoren series by powers $\omega^2$. That provides the validity of equalities

$$\left.\frac{d^iA^2(\omega)}{d\omega^{2i}}\right|_{\omega=0}=0,\ i=\overline{0,K-1}\ . \tag{5.31}$$

If $A^2(\omega)=|H_e(j\omega)|^2$ then the requirement (5.31) confirms the impossibility to derive the finite dispersion of dynamic error at indefinitely large dispersion of reference action in a system with no astatism.

**Interaction of approximating polynomial coefficients.** At $\beta\to\infty$ the correct choice of coefficients $\{c_i\}_K^N$ (or $\{q_i\}_K^N$) should ensure the contact of curves $A^2(\omega)$ and $C_{2N}(\omega)$ (or $Q_{2N}(\omega)$) at least in one point at $\omega\geq 0$. Otherwise formula (5.30) provides definitely not the strongest estimation $\overline{I}$ (or $\underline{I}$).

Let the abscissa of contact point be equal to $\chi$. Then, except for condition (5.29), the polynomial $C_{2N}(\omega)$ should satisfy the requirements (Fig. 5.4*a*)

$$C_{2N}(\chi)=A^2(\chi), \tag{5.32}$$

$$\left.\frac{dC_{2N}(\omega)}{d\omega^2}\right|_{\omega=\chi} = \left.\frac{dA^2(\omega)}{d\omega^2}\right|_{\omega=\chi}. \tag{5.33}$$

Similar requirements can be written for polynomial $Q_{2N}(\omega)$.

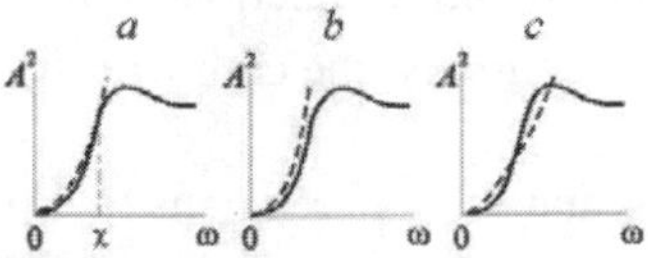


Fig. 5.4

If only one coefficient of polynomial $C_{2N}(\omega)$ is unknown, for example the coefficient $C_N$, then (5.32) and (5.33) can be considered as a system of equations with two unknowns; $\chi$ and $C_N$. The result of its solution is the expression for $C_N$ through other coefficients of polynomial $C_{2N}(\omega)$ and parameters of function

$$C_N = \Psi\left(\left\{c_i\right\}_K^{N-1}, A^2\right), \tag{5.34}$$

and also the appropriate expression for frequency $\chi$.

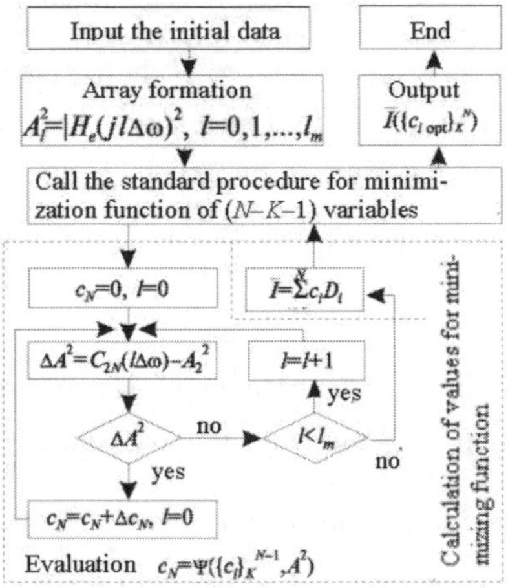


Fig. 5.5

If the coefficient $C_N$ has the greater value, than the value found by the expression (5.34), then the curve $C_{2N}(\omega)$ lies above the curve $A^2(\omega)$ in all frequencies $\omega > 0$ (Fig. 5.4*b*), i.e. equation (5.32) may have only one non-negative real root in the point $\omega = 0$. If the coefficient $C_N$ has a little bit of a smaller value than its value found by (5.34), then the curves $C_{2N}(\omega)$ and $A^2(\omega)$ are intersected twice at $\omega > 0$ (Fig. 5.4*c*), i.e. the equation (5.34) has two various non-negative real roots. At the exact realization of equality (5.34) these two roots are coincided, i.e. there is a non-negative real root of multiplicity 2 among the solutions of equation (5.32). The latter statement may be used to determine expression (5.34) by the analysis of the equation (5.32), avoiding the solution of combined equations (5.32) and (5.33). Examples of such research are given in the book [8] on pages 70-71.

**Coefficients optimization for approximating polynomial.** According to (5.30) and (5.34), write the criterion function for coefficient optimization in the polynomial $C_{2N}(\omega)$ as:

$$\bar{I} = \sum_{i=K}^{N-1} c_i D_i + \Psi\left(\left\{c_i\right\}_K^{N-1}, A^2\right) D_N \to \min . \qquad (5.35)$$

By the numerical optimization based on expressions (5.32) - (5.35) it is possible to use the algorithm which structure is shown in Fig. 5.5.

The standard procedure to extremize the function of several variables from the structure of universal computer software is used in an algorithm. The coefficient $C_N$ satisfying to (5.34) for each set of coefficient $\left\{c_i\right\}_K^{N-1}$ values can be determined by enumerative technique. Beginning from value $C_N = 0$ this factor increases with step $\Delta C_N$ until the realization of (5.29) for all frequencies $\omega \in \left[0, l_m \Delta\omega\right]$ are not be provided. Here $\Delta\omega$ is a step of frequency variation, $l_m$ is the number of increments showing the number by which the considered frequency range is divided. The bound frequency $l_m \Delta\omega$ is selected so that its value exceeds several times the width of the interval, where the intensive variation of function $A^2(\omega)$ derivatives takes place. The step value $\Delta C_N$ determines the accuracy of evaluation of the optimal value $c_N$. Initial conditions for calculation must include the parameters of function $A^2(\omega)$, initial values of coefficients $\left\{c_i\right\}_K^{N-1}$ and the required accuracies of their optimization.

For optimization of coefficients $\left\{c_i\right\}_K^N$ (or $\left\{q_i\right\}_K^N$) it is possible to use the linear programming methods, in particular, the simplex-method. If $N = K + 1$ then such a problem has the obvious geometrical interpretation. Thus it is possible to show, that the optimal value $c_K$ coincides with the value of $K$-th expansion coefficient of $A^2(\omega)$ in McLoren series by powers $\omega^2$ independent on values $\left\{D_i\right\}_K^N$ within the limits of some set of these values providing rather small width of action spectrum. The analytical solution of optimization problem [8] is possible in simple cases.

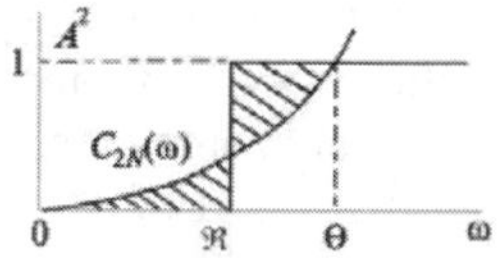


Fig. 5.6

**The account of additional information about spectrum of action.** One of the advantages of the approximative method is the possibility to account the additional information on the shape of the curve for spectral density of action, for example, on its unimodality. A procedure of such an account is explained in an example of rectangular function $A^2(\omega)$, shown in Fig. 5.6. It is known, that the function $S(\omega)$ is non-increasing in interval $\omega \in [0,\infty)$ or at least in interval $\omega \in [\Re,\infty)$ at $\int_0^{\Re} S(\omega)\, d\omega \geq S(\Re)\Re$.

When choosing the polynomial $C_{2N}(\omega)$, the condition (5.29) may be partially broken, which will cause the amplification of estimation $\overline{I}$.

In reality the curve $C_{2N}(\omega)$ moves so that the squares of two hatched figures in Fig. 5.6 are equal, i.e.

$$\int_0^{\Re} C_{2N}(\omega)\, d\omega = \int_{\Re}^{\theta} [1 - C_{2N}(\omega)]\, d\omega \tag{5.36}$$

or, due to condition for non-increasing function $S(\omega)$,

$$\int_0^{\Re} C_{2N}(\omega) S(\omega)\, d\omega \geq \int_{\Re}^{0} [1 - C_{2N}(\omega)] S(\omega)\, d\omega .$$

Thus the relations are fulfilled:

$$I = \int_{\Re}^{\infty} S(\omega)\, d\omega \leq \int_0^{\infty} C_{2N}(\omega) S(\omega)\, d\omega = \sum_{i=K}^{N} c_i D_i = \overline{I} \tag{5.37}$$

and estimation (5.30) is valid.

At $N = K$ from condition $C_{2N}(\Theta) = c_N \Theta^{2N} = 1$ follows $\Theta = \sqrt[2N]{\frac{1}{c_N}}$, and from equation (5.36) — $c_N = [2N/(2N+1)]^{2N} \Re^{-2N}$, and (5.37) becomes

$$\overline{I} = \left( \frac{2N}{2N+1} \right)^{2N} \frac{D_N}{\Re^{2N}} . \tag{5.38}$$

If $N = 1$ from (5.38) the widely used inequality developed by Gauss: $\overline{I} = 4D_1 / 9\Re^2$ is obtained. Its proof, which has been carried out here on the basis of the approximative method, differs with its simplicity and refinement. The used method allows development of a more common result [62].

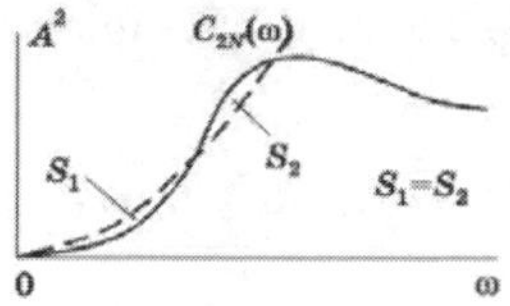


Fig. 5.7

In the case of the non-rectangular function $A^2(\omega)$ and non-increasing spectral density $S(\omega)$ the idea of the analysis is saved, and the approximating polynomial $C_{2N}(\omega)$ may be similar, appropriate to the graph in Fig. 5.7.

## §5.4. Upper estimate of dynamic error dispersion

**Singularities in application of approximative method.** An estimation of dynamic error dispersion from above is the main problem of accuracy analysis for robust systems by dispersions of action derivatives. Consider its solution by the approximative method for the system with unit feedback and fractional rational transfer function of the open loop like (5.1)

$$W(s)=\frac{1+b_1 s+\ldots+b_{n-1}s^{n-1}}{a_0+a_1 s+\ldots+a_n s^n};\ \{a_i\}_0^n,\ \{b_j\}_1^{n-1}\in[0,\infty).$$

Thus it is necessary to accept the following in expression (5.5):

$$I=D_{eg},\ A^2(\omega)=\left|H_e(j\omega)\right|^2=\left|1+W(j\omega)\right|^{-2},\ \omega_*=\beta\to\infty .$$

The approximate kind of function $\left|H_e(j\omega)\right|^2$ and variants of its approximation by polynomials $C_{2N}(\omega)=\sum_{i=K}^{N}c_i\omega^{2i}$ are shown in Fig. 5.8.

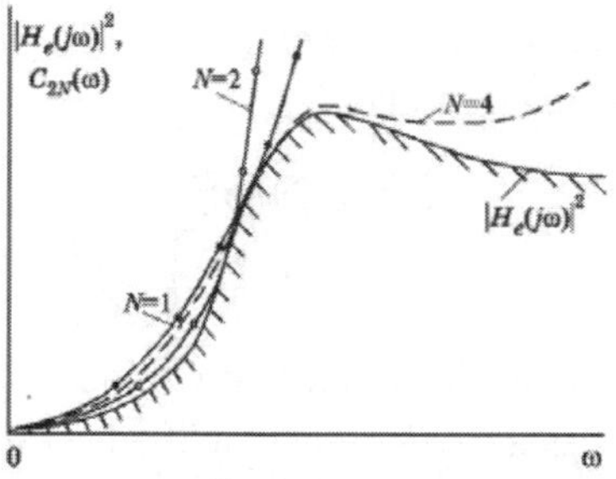


Fig. 5.8

The required coefficients $\{c_i\}_K^N$ should satisfy at least three obvious relations: $K\le k$ (see (5.1), (5.31)) $c_K>0$ and $c_N\ge 0$ (if $c_N=0$ then $c_{N-1}\ge 0$ etc.). The solution is possible even at one known dispersion $D_N$, when $K=N$. However, the increase of their number in common cases allows strengthening the estimation as (5.30):

$$\overline{D}_{eg} = \sum_{i=K}^{N} c_i D_i \, .$$

The strongest estimation is obtained at the optimization of the coefficients $\{c_i\}_K^N$ by criterion (5.27). Except for a case, when $n = 1$, considered in [8] on page 71, optimization is connected to the use of a computer (see § 5.3). However, the close to optimal values of coefficients can be analytically found rather easily. Consider two methods of such an investigation.

**Use the expansion of approximated function in power series.** For deriving the good estimate $\overline{D}_{eg}$ by the formula (5.30), the practical coincidence of curves $C_{2N}(\omega)$ and $|H_e(j\omega)|^2$ should be provided first of all on frequencies where the basic power of reference action is concentrated. When researching high-precision control systems, the characteristic case is the occurrence when these frequencies correspond to the initial interval of curve $|H_e(j\omega)|^2$, lying near the origin of coordinates. The strongest influence on function $C_{2N}(\omega)$ in a low-frequency area is provided by the coefficients $c_i$ with the younger indexes, the weak influence is provided by the coefficient $c_N$. Therefore, the optimal values of coefficients $c_i$ with younger indexes should be close to the appropriate coefficients for expansion of function $|H_e(j\omega)|^2$; in McLoren's series by the powers $\omega^2$, even if some increase of coefficient $c_N$ in comparison with coefficient at $\omega^{2N}$ in McLoren's series is required to utilize condition (5.29) $C_{2N}(\omega) \geq |H_e(j\omega)|^2$ on higher frequencies. Therefore, it is possible to refuse the strict optimization of coefficients $c_i$ with younger indexes without the essential turn for the worse of estimation $\overline{D}_{eg}$ and to determine them by the formula

$$c_i = \frac{1}{i!} \frac{d^i |H_e(j\omega)|^2}{d\omega^{2i}} \Bigg|_{\omega=0} . \qquad (5.42)$$

It causes a decreasing number of unknowns in the system of equations (5.32), (5.33), (5.35) and simplifies the determination of all coefficients $\{c_i\}_K^N$. In a limiting case the formula (5.42) can be used at $i = \overline{K, N-1}$, and the coefficient $c_N$ can be found as a type of (5.34), from the condition of actualization of inequality (5.29).

Let, for example, the transfer function for the open loop of a system be described by expression (5.1) at $n = N = 2$, $k = K = 1$, i.e. it looks like

$$W(s)=\frac{1+b_1 s}{a_1 s+a_2 s^2}=\frac{K_1(1+b_1 s)}{s(1+T_1 s)}, \tag{5.43}$$

and $D_2/D_1 << a_2^{-1}$, $b_1 < a_2/a_1$. Coefficients $c_0$ and $c_1$ are determined by formula (5.42):

$$|H_e(j\omega)|^2=\frac{a_1^2\omega^2+a_2^2\omega^4}{1+\left[(a_1+b_1)^2-2a_2\right]\omega^2+a_2^2\omega^4},$$

$$c_0=|H_e(j0)|=0, c_1=\left.\frac{d|H_e(j\omega)|^2}{d\omega^2}\right|_{\omega=0}=a_1^2.$$

Thus $C_{2N}(\omega)=a_1^2\omega^2+c_2\omega^4$, where coefficient $c_2$ must be found from the solution of combined equations (5.32), (5.33) whereby condition $C_{2N}(\omega)\geq|H_e(j\omega)|^2$ is satisfied. It is also possible to calculate it by the immediate solution of the indicated system of equations, or as the result of investigation of equation (5.32). In this case it looks like

$$\frac{a_1^2\chi^2+a_2^2\chi^4}{1+\left[(a_1+b_1)^2-2a_2\right]\chi^2+a_2^2\chi^4}=a_1^2\chi^2+c_2\chi^4$$

or, getting rid of the denominator in the left hand side and not taking into consideration the fourfold zero root,

$$\Theta_0\chi^4+\Theta_1\chi^2+\Theta_2=0, \tag{5.44}$$

where

$$\Theta_0=c_2a_2^2, \Theta_1=a_1^2a_2^2+c_2(a_1+b_1)^2-2c_2a_2,$$
$$\Theta_2=c_2+\left[(a_1+b_1)^2-2a_2\right]a_1^2-a_2^2.$$

At the correct choice of magnitude $c_2$ equation (5.44) should have double positive real roots or should have no nonzero real roots at all. In the first case $\chi>0$, in second one $\chi=0$. The enumerated requirements allow correlation of the coefficients of equation (5.44) by relations

$$\begin{cases}\Theta_1^2-4\Theta_0\Theta_2=0 & \text{at } \Theta_1<0,\\ -\Theta_1+\sqrt{\Theta_1^2-4\Theta_0\Theta_2}=0 & \text{at } \Theta_1\geq 0.\end{cases} \tag{5.45}$$

From here after the transformations, the required expression is obtained:

$$
c_2 = \begin{cases} a_2^2 \dfrac{2a_2\sqrt{\left(a_1^2 + a_2\right)^2 - a_1^2\left(a_1 + b_1\right)^2} + 2a_2\left(a_1^2 + a_2\right) - a_1^2\left(a_1 + b_1\right)^2}{\left(a_1 + b_1\right)^2\left[4a_2 - \left(a_1 + b_1\right)^2\right]} \\ \qquad\qquad \text{at } \left(a_1 + b_1\right)^2 < d, \\ a_2^2 + a_1^2\left[2a_2 - \left(a_1 + b_1\right)^2\right] \quad \text{at} \quad d \le \left(a_1 + b_1\right)^2 \le 2a_2 + a_2^2/a_1^2, \\ \qquad 0 \quad \text{at} \quad \left(a_1 + b_1\right)^2 > 2a_2 + a_2^2/a_1^2, \end{cases} \tag{5.46}
$$

where $d = 2a_2\left(3a_1^2 + 2a_2\right)\left[4a_1^2 + a_2 + \sqrt{4a_1^4 + a_2^2}\right]^{-1}$.

For the numerical values of system parameters $a_1 = b_1 = 1\,\text{sec}$, $a_2 = 4\text{sec}^2$ the expression (5.45) gives $c_2 = 24.2\text{sec}^4$. The graphs for functions $\left|H_e(j\omega)\right|^2$ and $C_{2N}(\omega) = C_4(\omega)$ in this case look like the one shown in Fig 5.9. The curve $Q_4(\omega)$, constructed at the determination of the lower estimate $\underline{D}_{eg}$ (see §5.5) at, $q_2 = -0.22\text{sec}^4$, is also figured there. If $D_1 = 1\,(\text{deg/sec})^2$, $D_2 = 10^{-2}\,(\text{deg/sec})^2$ then the found estimations equal $\overline{D}_{eg} = 1.24\,\text{deg}^2$, $\underline{D}_{eg} = 0.998\,\text{deg}^2$.

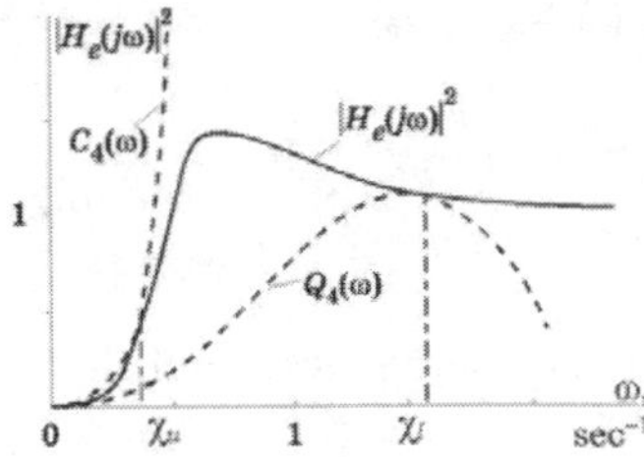


Fig. 5.9

Since the used mode to determine the coefficients $c_1$ and $c_2$ does not include the procedure for their optimization, the obtained estimation $\overline{D}_{eg} = C_1 D_1 + C_2 D_2$ is not necessarily the extreme strong, but may insignificantly differ from it in practice. As it was already noted, with the application of the simplex-method it is possible to show that the obtained estimate in some important cases coincides strictly with the extreme strong estimation.

If the analytical solution is impossible then the coefficient $c_N = \Psi\left(\left\{c_i\right\}_0^{N-1}, \left\{a_j\right\}_0^n, \left\{b_k\right\}_0^{n-1}\right)$ may be found by the numerical methods based on the algorithm, of which the block diagram is marked in Fig. 5.5 by the dotted line.

**Elementary mode to determine the coefficients of approximative polynomial.** If the order of system transfer functions coincides with the order of the higher restricted action derivative, and the order of system astatism coincides with the order of the lower restricted action derivative, i.e. $n = N$ and $k = K$, then it is possible to use the following elementary method to determine the coefficients $\left\{c_i\right\}_K^N$. The following expression can be written according to (5.1):

$$\left|H_e(j\omega)\right|^2 = \left|1+W(j\omega)\right|^{-2} = \left|\sum_{i=K}^{N} a_i (j\omega)^i\right|^2 [P(\omega)]^{-1}, \qquad (5.47)$$

where $P(\omega) = \left|1 + \sum_{j=1}^{N-1} b_j (j\omega)^j + \sum_{i=K}^{N} a_i (j\omega)^i\right|^2$.

Let $P_M = \min_{\omega \in [0,\infty)} P(\omega)$. Then, with the account of (5.47), the inequality

$$\left|H_e(j\omega)\right|^2 \le \left|a_K (j\omega)^K + a_{K+1}(j\omega)^{K+1} + \ldots + a_N (j\omega)^N\right|^2 P_M^{-1}, \omega \in [0,\infty)$$

is valid.

It provides the foundation for polynomial $C_{2N}(\omega)$ choice, satisfying condition (5.29), as

$$C_{2N}(\omega) = \left|\sum_{i=K}^{N} a_i (j\omega)^i\right|^2 P_M^{-1}$$

or

$$c_0 = a_0^2 / P_M, c_1 = \left(a_1^2 - 2a_0 a_2\right)/P_M, c_2 = \left(a_2^2 - 2a_1 a_3 + 2a_0 a_4\right)/P_M,$$
$$c_3 = \left(a_3^2 - 2a_2 a_4 + 2a_1 a_5 - 2a_0 a_6\right)/P_M, \ldots, c_n = a_n^2 / P_M. \qquad (5.48)$$

It is easy to show, that at real poles of function $\left|H_e(j\omega)\right|^2$, the obtained expression is $P_M = \left|H_e(j0)\right|^2 = (1+a_0)^2$, and at complex poles it is: $P_M \le (1+a_0)^2$. If $b_j = 0, j = \overline{1, N-1}$ then $P_M = M^{-2}$, where $M$ is an oscillation index for the closed loop system. In any case, the magnitude $P_M$ is the non-increasing function of the oscillation index, and $P_M \ge M^2$.

On frequencies, which are essentially less than the cutoff frequency, the function $P(\omega)$ varies rather poorly and practically coincides with its value on zero frequency $P(0) = (1+a_0)^2$. Therefore on a very low-frequency spectrum of reference action the use of formula (5.48) instead of the more accurate formula (5.42) for the choice of coefficients $c_i$ may cause the overestimation of $\overline{D}_{eg}$ not more than $P(0)/P_M$ times. At the extension of reference action spectrum, or at decrease the oscillation index of the closed loop system, the difference in accuracy of estimations $\overline{D}_{eg}$, obtained with use of formulas (5.48) and (5.42), decreases.

Note that at $k = n = K = N$, when the dispersion for only one derivative of action is known, the formula (5.48) offers the optimal values of coefficients $c_i$.

As an example consider once more the system with transfer function (5.43) at $N = 2$, $K = 1$. The following expression is therefore given:

$$P_M = \min_{\omega \in [0,\infty)} \left|1 + (a_1 + b_1) j\omega + a_2 (j\omega)\right|^2 =$$

$$= \begin{cases} (a_1 + b_1)^2 \left[4a_2 - (a_1 + b_1)^2\right] \left(4a_2^2\right)^{-1} & \text{at } b_1 \le \sqrt{2a_2} - a_1, \\ 1 & \text{at } b_1 > \sqrt{2a_2} - a_1, \end{cases} \qquad (5.49)$$

that allows the use of formulas (5.48) and (5.30), which looks like:

$$c_1 = a_1^2 / P_M,\ c_2 = a_2^2 / P_M,\ \overline{D}_{eg} = c_1 D_1 + c_2 D_2 .$$

The substitution of numerical values $a_1 = b_1 = 1 \text{ sec}$, $a_2 = 4 \text{ sec}^2$, $D_1 = 1 \text{ (deg/sec)}^2$, $D_2 = 10^2 \text{ (deg/sec}^2\text{)}^2$ equals $P_M = 0.75$, $c_1 = 1.33 \text{ sec}^2$, $c_2 = 21.3 \text{ sec}^4$, $\overline{D}_{eg} = 1.331 + 21.3 \cdot 10^{-2} = 1.54 \text{ deg}^2$. The found magnitude $\overline{D}_{eg}$ exceeds 24 % of the stronger upper estimation obtained earlier at $c_1 = a_1^2$ and with the use of formula (5.45) for coefficient $c_2$.

Detailed results of comparative analysis for estimations $\overline{D}_{eg}$, obtained using formula (5.45) and other modes explained above are also available in the book [8] (pages 75-76).

So, three modes to determine the coefficients of polynomial $C_{2N}(\omega)$, satisfying the condition $C_{2N}(\omega) \ge |H_e(j\omega)|^2$, are considered. The numerical analysis according to expressions (5.34) and (5.35) (mode 1) allows development of the strongest estimation $\overline{D}_{eg}$ such as (5.30). Formulas (5.42) and (5.34) (mode 2) and especially (5.48) (mode 3) in a general case present the less strong estimation $\overline{D}_{eg}$, however they are simpler.

**The case when spectrum of action width is finite.** Information regarding the spectrum of action can be considered as concentrated in finite interval of frequencies $\omega \in [0, \beta]$, and sometimes enables the strengthening of estimation $\overline{D}_{eg}$ at the expense of a greater freedom in choice of polynomial $C_{2N}(\omega)$, which corresponds to condition (5.29). Here two cases for mutual positioning of curves $|H_e(j\omega)|^2$ and $C_{2N}(\omega)$ appropriate to graphs in Fig. 5.10 are characteristic.

In a case when the bound frequency $\beta$ is small enough (Fig 5.10*a*), it is expedient to choose the coefficients $\{c_i\}_K^{N-1}$ by formula (5.42), and the coefficient $c_N$ from the condition $C_{2N}(\beta) = |H_e(j\beta)|^2$. In the case shown in Fig 5.10*b*, the coefficient $c_N$ is taken as negative. This causes the decrease of estimation $\overline{D}_{eg}$ at the increasing dispersion $D_N$.

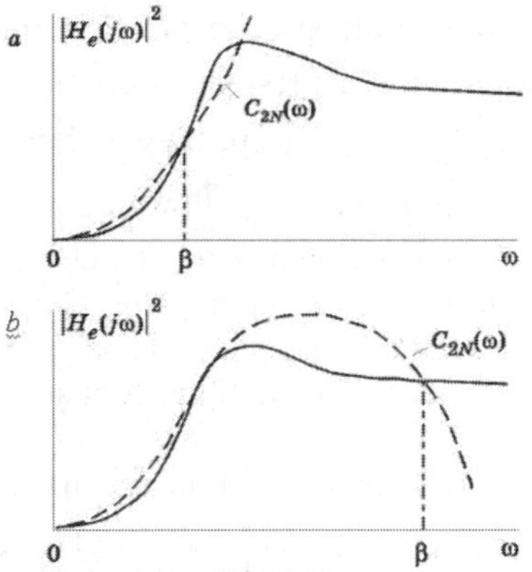


Fig. 5.10

However, it is necessary to note that it is possible only in a situation, when $D_N$ is the actual value for dispersion of the *N*-th derivative of action, but not for its greatest possible value which may not be reached in a considered system operation mode. Otherwise with multiplication by the coefficient $c_N < 0$ it would be necessary to take the smallest possible value of dispersion $D_N$ in the formula (5.30). If such a small value is equal to zero, then the account of finite width of the spectrum does not cause the amplification of estimation $\overline{D}_{eg}$.

This note is fair not only at $c_N < 0$, but also generally with the presence of negative coefficients with any indexes in the set $\{c_i\}_K^N$, independent of spectrum width β. If $\beta \to \infty$, then some coefficient $c_j$ may be negative at $K < j < N$, and is usually presented third among positive coefficients $c_{j-1}$ and $c_{j+1}$. If the dispersions of derivatives $\{D_i\}_K^N$ are restricted only by the problem situation from above, then it is necessary to substitute $D_j = 0$ into formula (5.30). If $K < j < N-1$, then at the accepted values $D_{j+1}$ and $D_{j+2}$ the minimum possible value according to (4.2) is nonzero and it satisfies condition $D_j \geq D_{j+1}^2 / D_{j+2}$.

Substitution of the different values of dispersions $D_i$, depending on the signs of coefficients $c_i$ into formula (5.30), makes the optimization problem of these coefficients a nonlinear programming problem and slightly complicates its solution with use of the algorithms shown in Fig. 5.5. It also forces the appropriate conditional transitions to be added into it and the more accurate search for the global extremums to be organized. However, in the majority of practical cases the necessity for such complication does not arise, because the optimal coefficients

$c_{K+1}$ are positive, although theoretically they might be negative. Nevertheless at $N = K$ or $N = K+1$, $\beta \to \infty$ all coefficients $\{c_i\}_K^N$ are positive.

Note, that at determination of the lower estimation of error dispersion $\underline{D}_{eg}$ it is necessary to take those values for dispersions of action derivatives, independent of the signs of coefficients $\{q_i\}_K^N$, which were used with the determination of the upper estimation $\overline{D}_{eg}$. Such estimation $\underline{D}_{eg}$ allows, as it was already shown in §5.1, insertion of the measure of indeterminacy for the characteristics of system accuracy, because a priori information is incomplete. Concerning the unconditional lower estimate of error dispersion, it has no practical interest, but at absence of restrictions from below on dispersions of actions derivatives it equals to zero in any case.

The additional a priori information about the spectral density of action $S_g(\omega)$ is a non-increasing function in interval $\omega \in [0, \infty)$. It always allows strengthening of the upper estimation of error dispersion $\overline{D}_{eg}$ due to the choice of polynomial $C_{2N}(\omega)$ with partial violation of condition (5.29). The procedure of such investigation is described in §5.3 (see Fig. 5.7).

## §5.5. The lower estimate for dispersion of dynamic error

**The basic variant of use of the approximate method.** For determination of the lower estimate of dynamic error dispersion $\underline{D}_{eg}$ by the approximate method it is necessary to find the polynomial $Q_{2N}(\omega) = \sum_{i=K}^{N} q_i \omega^{2i}$ coefficients, satisfying condition (5.29): $Q_{2N}(\omega) \le |H_e(j\omega)|^2$ on all frequencies where the spectral components of action are possible. At frequencies where their presence is most possible, it is necessary to find the polynomial coefficients that approximate the function $|H_e(j\omega)|^2$ well enough.

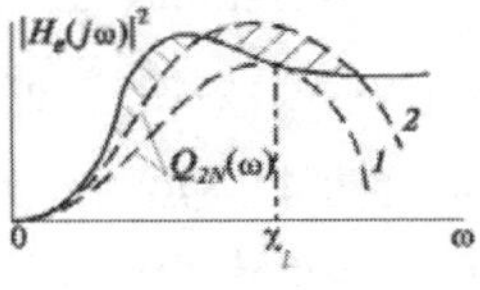


Fig. 5.11

In Fig. 5.11 the typical mutual position of curves $Q_{2N}(\omega)$ (curve 1) and $|H_e(j\omega)|^2$ at the unlimited spectrum of action width is shown. It is possible only at $K \ge k$ and $N \ge K+1$, i.e. at the restricted dispersions for at least two derivatives. The order of this younger derivative is not less than the order of system astatism.

It is possible to determine the set of coefficients $\{q_i\}_K^N$ in an optimal way from criterion (5.27) on the basis of an algorithm similar to the one described in §5.3, or with the use of approximate analysis. In particular, at $K = k$ it is admissible to take coefficients $q_i$ with the younger indexes $i = \overline{K, N-1}$ that are coincided with the appropriate coefficients of the function $|H_e(j\omega)|^2$ expansion in the McLoren series by powers $\omega^2$, and to find the coefficient $q_N < 0$ satisfying condition (5.29).

For example, as investigated in §5.4, the second order system with the first order astaticism (zero-constant-error system of type 1) is equal to transfer function (5.43). At $K = 1$, $N = 2$ the following is obtained:

$$q_1 = \left. \frac{d|H_e(j\omega)|^2}{d\omega^2} \right|_{\omega=0} = a_1^2 . \tag{5.50}$$

Determine the coefficient $q_2$, based on the condition, that inequality $Q_{2N}(\omega) < |H_e(j\omega)|^2$ is fulfilled on all frequencies $\omega \in (0, \infty)$ except for frequency $\omega = \chi_l$, where it turns into an equality (see Fig. 5.11). For this purpose it is necessary to investigate the expressions, which differ from (5.44) and (5.45) only by their parameters that depend not on coefficient $c_2 > 0$, but on coefficient $q_2 < 0$. All together it is possible to develop the expression

$$q_2 = \begin{cases} \dfrac{a_1^4 a_2^2}{a_1^2(a_1+b_1)^2 - 2a_2\left(a_1^2 + a_2\right) - 2a_2\sqrt{\left(a_1^2 + a_2\right)^2 - a_1^2(a_1+b_1)^2}} \\ \qquad \text{at } (a_1+b_1)^2 < d_1, \\ a_2^2 - a_1^2\left[(a_1+b_1)^2 - 2a_2\right] \ \text{at } (a_1+b_1)^2 \ge d_1, \end{cases} \tag{5.51}$$

where $d_1 = 2a_2\left(3a_1^2 + 2a_2\right)\left[4a_1^2 + a_2 - \sqrt{4a_1^2 + a_2^2}\,\right]^{-1}$.

Substitution of the numerical data $a_1 = b_1 = 1$ sec, $a_2 = 4$ sec$^2$ equals $q_1 = 1$ sec$^2$, $q_2 = -0.22$sec$^4$. From here at $D_1 = 1$ (deg/sec)$^2$, $D_2 = 10^{-2}$ (deg/sec$^2$)$^2$ $D_{eg} = q_1 D_1 + q_2 D_2 = 1 - 0.22 \cdot 10^{-2} = 0.998$ deg$^2$ is obtained. The graph of function $Q_4(\omega) = q_1\omega^2 + q_2\omega^4$ is shown in Fig. 5.9.

At the mentioned above mode to choose the coefficients close to optimal, as a rule, inequality $|q_N| << a_N^2$ is valid. Therefore at realization of condition

$\sqrt{D_N/D_{N-1}} \approx a_{N-1}/a_N$ or, especially, conditions $\sqrt{D_N/D_{N-1}} << a_{N-1}/a_N$ , it is possible to claim, that

$$\underline{D}_{eg} \cong \sum_{i=K}^{N-1} q_i D_i = \sum_{i=K}^{N-1} \frac{D_i}{i!} \frac{d^i |H_e(j\omega)|^2}{d\omega^{2i}} \bigg|_{\omega-0} . \tag{5.52}$$

In the example considered above the expression (5.52) results in $\underline{D}_{eg} = 1 \text{ deg}^2$, which has a relative error equal to 0.22 %.

If at determination of estimate $\underline{D}_{eg}$ formula (5.52) was used, and at determination of estimate $\overline{D}_{eg}$ — formulas (5.42) and (5.30), i.e. $q_i = c_i$, $i = \overline{K, N-1}$, $q_N < 0$, $c_N > 0$, $|q_N| << c_N$ then the difference between the upper and lower estimations on hand (5.48) makes:

$$\overline{D}_{eg} - \underline{D}_{eg} = (c_N - q_N) D_N \approx c_N D_N \approx a_N^2 P_M^{-1} D_N . \tag{5.53}$$

The information about the non-increasing function $S_g(\omega)$ in an interval of $\omega \in [0, \infty)$ allows for the strengthening of the estimate $\underline{D}_{eg}$ when choosing polynomial $Q_{2N}(\omega)$ with partial violation of condition (5.29). The appropriate graph is shown in Fig. 5.11 (curve 2) where the squares of two hatched figures are equal.

If the spectrum of action is restricted in width by an interval $\omega \in [0, \beta)$ then the realization of condition $Q_{2N}(\omega) \le |H_e(j\omega)|^2$ on frequency $\omega > \beta$ is not required. It allows the "higher" position of function $Q_{2N}(\omega)$ graph to be obtained and the estimate $\underline{D}_{eg}$ to be strengthened in comparison with the case $\beta \to \infty$ . See the example in Fig. 5.12.

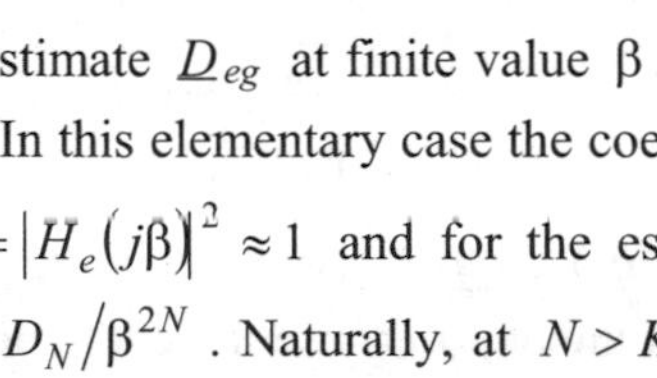

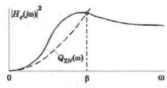

Fig. 5.12

For determination of estimate $\underline{D}_{eg}$ at finite value $\beta$, only one dispersion $D_N$ is enough if $N = K \ge k$. In this elementary case the coefficient $q_N$ is determined from condition $q_N \beta^{2N} = |H_e(j\beta)|^2 \approx 1$ and for the estimation it is possible to write the formula $\underline{D}_{eg} \approx D_N / \beta^{2N}$ . Naturally, at $N > K$ the estimation becomes stronger.

**Simplified variants of use of the approximative method.** If $N = n$, $K = k$ then the formula

$$Q_{2N}(\omega)=\left|a_k(j\omega)^k+a_{k+1}(j\omega)^{k+1}+...+a_n(j\omega)^n\right|^2 R^{-1}, \qquad (5.54)$$

where $R=\max\limits_{\omega\in[0,\beta]}\left|1+\sum_{i=1}^{n-1}b_i(j\omega)^i+\sum_{i=k}^{n}a_i(j\omega)^i\right|^2$,

gives the acceptable values of polynomial $Q_{2N}(\omega)$ coefficients.

One more way to determine the estimate $\underline{D}_{eg}$ at any (in particular — indefinitely large) value β consists of selection of polynomial $Q_{2N}(\omega)$, satisfying the condition $Q_{2N}(\omega)\le\left|H_e(j\omega)\right|^2$ in a finite interval of frequencies $\omega\in[0,\omega_*]$, $\omega_*<\beta$ with subsequent use of the formula

$$\underline{D}_{eg}=\sum_{i=K}^{N}q_iD_i^*, \qquad (5.55)$$

where $D_i^*$ is the lower (at $q_i\ge 0$) or upper (at $q_i<0$) estimation of integral similar to (5.5) at $A^2(\omega)=\omega^{2i}, S(\omega)=S_g(\omega)$.

The calculation of values $\{D_i^*\}_K^N$ can be made by the approximative method, or by the canonical presentation method. In particular, the simple formulas (5.11) and (5.12) can be used for this purpose. Such problems are considered in more detail in [53] in pages 79-82.

It is necessary to note that the indicated mode does not allow obtainment of the stronger estimation $\underline{D}_{eg}$ than the immediate use of the approximative method with the optimization of coefficients $\{q_i\}_K^N$. Its advantage is only the simplicity of calculations.

## §5.6. Design equations at known dispersions of action and two of its younger derivatives

At the estimation of dynamic error dispersion, the typical situation is one, when the dispersion of reference action is $D_0$ and the dispersion of two of its younger derivatives $D_1$ and $D_2$, i.e. $N=2$, are known. Consider the results obtained in §5.2-5.5 from the relations between upper and lower boundaries of error dispersions in this important special case.

The following cases may be formally reduced to the indicated modes. If only two dispersions $D_0$ and $D_1$ or $D_1$ and $D_2$ are known, then for the unknown third dispersion it is possible to accept the indefinitely large value and to assume, accordingly, $D_2\to\infty$ or $D_0\to\infty$. If $N>2$, but only the dispersions $D_{N-2}$, $D_{N-1}$ and $D_N$ are known, then is possible not to consider the reference action, but its $(N-2)$-th derivative and to transform the block diagram of a system in the appropriate way.

The upper and lower estimations of dynamic error dispersion in systems with transfer functions similar to (5.1) can be found by using the canonical presentations method (the canonical presentations at $K=0$, $N=2$ are expressed by the formulas (5.15) and (5.16)), or with use of the approximative method. The outcomes for such research of first, second and third orders systems are considered below.

**First order systems.** In a case

$$W(j\omega)=(a_0+a_1 s)^{-1}=K_0(1+T_1 s)^{-1},$$

where $a_0=K_0^{-1}\in[0,\infty), a_1=T_1K_0^{-1}\in[0,\infty)$, the following function is obtained:

$$|H_e(j\omega)|^2=\frac{a_0^2+a_1^2\omega^2}{(1+a_0)^2+a_1^2\omega^2}=\frac{1+T_1^2\omega^2}{(K_0+1)^2+T_1^2\omega^2}, \tag{5.56}$$

It has non-negative third derivative by $\omega^2$ and satisfy condition (5.13) in interval $\omega\in[0,\infty)$. Hence, this function is equivalent to the Chebyshev prolongation of power function system $\{\omega^{2i}\}_0^2$, and the accurate boundaries for error dispersion may be found by using the upper and lower main presentations. Such boundaries are expressed by formulas (5.19) — (5.22). Some special cases are particularly considered in the book [8].

**Second order systems.** In the case

$$W(j\omega)=(1+b_1 j\omega)\left[a_0+a_1 j\omega+a_2(j\omega)^2\right]^{-1},$$

where $a_0, a_1\in[0,\infty), a_2\in(0,\infty)$, the following function is obtained

$$|H_e(j\omega)|^2=\frac{a_0^2+\left(a_1^2-2a_0a_2\right)\omega^2+a_2^2\omega^4}{(1+a_0)^2+\left[(a_1+b_1)^2-2(1+a_0)a_2\right]\omega^2+a_2^2\omega^4}, \tag{5.57}$$

Its third derivative by $\omega^2$ has negative values in the low frequency area. Hence, condition (5.13) is not fulfilled and use of the canonical presentation method to determine the accurate boundaries of error dispersion is complicated.

For deriving the strongest estimation of error dispersion from above (5.30) it is necessary to choose the polynomial coefficients $C_4(\omega)$ in an optimal way. It is possible to solve the problem by using numerical methods (see Fig. 5.5). The estimate, which is rather close to the strongest one, can be founded by formula (5.42). In the case of $a_0=0$ it is possible to use formulas (5.43) — (5.46).

The simplest estimate from above, according to expressions (5.47) and (5.48) looks like:

$$\overline{D}_{eg}=\left[a_0^2D_0+\left(a_1^2-2a_0a_2\right)D_1+a_2^2D_2\right]P_M^{-1}, \tag{5.58}$$

where

$$P_M = \begin{cases} \dfrac{(a_1+b_1)^2\left[4a_2(1+a_0)-(a_1+b_1)^2\right]}{(4a_2^2)} & \text{at} \quad b_1 \le \sqrt{2a_2(1+a_0)}-a_1, \\ (1+a)^2 & \text{at} \quad b_1 \ge \sqrt{2a_2(1+a_0)}-a_1. \end{cases} \tag{5.59}$$

Formula (5.59) at $a_0 = 0$ turns into the simpler formula (5.49). The lower estimate at $a_0 = 0$ can be found by using formulas (5.50) and (5.51).

If condition $\sqrt{D_2/D_1} < a_1/a_2$ is satisfied, then the approximate expression (5.60) can be obtained from (5.52)

$$\underline{D}_{eg} \approx \frac{a_0^2 D_0}{(1+a_0)^2} + \frac{(a_1^2 - 2a_0a_2)(1+2a_0) - a_0^2 b_1(b_1+2a_1)}{(1+a_0)^4} D_1 . \tag{5.60}$$

The rather strong estimation from below by formula (5.54) is the most universal. It looks like

$$\underline{D}_{eg} = \left[a_0^2\rho_1 + (a_1^2 - 2a_0a_2)\rho_1\omega_1^2 + a_2^2\rho_1\omega_1^4\right]R^{-1}, \tag{5.61}$$

where

$$R = \max\left\{\left(1+a_0 - a_2\omega_{*0}^2\right)^2 + (a_1+b_1)^2\omega_{*0}^2, (1+a_0)^2\right\} \tag{5.62}$$

and the magnitudes $\rho_1$ and $\omega_1$ are determined by formula (5.15), and it is possible to accept a value

$$\omega_{*0} = \frac{1}{a_2}\sqrt{2a_2(1+a_0)-(a_1+b_1)^2} \approx \frac{1}{a_2}\sqrt{2a_2 - b_1^2}, \tag{5.63}$$

as the approximate value of frequency $\omega_{*0} \ge \sqrt{D_2/D_1}$. Note, that (5.63) is the root of equation

$$\left(1+a_0 - a_2\omega_{*0}^2\right)^2 + (a_1+b_1)^2\omega_{*0}^2 = (1+a_0)^2 .$$

With the use of more simple relations [8] the following is obtained:

$$\underline{D}_{eg} = \left[a_0^2\left(D_0 - \frac{D_2}{\omega_{*0}^4}\right) + (a_1^2 - 2a_0a_1)\left(D_1 - \frac{D_2}{\omega_{*0}^2}\right) + \frac{a_2^2}{D_0}\left(D_1 - \frac{D_2}{\omega_{*0}^2}\right)\right]R^{-1} . \tag{5.64}$$

**Third order systems.** In a case

$$W(j\omega) = \left[1 + b_1 j\omega + b_2 (j\omega)^2\right]\left[a_0 + a_1 j\omega + a_2 (j\omega)^2 + a_3 (j\omega)^3\right]^{-1}$$

the following function is obtained

$$|H_e(j\omega)|^2 = \left[a_0^2 + \left(a_1^2 - 2a_0a_2\right)\omega^2 + \left(a_2^2 - 2a_1a_3\right)\omega^4 + a_3^2\omega^6\right]\left\{(1+a_0)^2 + \right.$$
$$\left. + \left[(a_1+b_1)^2 - 2(1+a_0)(a_2+b_2)\right]\omega^2 + \left[(a_2+b_2)^2 - 2a_3(a_1+b_1)\right]\omega^4 + a_3^2\omega^6\right\}^{-1},$$
(5.65)

It has a non-negative third derivative by $\omega^2$ in the low frequencies area $\omega \in [0, \beta]$. The magnitude $\beta$ depends on the values of parameters $\{a_i\}_0^3, \{b_j\}_1^2$. In the indicated frequency area function (5.65) is the Chebyshev prolongation of power function system $\{\omega^{2i}\}_0^2$. Therefore, the strongest estimation of error dispersion from below in the account of (5.10), (5.16) and (5.65) can be found by the formula

$$D_{eg\min} = \sum_{j=1,2} \rho_j \left|H_e(j\omega_j)\right|^2 = \left(D_0 - \frac{D_1^2}{D_2}\right)\frac{a_0}{(1+a_0)^2} + \frac{D_1^2}{D_2}\left|H_e\left(j\sqrt{\frac{D_2}{D_1}}\right)\right|^2, \quad (5.66)$$

which is valid when condition $\omega_2 = \sqrt{D_2/D_1} \le \beta$ is realized. At values of parameters $a_i$ and $b_i$, providing high control accuracy, this condition is usually fulfilled. Otherwise, in order to determine the lower estimate it is necessary to use the approximative method.

The upper bound of the error dispersion, found by the use of upper main presentation and with formula (5.10), is accurate only when it is possible to accept the assumption about the absence of spectral components in reference action on frequencies $\omega > \beta$, where condition (5.13) is no longer fulfilled.

For the accurate determination of magnitude $\beta$ it is necessary to find the smallest real positive root of equation $d^3|H_e(j\omega)|^2/(d\omega^2)^3 = 0$. In a common case this can be done with use of numerical methods. Present the solution only for two characteristic, [8] special cases, such as at $W(j\omega) = 0.25(1+3aj\omega)^2/(aj\omega)^3$ the value is $\beta = 0.92/a$ and at $W(j\omega) = \left[1 + 2aj\omega + 2(aj\omega)^2\right]/(aj\omega)^3$ its value is $\beta = 0.71/a$. The first approximation value $\beta$ can be determined from the Bode diagram of the open loop system, and the right boundary of Bode diagram interval with inclinations – 60 dB/dec.

The estimation of error dispersion from above by the approximate method with the use of formulas (5.47) and (5.48) looks like

$$\overline{D}_{eg} = \left[a_0^2 D_0 + \left(a_1^2 - 2a_0a_2\right)D_1 + \left(a_2^2 - 2a_1a_3\right)D_2 + a_3^2 D_3\right]/P_M, \quad (5.67)$$

$$P_M = \begin{cases} (1+a_0)^2 - A_3 & \text{at } A_1 < 0, \\ (1+a_0)^2 & \text{at } A_1 \ge 0, A_2 > -a_3\sqrt{3A_1}, \\ (1+a_0)^2 - \max\{0, A_3\} & \text{at } A_1 < 0, \end{cases}$$

where $A_1 = (a_1+b_1)^2 - 2(1+a_0)(a_2+b_2)$, $A_2 = (a_2+b_2)^2 - 2a_3(a_1+b_1)$,

$$A_3 = \frac{2\left(A_2^2 - 3a_3^2 A_1\right)^{3/2} - 2A_2^3 + 9a_3^2 A_1 A_2}{27a_3^4} = \frac{A_1^2\left(A_2^2 - 4a_3^2 A_1\right)}{2\left(A_2^2 - 3a_3^2 A_1^2\right)^{3/2} + 2A_2^3 - 9a_3^2 A_1 A_2}.$$

Expression (5.67) may be used, if the dispersions $D_0$, $D_1$, $D_2$ and the dispersion of the third derivative of reference action $D_3$ are known. Note that at $a_3 = 0$, and $b_2 = 0$ it coincides with expression (5.55).

## §5.7. Estimations for dispersion of error from interference

**Use of the approximative method.** With the research the dispersion of error from interference $D_{ev}$ by dispersions of interference derivatives it is necessary to accept in (5.5) $A^2(\omega) = |H(j\omega)|^2$, $S(\omega) = S_v(\omega)$, and $\omega_* \to \infty$. The determination of good upper and lower estimations of dispersion $D_{ev}$ is a more difficult problem then the determination of estimations for dynamic error dispersion, because the width of the interference spectrum usually and essentially exceeds the width of the system bandpass. Recovery of the low-frequency part of the spectral density curve, or the curve for correlation functions of interference at large values of argument $\tau$, requires knowledge of a rather great number of expansion coefficients for those functions in a series such as (3.7). Actually it demands the knowledge of interference derivatives dispersions of $i = \overline{K, N}$ orders, at $K = 0$. The last requirement (i.e. the boundedness of interference dispersion) is basic. The minimal number of dispersions, which is necessary for the rough estimation of magnitude $D_{ev}$ at the unlimited width of interference spectrum, is equal to three, i.e. $N \geq 2$.

The typical shape of function $|H(j\omega)|^2$ and the variant of its approximation at $N = 2$ by functions $C_4(\omega)$ and $Q_4(\omega)$ like (5.28) are shown in Fig. 5.13.

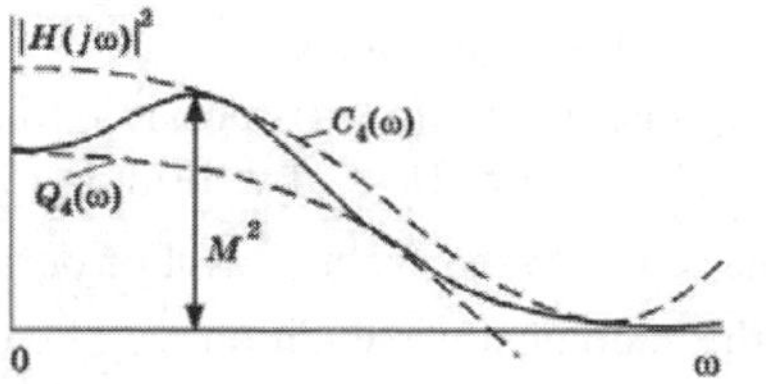


Fig. 5.13

It is easy to make sure that the optimal values of power polynomials coefficients satisfy the conditions

$$c_0 \geq M^2,\ c_1 < 0,\ c_2 > 0,$$

$$q_0 \leq |H(j\omega)|^2,\ q_2 > 0.$$

As the dispersion of the first derivative of interference at the calculation of the upper estimation $\overline{D}_{ev}$ by formula (5.30) is multiplied by the negative coefficient, the increase of this dispersion causes the decrease of the upper estimate for error dispersion, which corresponds to the physical sense of the problem.

The estimations $\overline{D}_{ev}$ and $\underline{D}_{ev}$ are strengthened while increasing the number of known derivatives of interference.

**Use of the canonical representation method.** The boundaries of error dispersion can be determined more simply, but not always accurately by using the canonical representations method. At some canonical representation

$$D_{\nu i} = \sum_{j=1}^{\nu} \rho_{\upsilon j} \omega_{\upsilon j}^{2i}, \ \omega_{\upsilon 1} < \omega_{\upsilon 2} < \ldots < \omega_{\upsilon \nu}, \ i = 0, 1, \ldots, N \tag{5.68}$$

the dispersion of error on account of disturbance makes:

$$D_{e\nu} = \sum_{j=1}^{\nu} \rho_{\upsilon j} \left| H\left(j\omega_{\upsilon j}\right) \right|^2. \tag{5.69}$$

If condition

$$\frac{d^{N+1}\left|H(j\omega)\right|^2}{d\omega^{2(N+1)}} \le 0, \omega \in [0, \infty), \tag{5.70}$$

equivalent to condition (5.13) for function $-\left|H(j\omega)\right|^2$ is satisfied, then the exact boundaries of error dispersion are calculated by formula (5.69) at main presentations. At the lower main representation formula (5.69) provides the upper boundary $D_{e\upsilon\max}$, and at the upper main presentation it provides the lower boundary $D_{e\upsilon\min}$, because the Chebyshev prolongation of function system $\left\{\omega^{2i}\right\}_0^N$ is the function $-\left|H(j\omega)\right|^2$, but not $\left|H(j\omega)\right|^2$. However, condition (5.70) is fulfilled only for $n = 1$ order systems and only at even $N$.

**Approximate formulas.** In practice it is possible to determine the approximate values for boundaries $D_{e\upsilon\max}$ and $D_{e\upsilon\min}$ by formula (5.69) for any order system. Those values are rather close to the exact ones. For deriving the upper boundary it is necessary to take the canonical presentation $\left\{\underline{\omega}_{\upsilon j}\right\}_1^{\nu}, \left\{\underline{\rho}_{\upsilon j}\right\}_1^{\nu}$, which has a component on resonance frequency of system $\underline{\omega}_{\upsilon 1} = \omega_r$, or on zero frequency $\underline{\omega}_{\upsilon 1} = 0$, if gain plot of the closed loop system has no resonance peak. For deriving the lower boundary it is necessary to take the canonical presentation $\left\{\overline{\omega}_{\upsilon j}\right\}_1^{\nu}, \left\{\overline{\rho}_{\upsilon j}\right\}_1^{\nu}$, where the frequency $\overline{\omega}_{\upsilon 1}$ has a maximum possible value $\overline{\omega}_{\upsilon 1} - \overline{\omega}_{\upsilon 1M}$ at realization of relations in (5.68). At odd $N$, when $\nu = (N+3)/2$, in order to provide the uniqueness of the solution it is possible to set the additional conditions $\underline{\rho}_{\upsilon 1} \Rightarrow \max, \overline{\rho}_{\upsilon 1} \Rightarrow \min$.

Really, the lowest-frequency component for canonical presentation with index $j=1$ has the greatest influence on error dispersion. Points $\{\omega_{\upsilon j}\}_1^{\nu}$ are more or less uniformly distributed over the interval on the frequency axis, appropriate to a spectrum width of interference. With a rather large spectrum width and small enough number $\nu$, components with index $j=2$ and the higher-frequency ones lay beyond the effective bandpass of the system and weakly affect the error dispersion.

The magnitudes $\overline{\omega}_{\upsilon 1M}$ and $\overline{\rho}_{\upsilon 1}$ can be roughly estimated as follows. Taking into account, that $\overline{\omega}_{\upsilon 1} < \overline{\omega}_{\upsilon 2} <\ldots< \overline{\omega}_{\upsilon\nu}$, from (5.68) it is possible to obtain the inequality:

$$\overline{\omega}_{\upsilon 1M}^{2i} \sum_{j=1}^{\nu} \overline{\rho}_{\upsilon 1} \le D_i, \overline{\omega}_{\upsilon 1M} \le \sqrt[2i]{D_{\upsilon i}/D_{\upsilon 0}}$$

or, with the account of (4.3),

$$\overline{\omega}_{\upsilon 1M} \le \sqrt{D_{\upsilon 1}/D_{\upsilon 0}} \,. \tag{5.71}$$

It is easy to be convinced that at $\nu=1$ the formulas $\overline{\omega}_{\upsilon 1M} = \sqrt{D_{\upsilon 1}/D_{\upsilon 0}}$, $\overline{\rho}_{\upsilon 1} = D_{\upsilon 0}$ are valid, at $\nu=2$ — $\overline{\omega}_{\upsilon 1M} = \sqrt{D_{\upsilon 1}/D_{\upsilon 0}}$, $\overline{\rho}_{\upsilon 1} = D_{\upsilon 0}$, $\overline{\omega}_{\upsilon 2} \to \infty$, $\overline{\rho}_{\upsilon 2}\overline{\omega}_{\upsilon 2}^4 = D_{\upsilon 2} - D_{\upsilon 1}^2/D_{\upsilon 0}$, and at $\nu>2$ — $\overline{\omega}_{\upsilon 1M} < \sqrt{D_{\upsilon 1}/D_{\upsilon 0}}$, $\overline{\rho}_{\upsilon 1} < D_{\upsilon 0}$. As the value for the lower boundary of error dispersion should not decrease with increasing $N$ and $\nu$, from (5.69) and (5.71) the inequality follows:

$$D_{e\upsilon\min} \ge D_{\upsilon 0}\left|H\left(j\sqrt{D_{\upsilon 1}/D_{\upsilon 0}}\right)\right|^2 . \tag{5.72}$$

For an estimation of value $\underline{\rho}_{\upsilon 1}$ it is necessary to take into account, that at elimination of a component with frequency $\underline{\omega}_{\upsilon 1}$ and power $\underline{\rho}_{\upsilon 1}$ out of the interference spectrum, the rest of the spectral components should have power moments, satisfying conditions (3.3). Therefore the relations

$$\frac{D_{\upsilon,i+1} - \underline{\rho}_{\upsilon 1}\underline{\omega}_{\upsilon 1}^{2i+2}}{D_{\upsilon i} - \underline{\rho}_{\upsilon 1}\underline{\omega}_{\upsilon 1}^{2i}} \le \frac{D_{\upsilon,i+2} - \underline{\rho}_{\upsilon 1}\underline{\omega}_{\upsilon 1}^{2i+4}}{D_{\upsilon,i+1} - \underline{\rho}_{\upsilon 1}\underline{\omega}_{\upsilon 1}^{2i+2}}, i = 0,1,\ldots,N,$$

are valid. Whence

$$\underline{\rho}_{\upsilon 1} \le \frac{D_{\upsilon i}D_{\upsilon,i+2} - D_{\upsilon,i+1}^2}{\underline{\omega}_{\upsilon 1}^{2i}\left(D_{\upsilon,i+2} - 2D_{\upsilon,i+1}\omega_{\upsilon 1}^2 + D_{\upsilon i}\omega_{\upsilon 1}^4\right)} . \tag{5.73}$$

At $\underline{\omega}_{\upsilon 1} = \omega_r << \sqrt[4]{D_{\upsilon 2}/D_{\upsilon 0}}$ from (5.73) and (3.3) the following is accepted

$$\underline{\rho}_{\upsilon 1} \le D_{\upsilon 0} - D_{\upsilon 1}^2/D_{\upsilon 2} \tag{5.74}$$

and according to (5.69),

$$D_{e\upsilon\max} \approx \left(D_{\upsilon 0} - D_{\upsilon 1}^2 / D_{\upsilon 2}\right) M^2 + \sum_{j=2}^{\nu} \rho_{\upsilon j} \left|H\left(j\omega_{\upsilon j}\right)\right|^2 \geq \left(D_{\upsilon 0} - D_{\upsilon 1}^2 / D_{\upsilon 2}\right) M^2 . \quad (5.75)$$

At $N = 2$, $\nu = (N+2)/2 = 2$ relation (5.75) equals

$$D_{e\upsilon\max} \approx \left(D_{\upsilon 0} - D_{\upsilon 1}^2 / D_{\upsilon 2}\right) M^2 + D_{\upsilon 1}^2 D_{\upsilon 2}^{-2} \left|H\left(j\sqrt{D_{\upsilon 2}/D_{\upsilon 1}}\right)\right|^2 . \quad (5.76)$$

Comparing formulas (5.72) and (5.75), it is possible to conclude that the upper and lower boundaries of error dispersion may practically coincide at $D_{\upsilon 2}/D_{\upsilon 1} \approx D_{\upsilon 1}/D_{\upsilon 0}$. However, generally the knowledge of small number of the power moments for spectral density of interference allows for the finding only rather rough estimates of error dispersion. In order to raise the accuracy of estimation it is necessary to find any additional information on properties of disturbing action (interference).

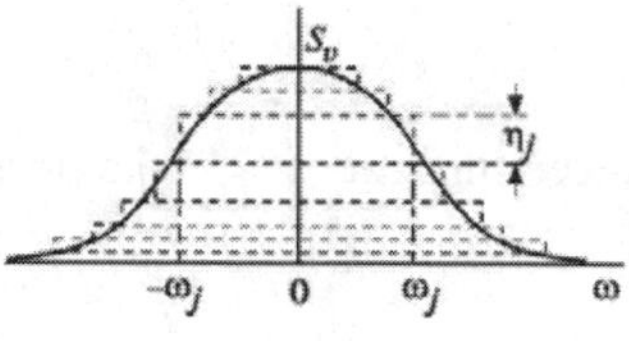


Fig. 5.14

**Use of additional a priori information.** If the statement is valid, that the spectral density of interference $S_\upsilon(\omega)$ is a continuous non-increasing function in interval $\omega \in [0,\infty)$, then at estimation of error dispersion it is possible to present it as the sum of $\upsilon$ rectangular components (Fig. 5.14)

$$S_v(\omega) = \sum_{j=1}^{\nu} S_{\upsilon j}(\omega), \quad (5.77)$$

where

$$S_{\upsilon j}(\omega) = \begin{cases} \eta_j \text{ at } |\omega| \leq \omega_j, \\ 0 \text{ at } |\omega| > \omega_j. \end{cases} \quad (5.78)$$

The values $\{\omega_j\}_1^\nu, \{\eta_j\}_1^\nu$ are determined from the solution of combined equation:

$$\sum_{j=1}^{\nu} \eta_j \omega_j^{2i+1} / (2i+1) = D_{\upsilon i},\ i = 0, 1, \ldots, N. \quad (5.79)$$

The necessary and sufficient condition for such solution to exist is the realization of inequalities [8]

$$\frac{D_{\upsilon,i+2}}{D_{\upsilon,i+1}} \geq \frac{(2i+3)^2}{(2i+5)(2i+1)} \frac{D_{\upsilon,i+1}}{D_{\upsilon,1}}, \ i = 0, 1, \ldots, N-2, \tag{5.80}$$

which are stronger, than inequalities (3.3).

Taking relations (5.77) - (5.80), into consideration by analogy with (5.76) and (5.72) it is possible to obtain the following formulae for boundaries of error dispersion:

$$D_{e\upsilon\max} \approx \left( D_{\upsilon 0} - \frac{9}{5} \frac{D_{\upsilon 1}^2}{D_{\upsilon 2}} \right) |H(j0)|^2 + \frac{9 D_{\upsilon 1}^2 \sqrt{3 D_{\upsilon 1}}}{5 D_{\upsilon 2} \sqrt{5 D_{\upsilon 2}}} \Delta f_{\text{eqv}}, \tag{5.81}$$

$$D_{e\upsilon\min} \geq \frac{D_{\upsilon 0} \sqrt{D_{\upsilon 0}}}{\sqrt{3 D_{\upsilon 1}}} \Delta f_{\text{eqv}}. \tag{5.82}$$

Here $\Delta f_{\text{eqv}} = \frac{1}{\pi} \int_0^\infty |H(j\omega)|^2 \, d\omega$ is the equivalent band pass of a closed-loop system.

More narrow boundaries for error dispersion may be found, if not power but generalized moments for spectral density of interference concerning the basis functions, appropriate to gain plot for low-pass filters, are given.

## §5.8. Use of equivalent harmonic actions

**Conditions for coincidence of strongest estimates with the exact boundaries.** Return to a common problem in order to estimate the extreme values of the functional $I(S(\omega))$ like (5.5) in a class of functions $S(\omega) \in R_1^+$, satisfying restrictions

$$\frac{1}{\pi} \int_0^{\omega_*} \omega^{2i} S(\omega) d\omega = D_i, \ i = \overline{K, N}, \ 0 \leq K \leq N, \ \omega_* \in (0, \infty). \tag{5.83}$$

The approximative method based on the approximation of function $A^2(\omega)$ by power polynomials $C_{2N}(\omega)$ and $Q_{2N}(\omega)$, satisfying conditions (5.29) $C_{2N}(\omega) \geq A^2(\omega) \geq Q_{2N}(\omega)$ in interval $\omega \in [0, \omega_*]$, allows for the obtainment of the strongest upper and lower estimations (5.30) $\overline{I} = \sum_{i=K}^{N} c_i D_i$ and $\underline{I} = \sum_{i=K}^{N} q_i D_i$ with an optimal choice of their coefficients.

On the other hand, with the realization of condition (5.13) $d^{N+1} A^2(\omega) / d\omega^{2(N+1)} \geq 0$ the method of canonical representations (see § 5.2), giving exact boundaries, is valid

$$I_{\max} = \sum_{j=1}^{\nu^*} \overline{\rho}_j A^2(\overline{\omega}_j), I_{\min} = \sum_{j=1}^{\nu_*} \underline{\rho}_j A^2(\underline{\omega}_j). \tag{5.84}$$

Note that the upper (or lower) canonical presentation of power moments (5.80) ensures realization of equalities

$$D_i = \sum_{j=1}^{\nu^*} \overline{\rho}_j \overline{\omega}_j^{2i} \left( \text{or } D_i = \sum_{j=1}^{\nu_*} \underline{\rho}_j \underline{\omega}_j^{2i} \right), i = \overline{K, N}. \tag{5.85}$$

Since (5.30), (5.81) and (5.82) conclude that for coincidence of strongest upper (lower) estimates of magnitude $I$ with the appropriate exact boundary, the following conditions should be satisfied

$$A^2(\overline{\omega}_j) = C_{2N}(\overline{\omega}_j), j = \overline{1, \nu^*} \left( A^2(\underline{\omega}_j) = Q_{2N}(\underline{\omega}_j), j = \overline{1, \nu_*} \right),$$

i.e. the values of function $A^2(\omega)$ and polynomial $C_{2N}(\omega)$ $(Q_{2N}(\omega))$, in which coefficients are optimal by criterion (5.27), and should coincide on the frequencies appropriate to the upper (lower) main presentation (Fig. 5.15).

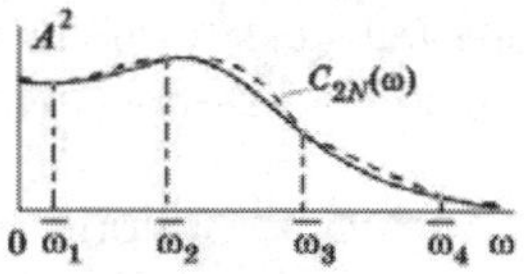


Fig. 5.15

The result obtained here can be spread on the wider class of functions $A^2(\omega)$, which was formally not restricted by condition (5.29), which is based on the following two theorems. Their proof is given in [8] on pages 95-96.

Theorem 1. If some polynomial $C_{2N}(\omega)$ $(Q_{2N}(\omega))$ converts the inequality (5.29) into an equality at $\omega = \omega_j, j = \overline{1, \nu}$ and there is also a set of coefficients $\rho_j$, making function

$$S(\omega) = \sum_{j=1}^{\nu} \rho_j \delta(\omega - \omega_j) \tag{5.86}$$

the solution for a set of equations (5.83), then the strongest estimation $\overline{I}(\underline{I})$ under the formula (5.30) coinciding with the exact boundary $I_{\max}(I_{\min})$ is reached at function $S(\omega)$ such as in (5.86).

Note. Abscissas $\{\omega_j\}_1^{\nu}$ correspond to tangent points of curves $C_{2N}(\omega)$ $(Q_{2N}(\omega))$ and $A^2(\omega)$, and also probably, to boundaries of actual integration intervals in (5.5).

In spite of the fact that condition (5.13) does not appear in theorem 1, the possibility of its practical application at determination of the exact boundaries $I_{max}$ and $I_{min}$ is immediately present. The point is that the problem in searching for magnitudes $\overline{\rho}_j$ and $\underline{\rho}_j$, satisfying requirements (5.85) may have no solution. The sufficient condition for existence of the solution to this problem is the realization of inequality (5.13). In this case the magnitudes $\left\{\overline{\omega}_j\right\}_1^{v^*}, \left\{\overline{\rho}_j\right\}_1^{v^*}$ and $\left\{\underline{\omega}_j\right\}_1^{v_*}, \left\{\underline{\rho}_j\right\}_1^{v_*}$ correspond to the upper and lower main canonical presentations.

If the conditions of theorem 1 are not fulfilled and there is no spectral density $S(\omega)$ similar to (5.86) with the preset power moments (5.83), then the formula (5.30) may present only the strongest estimate of magnitude $I$, but not of its exact boundary. It is important, that if such a spectral density does not exist at an optimal choice of approximating polynomial by criterion (5.27), it also does not exist at its non-optimal choice. Otherwise not even the strongest estimate would coincide with the exact boundary.

Theorem 2 (inverse). If some function $S(\omega)$ like in (5.86) satisfies the set of equations (5.85), and it is possible to select the polynomial coefficients $C_{2N}(\omega)$ $(Q_{2N}(\omega))$, of which values at points $\omega = \omega_j$ coincide with values of function $A^2(\omega)$ with the realization of condition (5.29), then the exact upper (lower) boundary of functional $I(S(\omega))$ is reached at function $S(\omega)$ of an indicated type and is expressed by formula (5.84).

Note. If the determination of the exact boundary for integral (5.5) is treated as dual to (5.27) the infinite-dimensional problem of linear programming then theorems 1 and 2 can be considered as the analog for duality and equilibrium theorems for infinite-dimensional cases.

**About the validity to use the equivalent harmonic actions.** The theorems mentioned above may actively be applied at practical estimation of error dispersion in arbitrary order systems. The most simple results are obtained by giving two dispersions $D_K$ and $D_{K+1}$, when $N = K+1$. In this case the canonical presentation turns into one spectral line with frequency $\omega_{K+1} = \sqrt{D_{K+1}/D_K}$ and with dispersion $\rho_{K+1} = D_K^{K+1}/D_{K+1}^K$ that corresponds to equivalent harmonic action

$$g_{eqv}(t) = \sqrt{2\rho_1}\sin(\omega_{K+1}t + \varphi), \tag{5.87}$$

where $\varphi$ is the casual initial phase. Such action is often viewed as typical to the analysis of dynamic systems accuracies. Theorem 2 answers the question in what case it causes the extreme value of magnitude $I$. For example, consider $K = 0$ and $K = 1$.

**Case of known dispersions of action and its first derivative.** At $K = 0$ for equivalent harmonic action the formula $\omega_1 = \sqrt{D_1/D_0}$, $\rho_1 = D_0$ is obtained. The formula

$$D_{eg\max} = \rho_1 \left| H_e(j\omega_1) \right|^2 = D_0 \left| H_e\left(j\sqrt{D_1/D_0}\right) \right|^2 \tag{5.88}$$

presents the exact upper bound of error dispersion only with the realization of the condition that it is possible to select coefficients $c_0$ and $c_1$, providing the curves tangency $C_{2N}(\omega) = c_0 + c_1\omega^2$ and $\left|H_e(j\omega)\right|^2$ on a point $\omega = \omega_1$. For this purpose condition $\omega_1 \in [\alpha_1, \gamma_1]$ should be satisfied, where $\alpha_1$ is the abscissa for the tangent point of curves $C_2(\omega) = \left|H_e(j\omega)\right|^2 + c_1\omega^2$ and $\left|H_e(j\omega)\right|^2$; $\gamma_1$ is the abscissa for the point of maximum for the function $\left|H_e(j\omega)\right|^2$. The position of interval $\omega \in [\alpha_1, \gamma_1]$ on a frequency axis is explained by the graphs in Fig. 5.16.

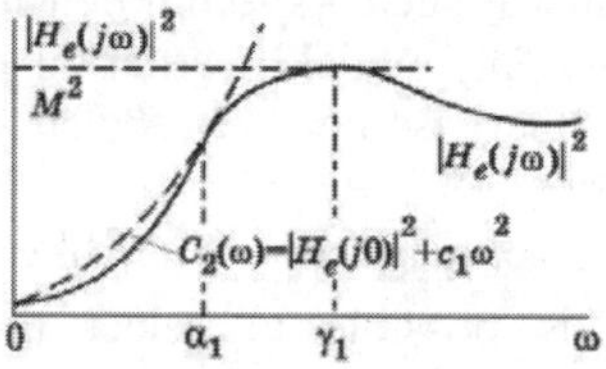


Fig. 5.16

In the case of a non-decreasing function $\left|H_e(j\omega)\right|^2$ the following is obtained: $\gamma_1 \to \infty$. If the order of system $n = 1$, then $\alpha_1 = 0$, $\gamma_1 \to \infty$ and the exact upper bound of error dispersion is reached in the processing of action (5.87) mode at any frequency $\omega_1$ (see also (5.22)).

In the case of $\omega_1 < \alpha_1$ the condition of action (5.87) processing cannot be considered as the heaviest, and formula (5.88) contains a little bit of a smaller value, than the exact upper bound. The maximum relative error for determination of upper bound by formula (5.88) in astatic systems is equal to the relative excess of value $C_2(\omega_1)$ above value $\left|H_e(j\omega_1)\right|^2$, because the upper estimation is $\overline{D}_{eg} = \rho_1 C_2(\omega_1) = c_1 D_1$.

The relative error increases with the rise of system order *n*, because the curvature of function $\left|H_e(j\omega)\right|^2$ graph increases on the interval $\omega \in [0, \alpha_1]$.

In a case where $\omega_1 > \gamma_1$ (which may hardly correspond to an admissible mode of system operation) it is accepted

$$\overline{D}_{eg} = D_0 \left|H_e(j\omega)\right|^2_{\max},$$

which also exceeds the value $D_{eg\max}$.

It is impossible to strengthen the upper estimate $\overline{D}_{eg}$ when assigning the restriction on spectrum of action width, because it does not allow for the decrease of the values of coefficients $c_0$ and $c_1$.

The exact lower boundary $D_{eg\min}$ in essence is not reached with action (5.87) processing, because it is impossible to choose the values of coefficients $q_0$ and $q_1$, providing the tangency of curves $Q_2(\omega)=q_0+q_1\omega^2$ and $|H_e(j\omega)|^2$ on point $\omega=\omega_1$ at $Q_2(\omega)\le|H_e(j\omega)|^2$.

Note that it is convenient to construct the graphs of functions $|H_e(j\omega)|^2$, $C_{2N}(\omega)$ and $Q_{2N}(\omega)$ on a square-law scale of frequency.

**Case of known first and second derivatives of action dispersions.** The following is obtained at $K=1$ for equivalent harmonic action $\omega_2=\sqrt{D_2/D_1}, \rho_2=D_1^2/D_2$. The formula

$$D_{eg\max}=\rho_2|H_e(j\omega_2)|^2=D_1^2D_2^{-1}\left|H_e\left(j\sqrt{D_2/D_1}\right)\right|^2 \qquad (5.89)$$

offers the exact upper bound of error dispersion only with the realization of the condition that it is possible to select the values of coefficients $c_1$ and $c_2$, providing the tangency of curves $C_4(\omega)=c_1\omega^2+c_2\omega^4$ and $|H_e(j\omega)|^2$ on point $\omega=\omega_2$. This is possible at $\omega_2\in[\alpha_2,\gamma_2]$, where $\alpha_2,\gamma_2$ are the abscissas for tangent points of curve $|H_e(j\omega)|^2$ with the curves $C_2(\omega)=c_1\omega^2$ and $C_4(\omega)=c_2\omega^4$ accordingly (Fig. 5.17).

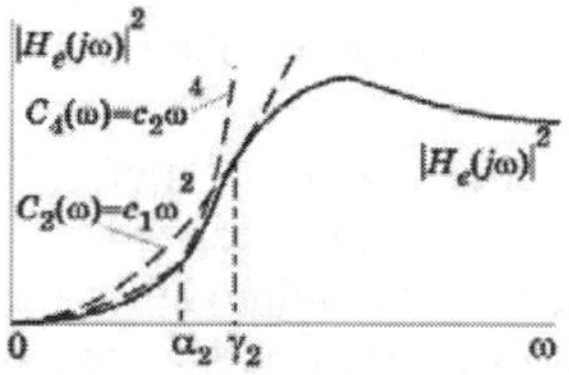


Fig. 5.17

At $\omega_2<\alpha_2$ and $\omega_2>\gamma_2$ formula (5.89) presents the underestimated value of upper boundary, and the processing of action (5.37) mode is impossible to be considered as the heaviest.

The upper estimation of error dispersion at $\omega_2<\alpha_2$ is determined by the formula $\overline{D}_{eg}=\rho_2c_2\omega_2^4=c_2D_2$. For its determination it is convenient to use the graph of function $|H_e(j\omega)|^2$, constructed at a scale appropriate $\omega^4$ on the abscissa axis. The straight line, tangent to a curve $|H_e(j\omega)|^2$ and passing through the origin of coordinates, is the graph of function $C_4(\omega) = c_2\omega^4$ in this case.

The lower boundary of error dispersion may be reached in the action (5.87) processing mode only in the first order systems (see (5.22)), because at $n > 1$ in low frequencies, the condition

$$Q_4(\omega) = q_1\omega^2 + q_2\omega^4 \leq |H_e(j\omega)|^2, \; q_2 < 0$$

is principally not satisfied.

*Questions*

1. What components does the resultant error of system include? What does each of them depend on?
2. What is the difference between the exact boundaries of error dispersion and its strongest upper and lower estimates?
3. Why is the determination of boundaries for system error dispersion by dispersions of action derivative the task for problems of moments in a mathematical sense?
4. What are the canonical presentations for a sequence of moments of action spectral density? How are they used with the determination of extreme values of system error dispersion?
5. Explain the idea of the approximative method for determination of the upper and lower estimations of error dispersion.
6. What are the advantages and disadvantages of the canonical representation method and the approximate method in determining the estimations of error dispersions?
7. In what condition does the estimation obtained by the approximative method appear the strongest?
8. How are the optimal and close to optimal modes for determination of the coefficients of approximating the polynomial realized?
9. How does the statement about the unimodality for the curve of action spectral density allow the estimations of error dispersion to strengthen?
10. How does the statement about the finite width of action spectrum allow the estimations of error dispersion to strengthen?
11. Why does knowledge on the dispersion for one derivative of action suffice to determine the upper estimation of system error dispersion, when determination of the lower estimation needs more numerical data?
12. What does the value of interval between the strongest upper and lower estimations of error control dispersion depend on? Do those estimations coincide?
13. In what cases can some coefficients of approximating polynomial be negative?
14. Why should the system in the case of unknown dispersion of reference action be necessarily astatic (zero-constant-error)?

15. Why is the knowledge of dispersion for derivatives of interference less useful for an estimation of error dispersion from interference than the knowledge of generalized moments for spectral density of interference with the use of non-power basis functions?
16. What additional information about the properties of interference is useful for the estimation of dispersion of the appropriate error?
17. In what case does the strongest estimate of error dispersion coincide with the exact boundary?
18. Under what conditions are the theoretically extreme values of dynamic error dispersion reached when processing the equivalent harmonic action?
19. How is it possible to define, whether the equivalent harmonic action corresponds to the most difficult operation of a system?
20. How can the reference action, providing the theoretical maximum of error dispersion of control, be selected?

# Chapter 6

## Analysis of maximum control error

### §6.1. Preliminary notes

**Problem definition.** Maximum (or practically maximum — see §2.4) error is the most objective characteristic of control accuracy in many cases. However, a rather long-life domination period of Wiener and Kalman methods of statistical synthesis with a full spectral-correlation description of actions in dynamic filtering regions cause the maximum error to be displaced by r.-m.-s. error as the main accuracy criterion.

Good conditions to revive the use of the maximum error when investigating control systems are created with the decision to use the concept of robust dynamic systems synthesis, with the refusal of the application of uncertain spectral models of actions and with their replacement, first of all by maximum values, by numerical characteristics of derivatives. Consider the analysis of maximum error over maximum values of action derivatives, when the execution of inequalities such as in (2.22)

$$\left|g^{(i)}(t)\right| \le g_M^i,\ i = \overline{K,N}, 0 \le K \le N. \tag{6.1}$$

are fulfilled.

Consider the evaluated dynamic error as an outcome from the passage of reference action through the filter with the transfer function

$$H_e(s) = [1 + W(s)]^{-1} = \sum_{i=r}^{n} a_i s^i \left(1 + \sum_{i=1}^{n-1} b_i s^i + \sum_{i=r}^{n} a_i s^i\right)^{-1}. \tag{6.2}$$

Such a transfer function corresponds to a closed-loop system with the transfer function of open loop similar to (5.1) at $a_i = 0$ for $i = \overline{0, r-1}$. That includes the presence of $r$-th order astatism property, $r \le n$.

For determination of the estimation of maximal dynamic error, basically only the execution of one inequality (6.1) at $i = k$ is sufficient, which restricts the derivative of the reference action with an order that coincides with the order of astatism for the considered system. For static systems only the value of maximum reference action can be preset. However, the additional information included in aggregate several magnitudes $g_M^{(i)}$ should improve the estimation of maximum error.

In fact, the maximum error is not a statistical characteristic of system accuracy, because it is reached at some determined action, most unfavorable in this sense, and only in particular instants. The determination of the most unfavorable reference action and maximum error is the interdependent problems, which should be solved jointly.

**Core of maximal error analysis in a time domain.** Having exposed the transfer function $H_e(s)$, expressed by formula (6.2), to inverse the Laplace transform, the pulse response of closed-loop system by the error $w_e(t) = L^{-1}\{H_e(s)\}$ is obtained, i.e. the law of error variation at $g(t) = \delta(t)$. Then the law of error variation at arbitrary action and zero initial conditions is determined by the formula of convolution such as $e_g(t) = w_e(t) * g(t)$, and for the maximum error the following expression can be written

$$e_{gm} = \int_0^{T_c} w_e(\tau) g_{mu}(T_c - \tau) d\tau . \qquad (6.3)$$

Here $\pm g_{mu}(t)$ is the most unfavorable action satisfying (6.1); $T_c$ is the duration of the control process. If the error can reach the maximum value already in some instant $t_0 < T_c$, then it is necessary to accept $g_{mu}(t) = 0$ at $t \in [0, T_c - t_0]$. This saves the validity of expression (6.3).

As it is shown below, the necessary condition of maximum error (6.3) finiteness is the validity of inequality $r \geq K$, where $K$ is the order of younger derivatives among those restricted by (6.1) $\left\{g^{(i)}\right\}_K^N$. If it is not fulfilled, then there is no reason to apply the problem of maximum error evaluation. Taking this into account, it is convenient to deal with transfer function

$$H_{eK}(s) = H_e(s) / s^K \qquad (6.4)$$

and with the appropriate pulse response $w_{eK}(t) = L^{-1}\{H_{eK}(p)\}$ in the analysis. Transfer function (6.4) connects the error image to the $K$-th derivative of action. Then expression (6.3) becomes:

$$e_{gm} = \max_{t,g} \left\{h_{eK}(t) * g^{(K)}(t)\right\} = \int_0^{T_c} w_{eK}(\tau) g_{mu}^{(K)}(T_c - \tau) d\tau . \qquad (6.5)$$

Formula (6.3) and, therefore, the equivalent formula (6.5) create the basis to determine the maximal error of the most unfavorable action. It was pointed out for the first time in a paper by B. V. Bulgakov [10] and was followed by a series of works on investigation of the maximum value of output magnitude in dynamic systems by the so-called perturbation accumulation method. Such a method is the most effective one in a case, when the inequalities in (6.1) restrict only one derivative of action, i.e. $N = K$. The singularities of its use with reference to functions $H_{eK}(s)$, typical for investigation of accuracy for closed-loop control systems, are considered in §6.2.

With an increased number of restricted derivatives of reference action, the complexity of the solution to the problem of determining the maximal error sharply increases, which generally forces the use of numerical methods. The analytical solution described in §6.3 is the solution to this particular case, when the higher order from restricted action derivatives no longer exceeds the order of system astatism, than one unit. In this case the possible forms of the most unfavorable action, which become periodic at the greater duration of control process execution, are analyzed.

**Other analysis methods.** The approximate method used to derive the upper-bound estimation of maximum error based on the reduction of several inequalities (6.1) to one is considered in §6.4. Such estimation is overstated and does not correspond to any particular action.

The approximate estimation of maximal error, as it is shown in §6.5, is possible when analyzing it in a frequency domain. The spectral presentations of reference action and errors are used for this purpose. Actually, the equivalent reference action must be found as the sum of harmonic functions or one harmonic function. This gives the approximate lower estimation of maximum error and its rather small fractional uncertainty can be guaranteed. Such an outcome is confirmed by a designed example in §6.6.

It is necessary to note that in principle the strict determination of maximum for original directly by the image, excluding the determination of original, is impossible. However, the suggested frequency method connected to the selection of action as the sum of several harmonic functions, satisfying (6.1) and maximizing the control error, gives the quite acceptable estimation for maximum error, where inaccuracy is well controlled. Such quality of estimation, as it is shown in §6.6, grows with the increasing number of restricted derivatives of action. It can be explained that the real most unfavorable action becomes more "smooth" and the above-stated model describes its properties better.

The problem in selecting the action with a line amplitude spectrum maximizing the control error is similar to the problem of selecting the line spectrum of action power maximizing the r.-m.-s. error described in §5.2. The passage to the dual problem and application of the approximative method explained in §6.3, are also useful here. Thus, the generality in analysis methods for r.-m.-s. error by the dispersions of action derivatives, and the frequency method of maximum error analysis by maximum values of derivatives are determined. This circumstance is used in chapter 7 when considering the uniform method for synthesis of robust systems by dispersions or by maximum values of action derivatives.

## §6.2. Calculation of maximum error with the restriction of one action derivative

**Rough estimation at the constant value of restricted derivative.** Assume that only one property of reference action is known, that its $K$-th derivative submits to the condition:

$$\left|g^{(K)}(t)\right| \le g_M^{(K)},\ K \ge 0\,. \tag{6.6}$$

It corresponds to common restriction (6.1) at $N = K$.

In particular, actions of a kind

$$g(t) = g_0 + \dot{g}_0 t + \ldots + \frac{g_0^{(K-1)} t^{K-1}}{(K-1)!} + \frac{g_M^{(K)} t^K}{K!}, \tag{6.7}$$

belong to the class of actions, restricted by an inequality (6.6).

Here $g_0, \dot{g}_0, \ldots, g_0^{(K-1)} \in (-\infty, \infty)$ are the values of action and both its younger derivatives in start time $t = 0$, and the higher derivative have the constant maximally acceptable value $g^{(K)}(t) = g_M^{(K)}$.

At reference action (6.7) the error in steady-state mode varies under the law

$$e_{ss}(t) = c_0 g(t) + c_1 \dot{g}(t) + \ldots + \frac{c_{K-1} g^{(K-1)}(t)}{(K-1)!} + \frac{c_K g_M^{(K)}}{K!}, \tag{6.8}$$

where

$$c_i = \left.\frac{d^i H_e(s)}{ds^i}\right|_{s=0},\ i - \overline{0, K} \tag{6.9}$$

are the error coefficients [9].

It is important, as the action described by formula (6.7) and its younger derivatives can have indefinitely large initial values, or they can unrestrictedly grow in time, therefore the necessary condition for finiteness of error is the equality of error coefficients $c_0, c_1, \ldots, c_{K-1}$ to zero. For this purpose the system should have an astatism of order not lower then $K$, i.e. the coefficients $a_0, a_1, \ldots, a_{K-1}$ in transfer functions (5.1) and (6.2) should be accepted equal to zero. Otherwise there is no sense in applying the problem of maximum error evaluation.

For a system with transfer function (6.2) at $r = K$ from (6.4) and (6.9) the following is obtained:

$$c_K = \left.\frac{d^K H_e(s)}{ds^K}\right|_{s=0} = H_{eK}(s)\big|_{s=0} = a_K\,,$$

whence, according to (6.8),

$$e_{ss}(t) = e_{ss} = a_K g_M^{(K)}. \tag{6.10}$$

Of course, it is impossible to consider the obtained value $e_{ss}$ as the maximum system error, because it is reached in processing the reference action of a narrower class, than selected by condition (6.6). This value can only be the rough estimation of maximum error from below.

**Derivation of the common expression of maximal error.** Research the formula (6.5) and discover the type of the most unfavorable reference action $g_{mu}(t)$ for accurate determination of maximum error. Such an action should ensure the maximization of the integrand in (6.5) if it is of a fixed sign. Therefore, according to (6.6) the $K$-th action derivative should vary by the law

$$g_{mu}^{(K)}(t) = g_M^{(K)} \operatorname{sign} w_{eK}(T_c - t). \tag{6.11}$$

Hence, the error reaches its maximal possible value.

$$e_{gM} = \int_0^{T_c} w_{eK}(\tau) g_{mu}^{(K)}(T_c - \tau) d\tau = g_M^{(K)} \int_0^{T_c} |w_{eK}(t)| dt. \tag{6.12}$$

The value of an integral in the right hand side of (6.12) is easily determined when reviewing the step response characteristic

$$h_{eK}(t) = \int_0^t w_{eK}(\tau) d\tau = L^{-1}\{H_{eK}(s)/s\}, \tag{6.13}$$

describing the error variation law for the action $g(t) = t^K 1(t)/K!$, of which the $K$-th derivative is the unit step-function, and at zero, initial conditions.

Note, that the initial value of step response characteristic $h_{eK}$, according to transfer function (6.2), can be determined as:

$$h_{eK}(0) = \lim_{s \to \infty} H_{eK}(s) = \begin{cases} 1 \text{ at } K = 0, \\ 0 \text{ at } K > 0, \end{cases} \tag{6.14}$$

and the final value

$$h_{eK}(\infty) = \lim_{s \to 0} H_{eK}(s) = \begin{cases} a_0/(1 + a_0) \approx a_0 & \text{at } K = 0 \\ a_K & \text{at } K > 0 \end{cases}. \tag{6.15}$$

It is not monotone in common case.

If some instants $t_1 < t_2 < \ldots < t_l$ correspond to the extreme points of step response characteristic $h_{eK}(t)$ on interval $t \in [0, T_c]$ or, that is the same, to sign reversal points of pulse response $w_{eK}(t)$, then taking (6.13) into consideration, the following expression is obtained:

$$\int_0^{T_c} |w_{eK}(t)| dt = \int_0^{t_1} w_{eK}(t) dt - \int_{t_1}^{t_2} w_{eK}(t) dt + \ldots + (-1)^{l-1} \int_{t_{l-1}}^{t_l} w_{eK}(t) dt +$$

$$+(-1)^l \int_{t_l}^{T_y} w_{eK}(t)dt = h_{eK}(t_1) - [h_{eK}(t_2) - h_{eK}(t_1)] + \ldots$$

$$\ldots + (-1)^{l-1}[h_{eK}(t_l) - h_{eK}(t_{l-1})] + (-1)^l [h_{eK}(T_c) - h_{eK}(t_l)] =$$
$$= 2[h_{eK}(t_1) - h_{eK}(t_2) + \ldots + (-1)^{l-1} h_{eK}(t_l)] + (-1)^l h_{eK}(T_c). \quad (6.16)$$

The approximate kind of pulse response $w_{eK}(t)$ and step response characteristic $h_{eK}(t)$ for a system with complex roots of characteristic equation at $K = 0$ are shown in Fig. 6.1*a*, and the most unfavorable reference action and error, caused by it, are shown in Fig. 6.1*b*. The appropriate curves for case $K > 0$ are shown in Fig. 6.2.

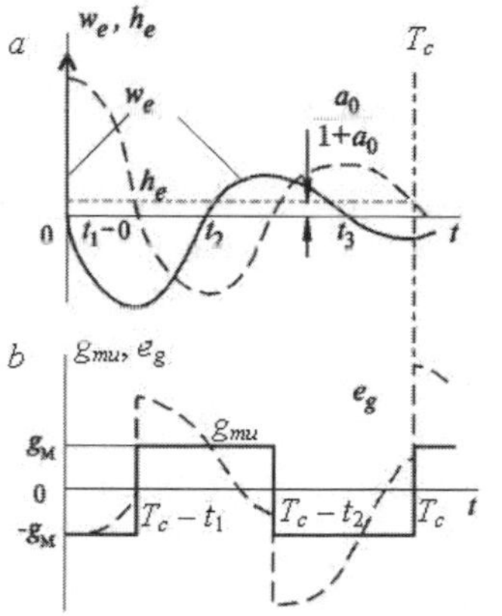


Fig. 6.1

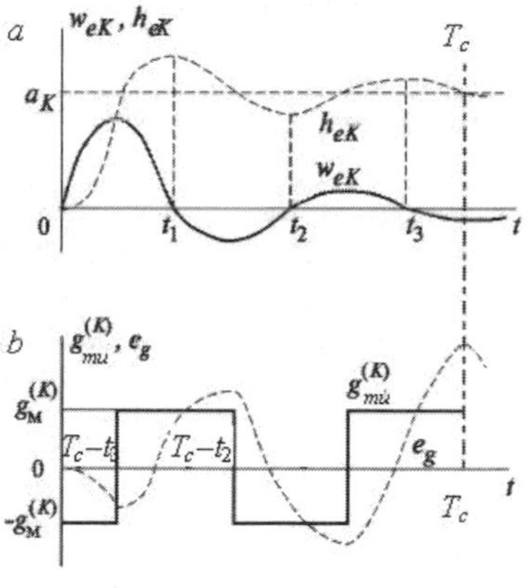


Fig. 6.2

From these graphs, in particular, it is visible, that function $g_{mu}(t)$ in case $K = 0$ has the step directly at the final instant $t = T_{c-0}$, and in a case $K > 0$ such step is absent. At $K = 0$ the maximum error can exceed the maximum value of reference action. Therefore, in practice, this case cannot be considered as a characteristic case.

**Error accumulation coefficient.** Enter the dimensionless error accumulation coefficient

$$k(T_c) = \frac{1}{a_K} \int_0^{T_c} |w_{eK}(t)| dt. \quad (6.17)$$

This shows, in how many times the maximum error at the most unfavorable reference action, defined by formula (6.11), exceeds the value of steady state error at a constant value of *K*-th derivative of reference action, determined by formula (6.10). Then the following expression for maximum error is valid.

$$e_M = ke_{ss} = ka_K g_M^{(K)}. \quad (6.18)$$

Curve $k(t)$, called the Bulgakovs' curve or accumulation diagram [56, 58], according to (6.16) and (6.17) can be obtained graphically from the step response characteristic $h_{eK}(t)$. This curve coincides with the normalized step response characteristic $h_{eK}(t)/a_K$ in interval $t \in [0, t_1]$. In interval $t \in [t_1, t_2]$ it is the mirror image of a $h_{eK}(t)/a_K$ characteristic concerning the horizontal line with the ordinate $h_{eK}(t_1)/a_K$, and in interval $t \in [t_2, t_3]$ it coincides with the characteristic $h_{eK}(t)/a_K$, displaced by value $h_{eK}(t_2)/a_K$, etc. In each interval $t \in [t_i, t_{i+1}]$, $i = 1, 2, \ldots$ the construction is made according to formula $k(t) = k(t_i) + (-1)^i h_{eK}(t)/a_K$. The whole curve $k(t)$ can be constructed by sequentially adding the characteristics $h_{eK}(t)/a_K$ to each other and interchanging them with the mirror imaged segments of this characteristic. The example shown in Fig. 6.3 illustrates the construction of curve $k(t)$ for the third order system with the real roots of the characteristic equation, when the step response characteristic $h_{eK}(t)$ has only two extremums.

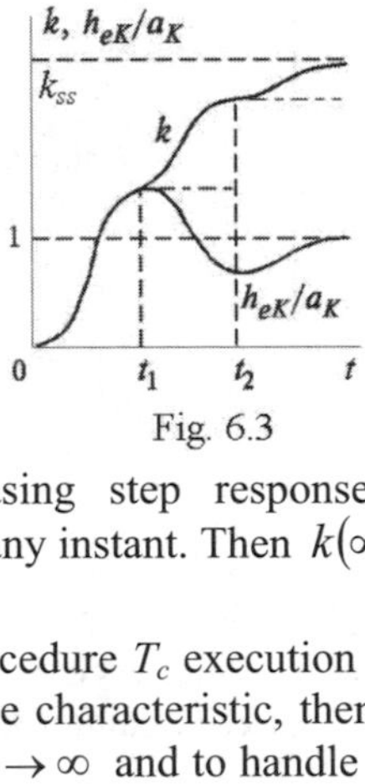


Fig. 6.3

At a monotonically increasing step response characteristic the relation $k(t) = h_{eK}(t)/a_K$ is fulfilled at any instant. Then $k(\infty) = 1$ and the maximum error cannot exceed the value $e_{ss}$.

If the duration of control procedure $T_c$ execution essentially exceeds the decay time of the system step response characteristic, then at the analysis of maximum error, it is possible to accept $T_c \to \infty$ and to handle only the steady state value of error accumulation coefficient $k_{ss} = k(\infty)$. Thus it is useful to define the following legitimacies.

If the step response characteristic has only one extremum, then according to (6.16) and (6.17) the relation $k_{ss} = 1 + 2\sigma$, where $\sigma = [h_{eK}(t_1) - h_{eK}(\infty)]a_K^{-1}$ is the overshoot determined over normalized step response characteristic, is fulfilled. With step response characteristic with two extremums the following is obtained:

$$k_{ss} = 1 + 2[h_{eK}(t_1) - h_{eK}(t_2)]a_K^{-1} < (1 + 4\sigma).$$

In common cases the error accumulation coefficient is more, than the system stability margin, which is less and the oscillatory of its stronger step response characteristic. At the approach to system oscillatory bound of stability, the error accumulation coefficient converges to infinity $k_{ss} \to \infty$.

The accumulation coefficient for a circuit with several sequentially joint dynamic units does not exceed the product of accumulation coefficients for these units. If the units in sequential circuit essentially differ on their response speed, then it is possible to take into account only the unit accumulation coefficient with the least speed of response. This strongly simplifies the estimation of maximum error. It is possible to consider also a parallel connection of units. Then the resulting value $k_{ss}$ does not exceed the sum of maximal output magnitudes of units.

**Design equation.** Analyze an error accumulation in systems with elementary transfer functions, with the use of relations (6.4), (6.13), (6.16) — (6.18).

*Static first order system.* At $W(s) = (a_0 + a_1 s)^{-1} = K_0/(1 + T_1 s)$, where $K_0 = 1/a_0$ is gain coefficient and $T_1 = a_1/a_0$ is an inertial time constant, $H_e(s) = \dfrac{a_0 + a_1 s}{1 + a_0 + a_1 s} = \dfrac{1 + T_1 s}{K_0 + 1 + T_1 s}$, $K = 0$ it is obtained:

$$h_{e1}(t) = \frac{a_0}{1 + a_0} + \frac{1}{1 + a_0}\exp\left(-\frac{1 + a_0}{a_1}t\right) = \frac{1}{K_0}\exp\left(-\frac{K_0 + 1}{T_1}t\right),$$

$$k_{ss} = a_0^{-1}(1 + a_0)\left[2h_e(0) - h_e(\infty)\right] = 2a_0^{-1}(1 + a_0) = 2K_0 + 1 \approx 2K_0,$$

whence

$$e_M = k_{ss} a_0 (1 + a_0)^{-1} g_M = k_{ss}(K_0 + 1)^{-1} g_M \approx 2 g_M .$$

*First order system with the first order astatism.* At $W(s) = K_1/s$, where $K_1$ is the gain coefficient over the velocity, $H_e(p) = p/K_1 + p$, $K = 1$, the following expressions are obtained

$$H_{e1}(s) = 1/(K_1 + s),\ h_{e1}(t) = K_1^{-1}\left[1 - \exp(-K_1 t)\right].$$

As the step response characteristic $h_{e1}(t)$ is monotone,

$$k_{ss} = K_1 h_{e1}(\infty) = 1 \text{ and } e_M = e_{ss} = \dot{g}_M / K_1 .$$

Note that this outcome can be obtained also by researching the differential equation $\dot{e}(t) = K_1 e(t) = g(t)$. As in the moment when the error reaches the maximal value, its derivative should be equal to zero, then $\max e(t) = \max \dot{g}(t)/K_1$.

*The second order system with the second order astatism.* At $W(s) = \dfrac{1 + b_1 s}{a_2 s^2} = \dfrac{K_2(1 + b_1 s)}{s^2}$, where $K_2$ is the gain coefficient by velocity, $H_e(s) = \dfrac{a_2 s^2}{1 + b_1 s + a_2 s^2}$, $K = 2$, the following expression is obtained

$$H_{e2}(p) = \frac{a_2}{1 + b_1 s + a_2 s^2} = \frac{1}{K_2 + K_2 b_1 s + s^2}.$$

In a case when the roots of the characteristic equation are real, i.e. at $a_2 \le b_1^2/4$, due to a monotonic property of step response characteristic, then

$$k_{ss} = a_2^{-1} h_{e2}(\infty) = a_2^{-1} \lim_{s\to 0} H_{e2}(s) = 1, \; e_M = e_{ss} = a_2 \ddot{g}_M = \dot{g}_M / K_2 .$$

At the complex roots of the characteristic equation, when $H_{e2}(s) = 1/\left(q^2 + 2\xi q s + s^2\right)$, where $q = \sqrt{K_2}$ is natural oscillations frequency, $\xi = b_1\sqrt{K_2}/2$ is the damping coefficient, the step response characteristic looks like:

$$h_{e2}(t) = q^{-2}\left[1 - \exp(-\gamma t)\cos(\lambda t) + \frac{\gamma}{\lambda}\sin(\lambda t)\right],$$

where $\gamma = \xi q$ is the damping coefficient, and $\lambda = q\sqrt{1-\xi^2}$ is damped frequency.

Extreme points of step response characteristic $t_i = i\pi/\lambda$, $i = 1, 2, \dots$, and $h_{e2}(t_i) = q^{-2}\left[1 - (-1)^i \exp(-i\pi\gamma/\lambda)\right]$. From here

$$k_{ss} = a_2^{-1}\left\{h_{e2}(\infty) + 2\sum_{i=1}^{\infty}\left[h_{e2}(t_{2i} - 1) - h_{e2}(t_{2i})\right]\right\} = 1 + 2\sum_{i=1}^{\infty}\exp(-i\pi\gamma/\lambda)$$

or, having calculated the sum of geometrical progression,

$$k_{ss} = 1 + \frac{2\exp(-\pi\gamma/\lambda)}{1 - \exp(-\pi\gamma/\lambda)} = \frac{1 + \exp\left(-\pi\xi/\sqrt{1-\xi^2}\right)}{1 - \exp\left(-\pi\xi/\sqrt{1-\xi^2}\right)} = \operatorname{cth}\frac{\pi\xi}{2\sqrt{1-\xi^2}},$$

$$e_M = k_{ss} a_2 \ddot{g}_M = k_{ss} \ddot{g}_M / K_2 .$$

The graph of function $k_{ss}(\xi)$ is shown in Fig. 6.4.

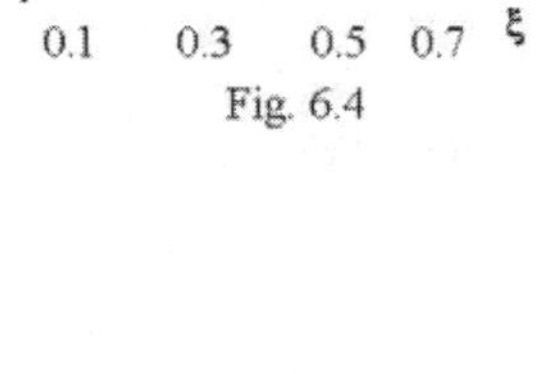

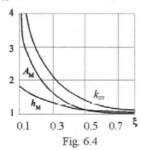

Fig. 6.4

The values of resonance peak for normalized gain plot $A_M = \max|K_2 H_{e2}(j\omega)| = \left(2\xi\sqrt{1-\xi^2}\right)^{-1}$, $\xi < 0.707$, and the value of maximum ejection for normalized step response characteristic $h_M = h_{e2}(t_1) = 1 + \exp\left(-\pi\xi/\sqrt{1-\xi^2}\right)$, are also shown there. The comparison of these curves allows for comparing the maximum error $e_M$ (achievable at the most unfavorable reference action, whose second derivative looks like meander) with errors at step and harmonic functions $\ddot{g}(t)$.

It is visible, that at $\xi \geq 0.707$ the accumulation effect is represented weakly, because $k_{ss} \leq 1.09$. At $\xi \geq 0.8$ it is practically absent because $k_{ss} \leq 1.03$. However, when decreasing the system stability margin, value $k_{ss}$ increases, noticeably exceeding a value $A_M$ and, especially, value $h_M$. In a limiting case at $\xi \to \infty$ the following is obtained: $k_{ss}/A_M = 4/\pi = 1.28$, and $k_{ss}/h_M \to \infty$. If $\xi < 0.5$, then with the use of harmonic action there is a stronger estimate for maximum error, than with the use of step action.

*The second order system with the first order astatism.* At $W(s) = \dfrac{1 + b_1 s}{a_1 s + a_2 s^2} = \dfrac{K_1(1 + b_1 s)}{s(1 + T_1 s)}$, where $K_1 = a_1^{-1}$ is the gain coefficient by velocity, $T_1 = a_2/a_1$ is the inertial time constant, $H_e(s) = \dfrac{s(a_1 + a_2 s)}{1 + (a_1 + b_1)s + a_2 s^2}$, $K = 1$, the following is obtained:

$$H_{e1}(s) = \frac{a_1 + a_2 s}{1 + (a_1 + b_1)s + a_2 s^2} = \frac{1 + T_1 s}{K_1 + (K_1 b_1 + 1)s + T_1 s^2}.$$

If $a_2 \leq (a_1 + b_1)^2/4$, then the characteristic equation has the real roots and

$$h_{e1}(t) = \{A_1[1 - \exp(-t/\tau_1)] + A_2[1 - \exp(-t/\tau_2)]\}K_1^{-1},$$

where

$$A_{1,2} = \frac{1}{2}\left[a_1 \pm \frac{a_1(a_1 + b_1)/2 - a_2}{\sqrt{(a_1 + b_1)^2/4 - a_2}}\right], \tau_{1,2} = \frac{(a_1 + b_1)}{2} \pm \sqrt{\frac{(a_1 + b_1)^2}{4} - a_2}.$$

At $a_2 \leq a_1 b_1$ the step response characteristic $h_{e1}(t)$ is monotone, $k_{ss} = K_1 h_{e1}(\infty) = 1$ and $e_M = e_{ss} = \dot{g}_M/K_1$.

At $a_1 b_1 < a_2 < (a_1 + b_1)^2/4$ the coefficient $A_2$ is negative. In this case the step response characteristic $h_{e2}(t)$ has one maximum in point $t_1 = \frac{\tau_1 \tau_2}{\tau_2 - \tau_1} \ln\left(-\frac{A_1 \tau_2}{A_2 \tau_2}\right)$, and therefore,

$$k_{ss} = K_1\left[2h_{e1}(t) - h_{e1}(\infty)\right] =$$

$$= 1 + \frac{K_1 A_1 \tau_2}{\tau_1}\left[\exp\left(-\frac{\tau_1}{\tau_2 - \tau_1}\right) + \frac{A_1}{A_2}\exp\left(-\frac{\tau_2}{\tau_2 - \tau_1}\right)\right],$$

$$e_M = k_{ss}\, \dot{g}_M / K_1 \text{, and } 1 < k_{ss} < 2 .$$

If $a_2 > (a_1 + b_1)^2/4$, the roots of characteristic equation are complex and

$$h_{e1}(t) = \left[1 - \exp(-\gamma t)\left(\cos(\lambda t) + \frac{\gamma - q^2 T_1}{\lambda}\sin(\lambda t)\right)\right] K_1^{-1},$$

where $\gamma = \xi q$, $\lambda = q\sqrt{1 - \xi^2}$, $q = \sqrt{K_1/T_1}$, $\xi = (K_1 b_1 + 1)/2\sqrt{K_1/T_1}$ .

In points $t_i = \frac{1}{\lambda}\left[\operatorname{arctg}\frac{\lambda T_1}{\gamma T_1 - 1} + (i-1)\pi\right], i = 1,2,\ldots$ the step response characteristic $h_{e1}(t)$ has extremums

$$h_{e2}(t) = \left\{1 - (-1)^i \exp\left[-(i-1)\frac{\pi\gamma}{\lambda}\right]\exp\left(-\frac{\gamma}{\lambda}\operatorname{arctg}\frac{\lambda T_1}{\gamma T_1 - 1}\right)\sqrt{q^2 T_1^2 - 2\gamma T_1 + 1}\right\} K_1^{-1},$$

therefore $k_{ss} = K_1\left\{h_{e1}(\infty) + 2\sum_{i=1}^{\infty}\left[h_{e1}(t_{2i-1}) - h_{e1}(t_{2i})\right]\right\} =$

$$= 1 + \frac{2\sqrt{q^2 \quad T_1^2 - 2\gamma T_1 + 1}}{1 - \exp(-\pi\gamma/\lambda)}\exp\left(-\frac{\gamma}{\lambda}\operatorname{arctg}\frac{\lambda T_1}{\gamma T_1 - 1}\right)$$

and $e_M = k_{ss}\dot{g}_M / K_1$ .

Assuming $a_1 = 0$, the obtained formulas can be used also to calculate the maximum error in the second order astatism system at $K = 1$, i.e. at the known maximum value of the first derivative of action.

## §6.3. Calculation of maximum error with the restriction of several action derivatives

**Derivation of common expression.** If the maximum values of two or several derivatives of reference action $\left\{ \overset{i}{g}_M \right\}_K^N$, $N > K$ (and $K \le r$, where $r$ is the system astatism order) are known, then it is possible to calculate the maximum dynamic error by researching the functional

$$e_g\left(t, g^{(K)}\right) = \int_0^t w_{eK}(\tau) g^{(K)}(t-\tau)\, d\tau, \tag{6.19}$$

written according to (6.5). Here the $K$-th action derivative $g^{(K)}(t)$ belongs to a class of functions $\Lambda$, satisfies the conditions:

$$\left|g^{(K)}(t)\right| \le g_M^{(K)}, \left|g^{(K)}(t)/dt\right| \le g_M^{(K+1)}, \ldots, \left|d^{N-K} g^{(K)}(t)/dt^{N-K}\right| \le g_M^{(N)}. \tag{6.20}$$

The maximum error is reached at an action, where the $K$-th derivative maximizes the functional (6.19), i.e.

$$e_{gM} = \max_{t\in[0,T_y]} \max_{g^{(K)}\in\Lambda} e\left(t, g^{(K)}\right) = e_g\left(T_y, g_{mu}^{(K)}\right) \approx \lim_{t\to\infty} e_g\left(t, g_{mu}^{(K)}\right).$$

The properties of function $g_{mu}^{(K)}(t)$ essentially depend on a kind of pulse response $w_{eK}(t)$. If it has the fixed sign (possible only at $K = r$), i.e. the step response characteristic $h_{eK}(t)$ is monotone, then the function $g_{mu}^{(K)}(t)$ is also monotone. In this case the maximum error appears when the function $g_{mu}^{(K)}(t)$ holds the maximum value $g_M^{(K)}$ long enough, irrespective of speed and law according to which it has reached this value. Thus the restriction of higher derivatives of action has no influence on the maximum error, for which the formula

$$e_{gM} = e_{gss} = g_M^{(K)} h_{eK}(\infty) = a_K g_M^{(K)}. \tag{6.21}$$

is valid.

The determination of $e_{gM}$ at an alternating-sign pulse response $w_{eK}(t)$ is a more difficult problem. In common case it has no analytical solution. Therefore some particular cases will be considered. From the relations

$$w_{ei}(t) = L^{-1}\left\{H_e(s)/s^i\right\}, h_{ei}(t) = \int_0^t w_{ei}(t)dt = L^{-1}\left\{H_e(s)/s^{i+1}\right\}$$

therefore if $r$ is the order of system astatism, $r \ge K$, then

$$\lim_{t\to\infty} h_{ei}(t) = \begin{cases} 0 & \text{at } i < r, \\ h_{er}(\infty) = \text{const} & \text{at } i = r, 0 < h_{er}(\infty) < \infty, \\ \infty & \text{at } i > r. \end{cases} \tag{6.22}$$

The approximate kind of functions $w_{er}(t), h_{er}(t)$ and $h_{e,r+1}(t)$ is shown in Fig. 6.5.

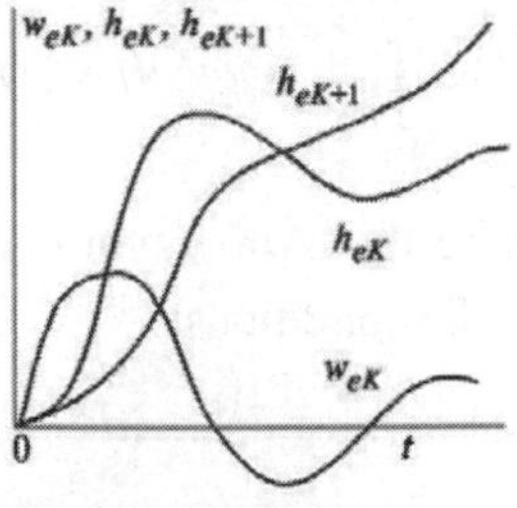


Fig. 6.5

By using a rule of partial integration at $K > r$, (6.19) can be written as:

$$e_g\left(t, g^{(K)}\right) = h_{eK}(\tau) g^{(K)}(t-\tau)\Big|_0^t + \int_0^t h_{eK}(\tau) g^{(K+1)}(t-\tau)\,d\tau =$$
$$= h_{eK}(t) g^{(K)}(0) + \int_0^t h_{eK}(\tau) g^{(K+1)}(t-\tau)\,d\tau = \tag{6.23}$$
$$= \sum_{i=K}^{r-1} h_{eK}(t) g^{(i)}(0) + \int_0^t w_{er}(\tau) g^{(r)}(t-\tau)\,d\tau.$$

Accepting $t \to \infty$, from (6.23) with the account of (6.22) the following expression is obtained:

$$e_g\left(t, g^{(K)}\right) = \int_0^t w_{er}(\tau) g^{(r)}(t-\tau)\,d\tau. \tag{6.24}$$

It is valid with this condition, that the duration of process execution essentially exceeds the decay time of system pulse response. This expression is convenient for determining the most unfavorable action at $N = r$, when the function $g^{(r)}(t-\tau)$ can vary in discrete steps.

With reference to a case $N = r + 1$, it is possible to convert expression (6.24) with use of the partial integration formula and having entered the step response function $h_{e(r)}(\tau) - h_{e(r)}(t)\Big|_{t\to\infty}$ displaced on the ordinate axis, which represents a free component for step response function $h_{e(r)}(\tau)$. Then the following is obtained:

$$e_g\left(t, g^{(K)}\right) = \int_0^t g^{(r)}(t-\tau)\,d\left[h_{e(r)}(\tau) - h_{e(r)}(t)\right] =$$
$$= h_{e(r)}(t) g^{(N-1)}(t) + \int_0^t \left[w_{eN}(\tau) - w_{eN}(t)\right] g^{(N)}(t-\tau)\,d\tau. \tag{6.25}$$

**Most unfavorable reference action.** The analysis of expression (6.25) allows us to find the shape of the most unfavorable action.

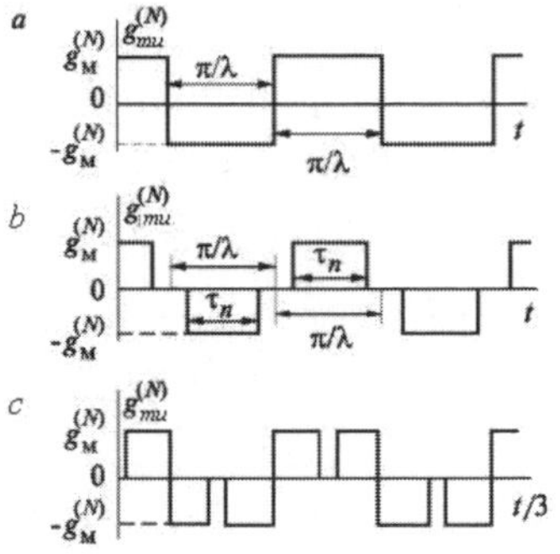

Fig. 6.6

At an alternating-sign pulse response $w_{eN}(t)$, in functions $w_{eN}(\tau) - w_{eN}(t)|_{t\to\infty}$, standing under the integral (6.25) is also an alternating-sign and it describes the damped oscillations concerning a zero level. Consider the characteristic case, when such oscillations are close in their shape to a damping harmonic function with frequency $\lambda_*$. Thus an integral value in (6.25) is maximum if the function $g^{(N)}(t-\tau)$ is periodic and it ensures the integrand of the fixed sign. Such a function must look like that shown in Fig. 6.6*a* or 6.6*b* depending on the following circumstances.

If the function $g^{(N)}(t)$ is meander with period $2\pi/\lambda$ (in that particular case $\lambda = \lambda_*$) and it varies in discrete steps between values $\pm g_M^{(N)}$, then each step has the maximum possible value $2g_M^{(N)}$ (Fig. 6.6*a*). The $(N-1)$-th derivative of such action varies by the saw-tooth law, and the $(N-2)$-th derivative consists of quadratic parabola segments, and the $K$-th derivative — of segments of $(N-K)$-th power parabola.

In order to the reach the coincidence between the turndown of the $i$-th derivative of described action and restrictions (6.20), the inequalities

$$\lambda \geq \eta_{N-1}^* \sqrt[N-i]{g_M^{(N)} / g_M^{(i)}}, \tag{6.26}$$

where $\eta_1^* = 1.57$, $\eta_2^* = 1.11$, $\eta_3^* = 1.09$, $\eta_4^* = 1.06$, $\eta_5^* = 1.05$, $\eta_6^* = 1.04, \ldots$ should be fulfilled [8].

If at $i = K$ the inequality (6.26) is strict, then the $K$-th derivative of reference action, except for the described variable component, can have a constant component $g_M^{(K)} - g_M^{(N)}\left(\eta_{N-K}^*/\lambda\right)^{N-K}$. Then the maximum value of the $K$-th derivative is equal to the extreme possible magnitude $g_M^{(K)}$.

The execution of inequalities in (6.26) allows acceptation the hypothesis that the most unfavorable action is that action, which derivatives are shown in Fig.

6.6*a*. Further, the maximum value of an error is found without any basic difficulties, for example, by numerical methods with solving the appropriate differential equation on a computer. The use of analytical method explained below is also possible.

Note, that the pointed out case is especially characteristic with strongly oscillatory pulse response $w_{eK}(t)$ and at large enough magnitude $g_M^{(N)}/g_M^{(N-1)}$, commensurable with frequency $\lambda_*$ (for example, at $0.1<\lambda_* g_M^{(N-1)}/g_M^{(N)}<1$).

If the value $g_M^{(N)}/g_M^{(N-1)}$ exceeds $\lambda_*$ then it is impossible to handle with Fig. 6.6*a* at $\lambda=\lambda_*$, because it can cause the violation of inequalities in (6.26). Then the *N*-th derivative of the most unfavorable reference action should look like that shown in Fig. 6.6*b*. Zero values of this function during the part of the oscillation period for pulse response $\lambda=\lambda_*$ ensure such decrease of ranges for variation of the lowest derivatives of the reference action, when the conditions (6.26) were fulfilled. For maximization of the integral in (6.25), the nonzero values of function $g^{(N)}(t-\tau)$ should correspond in time to the maximum deviations of function $w_{eN}(\tau)-w_{eN}(t)$ from zero in each period of its damped oscillations. The most difficult problem is to determine the most unfavorable action at $g_M^{(N)}/g_M^{(N-1)} << \lambda_*$, when the first addend in expression (6.25) essentially exceeds the second one.

If the pulse response $w_{eN}(t)$ has weak oscillations, then the variation periodicity of the most unfavorable action can have no connection with the oscillations period of function $w_{eN}(t)$. Thus, it is possible to consider the function, of which the *N*-th derivative is shown in Fig. 6.6*a* or 6.6*b*, but at $\lambda_*<\lambda$, or at the sinusoidal shape of action, as a characteristic action.

If pulse response $h_{eN}(t)$ has weak oscillations, then the system resonant properties can affect the variation period of unfavorable action. The period may turn out to be a multiple to the oscillations period in function $h_{eN}(t)$ in this case. For maximization of the integral in (6.25), the ratio of the first period to the second one should be the odd number or $\lambda_*/\lambda$ = 3, 5, 7... The typical form of curve $g_{mu}^{(N)}(t)$ at $\lambda_*/\lambda=3$ is shown in Fig. 6.6*c*. Zero values for this function in the middle part of each of its positive (negative) rectangular impulses ensure the decrease of variation ranges for lowest action derivatives, necessary for execution of inequalities in (6.26), and corresponding in time to the greatest negative (positive) deviation of function $h_{eN}(t)-h_{eN}(\infty)$ from zero.

The variation period for the most unfavorable action $2\pi/\lambda$ can be estimated rather simply on the basis of frequency analysis methods, which will be considered in §6.5.

**Analytical method at *N* = *r* + 1.** If the inequalities in (6.26) are fulfilled and the hypothesis that the *N*-th derivative of the most unfavorable action similar to that shown in Fig. 6.6*a* is lawful, compared to the formula in (6.25) for maximum error, then the following can be obtained:

$$e_{gM} = \lim_{t\to\infty} e_g\left(t, g_{mu}^{(K)}\right) \le h_{er}(\infty) g_N^{(N-1)} + g_M^{(N)} \int_0^\infty \left| w_{eN}(\tau) - w_{eN}(\infty) \right| d\tau . \quad (6.27)$$

In order to calculate an integral in (6.27), it is necessary to consider the function

$$\Delta h_{eN}(t) = \int_0^t \left[ w_{eN}(\tau) - w_{eN}(\infty) \right] d\tau .$$

Let the instants $t_1$, $t_2$, $t_3$ ... be the extreme points of function $\Delta h_{eN}(t)$, i.e. the sign reversal points of function $w_{eN}(t) - w_{eN}(\infty)$, according to the accepted assumption about the character of system pulse response $t_i = t_1 + i\pi/\lambda_*$, $i = 2, 3...$, $t_1 \in (0, \pi/\lambda_*)$. It is also essential, that as $w_{eN}(0) = 0$, $w_{eN}(\infty) > 0$, then at $t \in (0, t_1)$ the condition $w_{eN}(t) - w_{eN}(\infty) < 0$ is satisfied and, therefore, $\Delta h_{eN}(t_1) < 0$. It becomes easy to make sure of the validity of formula (intermediate calculations see in [8] in page 123)

$$\int_0^\infty \left| w_{eN}(\tau) - w_{eN}(\infty) \right| d\tau = 2\sum_{i=1}^\infty \left[ \Delta h_{eN}(t_{2i}) - \Delta h_{eN}(t_{2i-1}) \right] - \Delta h_{eN}(\infty). \quad (6.28)$$

The maximum error estimation by formula (6.27) is overstated, because according to Fig. 6.6*a* the function $g_{mu}^{(N-1)}(t)$ accepts the maximum positive value only in instants coinciding with negative steps of function $g_{mu}^{(N)}(t)$. However, the maximum positive values of addend in the right hand side of (6.25), fixed in (6.27), are reached in instants lagging concerning the negative steps of function $g_{mu}^{(N)}(t)$ on value $t_1$. Thus, estimation (6.27) is formed as the sum of the greatest possible values of the first and second addend in (6.25), actually reached in different instants biased of value $t_1$.

When observing the error value in an instant, which appropriates to the maximum of the second addend in (6.25), in addition to estimate (6.27) the underestimation looks like

$$e_{gM} \ge h_{eN}(\infty)\left[\max g^{(N-1)}(t) - g_M^{(N)} t_1\right] + g_M^{(N)} \int_0^\infty \left| h_{eN}(\tau) - h_{eN}(\infty) \right| d\tau , \quad (6.29)$$

where $\max g^{(N-1)}(t) = \begin{cases} g_M^{(N-1)} & \text{at } N-1 = K, \\ \pi g_M^{(N)}/(2\lambda_*) & \text{at } N-1 > K. \end{cases}$

If a difference between estimations (6.27) and (6.29) is so small, that they allow us to practically find the exact value of the maximal error, then the accepted hypothesis that the action, of which the $N$-th derivative shown in a Fig. 6.6*a* is the most unfavorable, is confirmed.

## §6.4. Approximate upper estimate of maximum error

**Estimation principle.** Explain the principle of approximate estimation for maximum error from above, first with reference to a case, when the system order is equal to the order of a higher restricted derivative for reference action, i.e. $n = N$.

Consider the transfer function $H_{eK}(s)$, expressed by formula (6.4) according to (6.2) at $r = K$. Having separated the numerator of this transfer function from the denominator, write it as product

$$H_{eK}(s) = H_s(s) H_p(s), \tag{6.30}$$

where $H_s(s) = a_K + a_{K+1}s + \ldots + a_n s^{n-K}$,

$$H_p(s) = \left(1 + \sum_{i=1}^{n-1} b_i s^i + \sum_{j=K}^{n} a_j s^j\right)^{-1}.$$

Expression (6.30) can be compared with the equivalent block diagram of the dynamic error formation represented in Fig. 6.7.

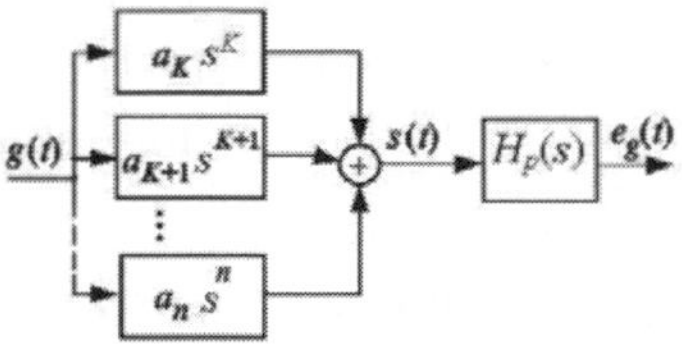


Fig. 6.7

The error is shown as an outcome from the passing of some signal $s(t)$, connected to the reference action derivatives by relation

$$s(t) = a_K g^{(K)}(t) + a_{K+1} g^{(K+1)}(t) + \ldots + a_n g^{(n)}(t), \tag{6.31}$$

through the positional unit with transfer function $H_p(s)$.

As each of addends in the right hand side of (6.31) is restricted on absolute value by inequalities (6.1), the signal $s(t)$ can be restricted by its absolute value by some magnitude $\max|s(t)| = s_M < \infty$, whose exact evaluation is an independent problem considered below. Then the estimation of maximum error is reduced to the estimation of the maximum value of the signal on the output of the positional unit with transfer function $H_p(s)$ at the input signal restricted at its absolute value. As it is shown in §6.2, such estimation looks like

$$e_{gM} \leq k_p s_M\,, \tag{6.32}$$

where $k = \int\limits_0^\infty |w_p(t)| dt$ is the accumulation coefficient; $w_p(t) = L^{-1}\{H_p(s)\}$.

The real roots of the system characteristic equation $k_p = 1$, at complex roots the value of $k_p$ exceeds one, but is close to it with a good system stability margin.

It is necessary to take into account, that the inequality (6.32) could be converted to the equality only when the module of signal $s(t)$ does not simply reach the value of $s_M$ but remains equal to it long enough. The periodicity of this signal sign variation corresponds to resonant properties of the unit with transfer function $H_p(s)$. As at $N > K$ this condition is not fulfilled, and estimation (6.29) cannot correspond to the exact upper bound of error and is always a little overestimated.

With the determination of magnitude $s_M$ it is necessary to distinguish two possible versions of problem definition. If the execution of conditions (6.1) is guaranteed only in a time interval, which is the definition interval for process $e_g(t)$, i.e. during the duration of control process execution, then all inequalities (6.1) can be converted to equalities at once on boundaries of this time interval. That results in

$$s_M = a_K g_M^{(K)} + a_{K+1} g_M^{(K+1)} + \ldots + a_n g_M^{(n)}. \tag{6.33}$$

If conditions (6.1) are fulfilled indefinitely long, then reaching the maximum possible positive values by all derivatives $g_i(t)$, $i = \overline{K,n}$ at the same instant is impossible, because the violation of conditions (6.1) would follow immediately after that. Therefore, $s_M$ has a smaller value, than the value found by (6.33) (except for cases $n = K$ and $n = K + 1$).

**Case of three restricted derivatives of action.** Obtain the expression for magnitude $s_M$ with reference to a practical and important case, when $n \leq K + 2$, i.e. the right hand side of (6.31) contains not more than three nonzero addends by using inequality (3.8), which is presented as

$$g^{(n-1)}(t) \leq \sqrt{2 g_M^{(n)} \left[g_M^{(n-2)} - g^{(n-2)}(t)\right]}. \tag{6.34}$$

At $g^{(n-2)}(t) \geq g_M^{(n-2)} - \left(g_M^{(n-1)}(t) / g_M^{(n)}\right)^2 / 2$ the inequality (6.34) is stronger, than the inequality $g^{(n-1)}(t) \leq g_M^{(n-1)}$.

According to (6.34), for the right hand side of (6.31), it is possible to write:

$$s(t) \leq a_{n-2} g^{(n-2)} + a_{n-1} \sqrt{2 g_M^{(n)} \left[g_M^{(n-2)} - g^{(n-2)}(t)\right]} + a_n g_M^{(n)}. \tag{6.35}$$

Having investigated the right hand side of inequality (6.32) on the extremum with the account of (3.8), it is possible to conclude that it accepts the maximum in a case, when

$$g^{(n-2)}(t)=g_*^{(n-2)}=\begin{cases} g_M^{(n-2)}-g_M^{(n)}/(2\alpha_n) & \text{at } \omega_{ng}\le\alpha_n, \\ g_M^{(n-2)}-g_M^{(n-1)}/(2\omega_{ng}) & \text{at } \omega_{ng}>\alpha_n, \end{cases}$$

where $\omega_{ng}=g_M^{(n)}/g_M^{(n-1)}, \alpha_n=a_{n-2}/a_{n-1}$. This maximum value is:

$$s_M=\begin{cases} a_{n-2}\quad g_M^{(n-2)}+[a_{n-1}/(2\alpha_n)+a_n]g_M^{(n)} & \text{at } \omega_{ng}\le\alpha_n, \\ a_{n-2}\left[g_M^{(n-2)}-g_M^{(n-1)}/(2\omega_{ng})\right]+a_{n-1}g_M^{(n-1)}+a_n g_M^{(n)} & \text{at } \omega_{ng}>\alpha_n. \end{cases} \tag{6.36}$$

Further clarify, at what type of action function $s(t)$ accepts the extreme large value $s_M$, i.e. determine the action maximizing the estimation of control error (6.32) and in this sense being the most unfavorable. Make the supposition, that of the $(n+1)$-th derivative of such an action, which is not restricted by inequalities in (6.1), and should accept indefinitely large values. As its $n$-th derivative fulfills condition $\left|g^{(n)}(t)\right|\le g_M^{(n)}$, function $g^{(n+1)}(t)$ can accept indefinitely large values only within the infinitesimal time intervals, and the positive and negative values of this function should be interchanged. The action, which $n$-th derivative looks like meander

$$g_{mu}(t)=(-1)^{[t/T_*+1/2]}g_M^{(n)}, \tag{6.37}$$

possess the indicated properties. Here $T_*<2/\omega_{ng}, [\ldots]$ is an integer part of the number in brackets. Thus, the $(n-1)$-th derivative varies by the saw-tooth law in range $\pm g_M^{(n)}T_*/2$, and the $(n+1)$-th derivative consists of parabola segments and varies in range $\pm g_M^{(n)}T_*/8$, so that at constant component $g_M^{(n-2)}-g_M^{(n)}T_*^2/8$ its maximum value $g_M^{(n-2)}$ is ensured to be reached. The described function system, and approximate shape of function $s_M$ are shown in Fig. 6.8.

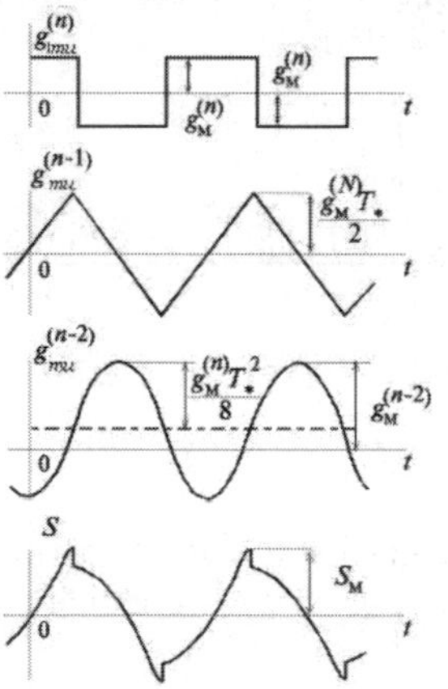

Fig 6.8

The maximum of function $s_M$, appropriate to the described above action, depends on magnitude $T_*$ and on parameters $a_{n-2}, a_{n-1}, a_n$. After research of this function's maximum it is possible to conclude that at $\alpha_n \geq \omega_{ng}$ and $T_* = 2/\alpha_n$

$$s_M = s_{M1} = a_{n-2} g_M^{(n-2)} + \left[a_{n-1}/(2\alpha_n) + a_n\right] g_M^{(n)}, \tag{6.38}$$

at $\alpha_n \leq \omega_{ng}$ and $T_* = 2/\omega_{ng}$

$$s_M = s_{M2} = a_{n-2}\left[g^{(n-2)} - g_M^{(n-1)}/(2\omega_{ng})\right] + a_{n-1} g_M^{(n-1)} + a_n g_M^{(n)}. \tag{6.39}$$

When comparing (6.35) and (6.39) with (6.36) it becomes clear, that these formulas provide an identical value of magnitude $s_M$. Thus, the stated supposition has been confirmed, i.e. the action, of which the $n$-th derivative looks like (6.37), is really the most unfavorable one.

In a case, when the system order is lower than the order of higher restricted derivative of action, i.e. $n < N$, the formulas (6.32), (6.33) and (6.36) are valid if the coefficients $\{a_i\}_{n+1}^{N}$ are accepted to be zero.

**Common case.** If $n > N$, the polynomial $H_s(s)$ should be written as the product of two polynomials $H_s(s) = H_{s1}(s) H_{s2}(s)$. The first of them has the order $N - K$, and second one — the order $n - N$. The roots of the first polynomial lie closer to an imaginary axis of a complex plane, than the roots of the second one.

Let $s_1, s_2, \ldots, s_{n-K}$ be the roots of polynomial $H_s(s)$, written in increasing order of real part absolute value. Then it is possible to use the formulas

$$H_{s1}(s) = a_n a_{n-N+K}^{-1} (s - s_1)(s - s_2)\ldots(s - s_{N-K}),$$

$$H_{s2}(s) = a_{n-N+K} (s - s_{N-K+1})(s - s_{N-K+2})\ldots(s - s_{n-K}),$$

These formulas are valid only if the roots $s_{N-K}$ and $s_{N-K+1}$ are not complex conjugates. In the opposite case, the polynomial $H_{s1}(s)$ should be taken with order $N - K - 1$, and the polynomial $H_{s2}(s)$ — with order $n - N + 1$.

After deriving polynomials $H_{s1}(s)$ and $H_{s2}(s)$, the magnitude $s_M$ is determined by the coefficients of polynomial $H_{s1}(s)$. The polynomial $H_{s2}(s)$ is then combined with transfer function $H_p(s)$ to determine the resulting accumulation coefficient.

## §6.5. Frequency method for maximum error estimation

**Use of spectral decomposition of action.** Go back again to expression (6.5). Taking into account the condition of physical realizability of system $h_{eK}(\tau) = 0$ at $\tau < 0$, enter the formally infinite limits of integration at error evaluation:

$$e_g(t) = \int_{-\infty}^{\infty} h_{eK}(\tau) g^{(K)}(t-\tau) d\tau . \qquad (6.40)$$

When taking into account the finiteness of the control process execution duration and boundedness of absolute values of functions $g^{(K)}(t)$ and $e_g(t)$, consider the conditions for their absolute integrity

$$\int_{-\infty}^{\infty} \left| g^{(K)}(t) \right| dt < \infty, \ \int_{-\infty}^{\infty} \left| e_g(t) \right| dt < \infty.$$

as fulfilled.

Having transformed the right and left hand sides of formula (6.40) with the Fourier transformation and having changed the integration order, the following is obtained:

$$\int_{-\infty}^{\infty} e_g(t) \exp(-j\omega t) dt = \int_{-\infty}^{\infty} \int_{-\infty}^{\infty} h_{eK}(\tau) g^{(K)}(t-\tau) \exp(-j\omega t) d\tau dt =$$

$$= \int_{-\infty}^{\infty} h_{eK}(\tau) \exp(-j\omega\tau) d\tau \int_{-\infty}^{\infty} g^{(K)}(t-\tau) \exp[-j\omega(t-\tau)] d(t-\tau)$$

or

$$E_g(j\omega) = H_{eK}(j\omega) G_K(j\omega), \qquad (6.41)$$

where $E_g(j\omega)$ and $G_K(j\omega)$ are the images of error by Fourier and the $K$-th derivative of reference action containing a sense of spectral densities for these process amplitudes, and $H_{eK}(j\omega)$ is the frequency transfer function appropriate to transfer function (6.4).

Having applied the Fourier reconversion to images $E_g(j\omega)$ and $G_K(j\omega)$, the initial functions are fulfilled:

$$e_g(t) = \frac{1}{2\pi} \int_{-\infty}^{\infty} E_g(j\omega) \exp(j\omega t) d\omega , \qquad (6.42)$$

$$g^{(K)}(t) = \frac{1}{2\pi} \int_{-\infty}^{\infty} G_K(j\omega) \exp(j\omega t) d\omega . \qquad (6.43)$$

The following formulas for derivative of action:

$$g^{(K+i)}(t) = \frac{1}{2\pi} \int_{-\infty}^{\infty} (j\omega)^i G_K(j\omega) \exp(j\omega t) d\omega, \ i = \overline{0, N-K} \qquad (6.44)$$

are also valid.

Taking into account the relations of (6.31) - (6.34) for maximum absolute values of originals, it is possible to write the following expressions:

$$\max\left|e_g(t)\right| \leq \frac{1}{\pi}\int_0^\infty \left|E_g(j\omega)\right| d\omega = \frac{1}{\pi}\int_0^\infty \left|H_{eK}(j\omega) G_K(j\omega)\right| d\omega\,, \tag{6.45}$$

$$\max\left|g^{(K+i)}(t)\right| \leq \frac{1}{\pi}\int_0^\infty \omega^i \left|G_K(j\omega)\right| d\omega,\; i = \overline{0, N-K}\,. \tag{6.46}$$

**Narrowing the class of considered action.** Select conditionally the class $\Xi$ of functions $g^{(K)}(t)$, for which inequality (6.46) can be converted into the equality irrespective complex function $G_K(j\omega)$ argument. This means, that inside the domain of function $g^{(K)}(t) \in \Xi$ there is always an instant, where all elementary harmonic components, in their population generating the function $g^{(K)}(t)$, have the phase, multiple to $2\pi$ or even close to it. In this instant the action $g^{(K)}(t)$ reaches the maximum value expressed by formula (6.46) at $i = 0$. As a discrete character of function $G_K(j\omega)$ the integral in (6.46) accepts the kind of total amplitudes of all harmonics in the function $g^{(K)}(t)$.

The segment of the harmonic function and also the sum of the constant component and segment of harmonic functions are the elementary example for functions possessing the indicated properties. Practically it is possible to refer the segment of the function that is the sum of arbitrary final number of harmonics to different frequencies $\omega_1 \ll \omega_2 \ll \ldots \ll \omega_\nu$. The segment of the function, which is the sum of several harmonic functions with non-multiple frequencies, also belongs to the same class. When reviewing such functions on a large enough time interval, there is always a point, where the phases of all harmonic components lie, for example, within the limits $\pm 10^\circ$. In this point the value of the function is not less than 98.5 % from the sum of harmonics amplitudes, as $\cos 10^\circ = 0.985$, and inequality (6.46) is practically converted to equality.

All enumerated functions posses the property that any of their linear transformations provides the function, which also belongs to the $\Xi$ class. In particular, from the assertion $g^{(K)}(t) \in \Xi$, that $g^{(K+i)}(t) \in \Xi,\; i = \overline{0, N-K}$ and $e_g(t) \in \Xi$ follows.

Assume that the most unfavorable action $g_{mu}^{(K)}(t)$ belongs to the class $\Xi$. Then the problem for maximum error estimation can be defined as follows.

$N - K + 1$ power moments of unknown function $\left|G_K(j\omega)\right|$ are given:

$$g_M^{(K+i)} = \frac{1}{\pi} \int_0^{\infty} \omega^i |G_K(j\omega)| d\omega, i = \overline{0, N-K}. \qquad (6.47)$$

It is required, to estimate the value of integral

$$e_* = \frac{1}{\pi} \int_0^{\infty} |H_{eK}(j\omega) G_K(j\omega)| d\omega, \qquad (6.48)$$

where $|H_{eK}(j\omega)|$ is the known gain plot of the filter with transfer function (6.4).

**Main relations.** The defined problem in essence does not differ from the problem for estimation of error dispersion by dispersions of action derivatives, considered in chapter 5. When solving this problem, it is possible to use the canonical presentations for sequence of power moments $\left\{g_M^{(j)}\right\}_K^N$, or the approximation of function $|H_{eK}(j\omega)|$ by power polynomials. In the latter case the analysis is carried out on the basis of relation

$$\frac{1}{\pi} \int_0^{\infty} \left(c_K + c_{K+1}\omega + \ldots + c_N \omega^{N-K}\right) G_K(j\omega) | d\omega = c_K g_M^{(K)} + \ldots + c_N g_M^{(N)}. \qquad (6.49)$$

If the polynomial $C_{N-K}(\omega) = \sum_{i=K}^{N} c_i \omega^{i-K}$ coefficients ensure the execution of condition $C_{N-K}(\omega) \geq |H_{eK}(j\omega)|$ in frequency domain $\omega \in [0, \beta]$, $\beta \in (0, \infty)$, where the spectral components of action are possible, from (6.47) — (6.49) then the following are obtained:

$$e_* \leq c_K g_M^{(K)} + c_{K+1} g_M^{(K+1)} + \ldots + c_N g_M^{(N)}. \qquad (6.50)$$

Estimate (6.50) becomes strongest, if the coefficients $\{c_i\}_K^N$ are optimal by criterion

$$\sum_{i=K}^{N} c_i g_M^{(i)} \to \min_c .$$

As the magnitude $e_*$, expressed by formula (6.50), has a sense of maximum error only by actions that belong to the class $\Xi$, then $e_* \leq e_M$. Due to this relation (6.50) generally does not allow a strict estimation of maximum error $e_M$ to be given and it is possible to write only the approximated formula

$$e_M \approx c_K g_M^{(K)} + c_{K+1} g_M^{(K+1)} + \ldots + c_N g_M^{(N)}. \qquad (6.51)$$

A stronger outcome is obtained in the following case. Let the canonical presentation for sequence of power moments $\left\{g_M^{(i)}\right\}_K^N$, such as $g_M^{(i)} = \sum_{j=1}^{\nu} u_j \omega_j^{i-K}$, and $i = \overline{K, N}$, where $u_j$ is the amplitude of potential function with frequency $\omega_j$, be found. Allow the possibility to ensure the execution of conditions $C_{N-K}(\omega_j) = |H_e(j\omega)|$, and $j = \overline{1, \nu}$ i.e. the abscissas of tangency points for polynomial $C_{N-K}(\omega)$ with the function $|H_{eK}(j\omega)|$ approximated by it, and to correspond to the points of canonical presentations $\{\omega_j\}_1^{\nu}$. Then inequality (6.47) is converted into the equality and it can obtain the strict estimation

$$e_M \geq e_* = \sum_{i=K}^{N} c_i g_M^{(i)} = \sum_{j=1}^{\nu} \left|H_{eK}(j\omega_j)\right| u_j . \tag{6.52}$$

Thus, the restriction of the considered set of reference actions by acceptance of assumption $g_{mu}^{(K)}(t) \in \Xi$ can use the frequency methods and simplify the procedure for the determination of the maximum error estimate. The fee for such simplification is a theoretical approximation of an obtained estimation. However, in practice the error of estimation can be rather small (see §6.6).

**Maximum error at harmonic reference action**. Make one more step in the direction of estimations simplification. Consider the extreme simple and rough frequency method to analyze the maximum error based on the acceptance of the assumption that the $K$-th derivative of the most unfavorable reference action is the harmonic function with frequency $\omega_1$ and amplitude $u_1$.

The maximum values for higher derivatives of such actions are: $\max g^{(i)}(t) = u_1 \omega_1^{i-K}, i = \overline{K, N}$. In this case the inequality (6.1) can be carried out only, when the amplitude and frequency of action are in the ratio as

$$u_1 \leq u_{1M}(\omega_1) = \min_i g_M^{(i)} / \omega_1^{i-K}, \; i = \overline{K, N} . \tag{6.53}$$

The error amplitude of action processing having only one spectral line on frequency $\omega_1$, according to expression (6.48) equals to

$$e_{*1} = \left|H_{eK}(j\omega_1)\right| u_1 . \tag{6.54}$$

It also depends on magnitude $\omega_1$.

From (6.53) and (6.54) obtain the expression for maximum error at harmonic reference action

$$e_{*1M} = \max e_{*1} = \max_{\omega_1} \left\{ \left|H_{eK}(j\omega_1)\right| \min_i g_M^{(i)} / \omega_1^{i-K} \right\}. \tag{6.55}$$

In order to determine the frequency, that ensures the maximum value of the right hand side of equality (6.52), for the most unfavorable harmonic action $\omega_{1mu}$, it is possible to consider the following. With the account of (6.53) at $u_1 = u_{1M}(\omega_1)$ rewrite (6.55) as:

$$e_{*1M} = \max_{\omega_1}\{H_{eK}(j\omega)u_1(\omega_1)\} - \max_{\omega_1}\{u_1(\omega_1)/|H_{eK}(j\omega_1)^{-1}|\}.$$

As in point $\omega_1 = \omega_{1mu}$ the derivative of maximized function equals zero, and the following equality should be valid

$$\left.\frac{du_1(\omega_1)}{d\omega_1}\right|_{\omega_1=\omega_{1mu}} |H_{eK}^{-1}(j\omega_{1mu})| - u_1(\omega_{1mu})\frac{d}{d\omega_1}\left.|H_{eK}^{-1}(j\omega_1)|\right|_{\omega_1=\omega_{1mu}} = 0$$

or

$$\left.\frac{d[\mu u_1(\omega_1)]}{d\omega_1}\right|_{\omega_1=\omega_{1mu}} |H_{eK}^{-1}(j\omega_{1mu})| = \mu u_1(\omega_{1mu})\frac{d}{d\omega_1}\left.|H_{eK}^{-1}(j\omega_1)|\right|_{\omega_1=\omega_{1mu}}, \tag{6.56}$$

where $\mu$ is a arbitrary scaling coefficient, $\mu\in(0,\infty)$.

Write the sufficient conditions for execution of equality (6.56) as

$$\begin{gathered} \mu u_1(\omega_{1mu}) = |H_{eK}^{-1}(j\omega_{1mu})|, \\ \left.\frac{d[\mu u_1(\omega_1)]}{d\omega_1}\right|_{\omega_1=\omega_{1mu}} = \frac{d}{d\omega_1}\left.|H_{eK}^{-1}(j\omega_1)|\right|_{\omega_1=\omega_{1mu}}. \end{gathered} \tag{6.57}$$

It is lawful to give the dimension to the coefficient $\mu$, which levels the dimension of the right and left-hand sides of equalities in (6.57).

The equalities in (6.57) can be treated as an analytical entry of the condition for the tangency of curves $\mu u_1(\omega_1)$ and $|H_{eK}^{-1}(j\omega_1)|$ at point $\omega_1 = \omega_{1mu}$. Therefore the frequency $\omega_{1mu}$ can be determined as the result of a graphical construction of these curves. Such construction is convenient to execute by using the logarithmic scale on axes of abscissas and ordinates, for example, in the gain plot shown in Fig. 6.9. Thus the graph for function $20\lg\mu u_1(\omega) = 20\lg\mu + 20\lg\left(\min_i g_M^i/\omega^{i-K}\right)$ consists of straight-line segments with inclinations $-(i-K)20$ dB/dec. Consider also, that $20\lg|H_{eK}^{-1}(j\omega)| = -20\lg|H_{eK}(j\omega)|$. By vertical self-parallel transitions of curve $20\lg u_1(\omega)$ (which corresponds to the variation of scale coefficient) it is necessary to achieve its tangency with curve $-20\lg|H_{eK}(j\omega)|$. The abscissa of a tangency point is the required frequency $\omega_{1mu}$.

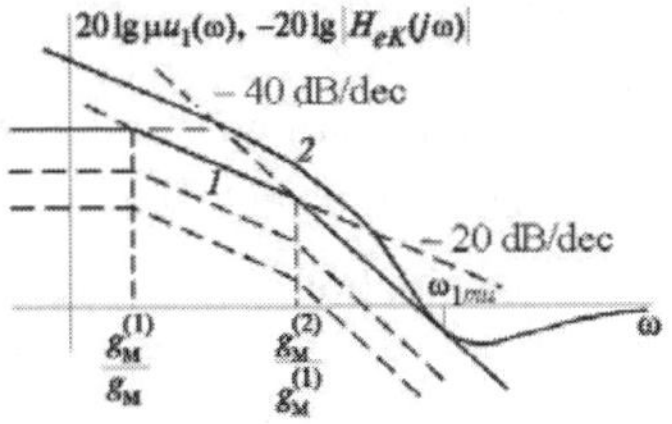


Fig. 6.9

From the graphs in Fig. 6.9 it is visible, that with the presence of a significant resonance peak for curve $\left|H_{eK}(j\omega)\right|$, the point $\omega_{1mu}$ lies slightly more to the left from the resonance frequency. At the monotone character of this curve the point $\omega_{1mu}$, as a rule, coincides with one of the kinks in the curve $20\lg\mu u_1(\omega)$, which has the ordinate $g_M^{(i+1)}/g_M^{(i)}$.

It is possible to construct the curves — $20\lg\mu u_1(\omega)$ and $20\lg\left|H_{eK}(j\omega)\right|$ instead of curves $20\lg\mu u_1(\omega)$ and $-20\lg\left|H_{eK}(j\omega)\right|$ to determine value $\omega_{1mu}$. Some standard examples of such construction are shown in Fig. 6.10.

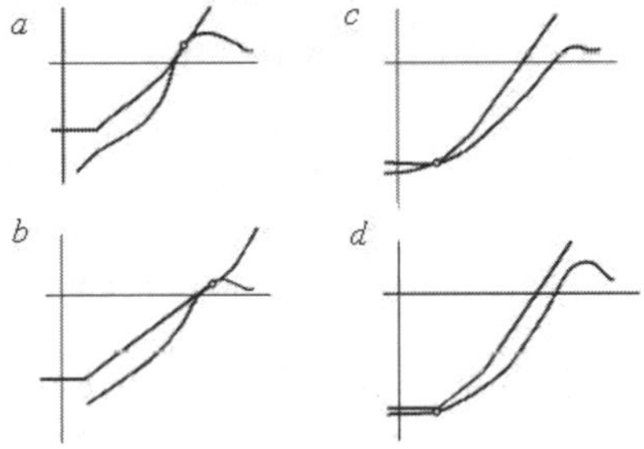


Fig. 6.10

It is possible to determine not only frequency $\omega_{1mu}$ by these graphs but to estimate roughly the shape of the most unfavorable action (select one of curves, shown in Fig. 6.6) and also its periodicity (accepting $\lambda \approx \omega_{1mu}$).

In the case shown in Fig. 6.10*a*, the function $g_{mu}^{(N)}(t)$ should have a shape as shown in Fig. 6.6*a* at $\lambda = \lambda_*$. The graphs in Fig. 6.10*b* and 6.10*c* are characteristic in a case, when the shape of function $g_{mu}^{(N)}(t)$ corresponds to Fig. 6.6*b* at $\lambda = \lambda_*$ and at $\lambda < \lambda_*$, and in Fig. 6.10*d* to Fig. 6.6*c* at $\lambda = \lambda_*/3$.

Let the frequency $\omega_{1mu}$ correspond to the segment of the broken line $20\lg\mu u_1(\omega)$ with inclination $-(I-K)\cdot 20$ dB/dec. Then, according to the determined frequency $\omega_{1mu}$ and index *I*, the expression (6.55) for maximum error at harmonic reference action can be concreted as

$$e_{*1m} = \left|H_{eK}\left(j\omega_{1mu}\right)\right| g_M^{(I)} / \omega_{1mu} = \left|H_{el}\left(j\omega_{1mu}\right)\right| g_M^{(I)}. \tag{6.58}$$

Underline that the most unfavorable harmonic action with frequency $\omega_{1mu}$ used here, in essence differs from the "equivalent" harmonic action with frequency $\omega_N = \sqrt{g_M^{(N)} / g_M^{(N-1)}}$. At its determination, the properties of the system are not taken into consideration.

## §6.6. Precision of the frequency domain method

**Purpose of investigations.** The magnitude $e_{*1M}$ appropriate to the harmonic action and expressed by formula (6.55), is the strict lower estimation of maximum error $e_M$, i.e. $e_M \geq e_{*1M}$. The purpose of investigation on precision shown below consists of determination of the greatest possible underestimation for this estimate, which is characterized by the coefficient $\Re = e_M / e_{*1M}$.

It is essential, that the magnitude $e_*$ expressed by formula (6.48) appropriate to restriction of considered actions by entered in §6.5 class $\Xi$, is the maximum error estimation, which is not weaker than $e_{*1M}$, because class $\Xi$ is wider than the class of harmonic functions. Thus, the relations $e_* \geq e_{*1M} = e_M / \Re$ are valid, whence $e_M \leq \Re\, e_{*1M}$. Taking into account (6.50),

$$e_M \leq \Re\left(c_K g_M^{(K)} + c_{K+1} g_M^{(K+1)} + \ldots + c_N g_M^{(N)}\right). \tag{6.59}$$

Therefore, after determination of coefficient $\Re$ it becomes possible to strictly estimate the maximum error from above with analysis in the frequency area by using the approximative method.

**Case of essentially restricted *N*-th derivative of action.** At the determination of magnitude $\Re$ consider at first, that the frequency $\omega_{1mu}$ corresponds to a segment of a broken line $20\lg \mu u_1(\omega)$ with inclination $-(N-K)\cdot 20$ dB/dec and the most unfavorable action looks meander, shown in Fig. 6.6*a*. Having decomposed this function into the Fourier series, the following is obtained:

$$g_{mu}^{(N)}(t) = \frac{4}{\pi}\left[\cos(\lambda t) - \frac{1}{3}\cos(3\lambda t) + \frac{1}{5}\cos(5\lambda t) - \ldots\right].$$

The error of such action processing (excluding the possible constant component of its *K*-th derivative) is:

$$e(t) = \frac{4}{\pi} g_M^{(N)} \sum_{l=1}^{\infty} \frac{(-1)^{l-1}}{2l-1} \left|H_{eN}\left[j(2l-1)\lambda\right]\right| \times \\ \times \cos\{(2l-1)\lambda t + \arg H_{eN}\left[j(2l-1)\lambda\right]\}, \tag{6.60}$$

where according to (6.4) $H_{eN}(j\omega) = H_e(j\omega)/(j\omega)^N$.

If $\lambda \approx \omega_{1mu}$, according to Fig. 6.9 the following inequality is fulfilled

$$|H_{eN}(j\lambda)| \geq |H_{eN}(j3\lambda)| \geq |H_{eN}(j5\lambda)| \geq \ldots$$

The function $|H_{e2}(j\omega)|$ can usually be well approximated by the gain plot of an oscillatory unit with damping coefficient $\xi \geq 0.6$ near the resonance peak. Then, analyzing the phase plot of an oscillatory unit [9], the following inequality can be developed

$$|\arg H_{eN}(j\lambda) - \arg H_{eN}(j3\lambda)| \leq \pi/2 .$$

Taking this into account, estimate the maximum value for the sum in expression (6.60) over the greatest possible values for three of its first members

$$\max\left\{\cos(\lambda t) - \frac{1}{3}\cos\left(3\lambda t + \frac{\pi}{2}\right) + \frac{1}{5}\cos\left(5\lambda t + \frac{\pi}{2}\right)\right\} =$$

$$= \max\left\{\cos(\lambda t) + \frac{1}{3}\sin(3\lambda t) - \frac{1}{3}\sin(5\lambda t)\right\} \approx 1.16 .$$

Then expression (6.60) becomes: $e_M = \max e(t) < 1.16\frac{4}{\pi}|H_{eN}(j\lambda)| g_M^{(N)}$.

According to (6.58) for $I = N$ and for inequality $|H_{eN}(j\omega_{1mu})| \geq |H_{eN}(j\lambda)|$ the following is developed:

$$\frac{e_M}{e_{*1M}} < 1.16\frac{4}{\pi} = 1.48 . \qquad (6.61)$$

If the by the gain plot of an oscillatory unit with the small damping coefficient approximates the resonance peak of function $|H_{e2}(j\omega)|$, and only the harmonic can be practically recovered from the action with resonance frequency $\lambda \approx \omega_{1mu}$, then the stronger inequality

$$\frac{e_M}{e_{*1M}} \leq \frac{4}{\pi} = 1.27 \qquad (6.62)$$

is obvious.

If *N*-th derivative of the most unfavorable action looks like that shown in Fig. 6.6*b*, and the expansion of such function into Fourier series equals

$$g_{mu}^{(N)}(t) = \frac{4}{\pi} g_M^{(N)}\left(\sin\frac{\lambda\tau_{int}}{2}\cos\lambda t + \frac{1}{3}\sin\frac{3\lambda\tau_{int}}{2}\cos 3\lambda t + \right.$$
$$\left. + \frac{1}{5}\sin\frac{5\lambda\tau_{int}}{2}\cos 5\lambda t + \ldots\right). \qquad (6.63)$$

The maximum error can be estimated by using the first three members of the sum in (6.63). Then by analogy with inequality (6.61) it is possible to write the expression

$$\frac{e_M}{e_{*M}} \leq 1.16 \frac{4}{\pi} \sin \frac{\lambda \tau_{\text{int}}}{2}. \tag{6.64}$$

At $\tau_{\text{int}} = \pi/\lambda$, when function $g_{mu}^{(N)}(t)$ has the shape of the meander, the relation (6.64) turns into inequality (6.61).

If the system stability margin is not large and the function $|H_{eN}(j\omega)|$ has the large resonance peak then the magnitude $e_M$ can be estimated with use of only the first member of the sum in expression (6.63). It equals

$$\frac{e_M}{e_{*M}} \leq \frac{4}{\pi} \sin \frac{\lambda \tau_{\text{int}}}{2}.$$

**Case of essentially restricted (*N* - 1)-th derivative of action.** Consider the case, when the frequency $\omega_{1mu}$ corresponds to the segment of broken line $20 \lg \mu u_1(\omega)$ with inclination $-(N-K-1) \cdot 20$ dB/dec. Thus the magnitude $\tau_{\text{int}}$ is determined first of all. by the necessity of inequality (6.4) at the execution at $i = N-1$ Integrating expression (6.63) obtain the Fourier series for the function $g_{mu}^{(N-1)}(t)$

$$g_{mu}^{(N-1)}(t) = \frac{4 g_M^{(N)}}{\pi \lambda} \sum_{l=1}^{\infty} \frac{1}{(2l-1)^2} \sin \frac{(2l-1)\lambda \tau_{int}}{2} \sin[(2l-1)\lambda t], \tag{6.65}$$

whence

$$\max g_{mu}^{(N-1)}(t) = \frac{4 g_M^{(N)}}{\pi \lambda} \left( \sin \frac{\lambda \tau_{int}}{2} - \frac{1}{9} \sin \frac{3\lambda \tau_{int}}{2} + \frac{1}{25} \sin \frac{5\lambda \tau_{int}}{2} - \ldots \right) = \frac{g_M^{(N)} \tau_{int}}{2}$$

and the execution of the condition $\max g_{mu}^{(N-1)}(t) = g_M^{(N-1)}$ of the magnitude $\tau_{\text{int}}$ should be equal to

$$\tau_{\text{int}} = 2 g_M^{(N-1)} / g_M^{(N)}. \tag{6.66}$$

With use of (6.66) and first member of series (6.65), obtain the estimation for the error $e_M$

$$e_M \leq \frac{4 g_M^{(N)}}{\pi \lambda} \sin \frac{\lambda \tau_{\text{int}}}{2} \left| H_{e(N-1)}(j\lambda) \right| = \frac{4 g_M^{(N)}}{\pi \lambda} \sin \frac{\lambda g_M^{(N-1)}}{g_M^{(N)}} \left| H_{e(N-1)}(j\lambda) \right|. \tag{6.67}$$

Comparing expressions (6.67) and (6.58) at $I = N-1$, according to inequality $\left| H_{e(N-1)}(j\omega_{1mu}) \right| \geq \left| H_{e(N-1)}(j\lambda) \right|$, equals

$$\frac{e_M}{e_{*1M}} \leq \frac{4}{\pi} \frac{g_M^{(N)}}{\lambda g_M^{(N-1)}} \sin \frac{\lambda g_M^{(N-1)}}{g_M^{(N)}}. \tag{6.68}$$

The inequality (6.68) is stronger than (6.62), because at $0 < \lambda g_M^{(N-1)} / g_M^{(N)} \le 1$ the following condition is satisfied

$$0.79 \le \frac{g_M^{(N)}}{\lambda g_M^{(N-1)}} \sin \frac{\lambda g_M^{(N-1)}}{g_M^{(N)}} < 1 .$$

For the characteristic case $\lambda = g_M^{(N)} / g_M^{(N-1)}$ from (6.68) $e_M / e_{*M} \le 1.07$ is obtained.

**Case of essentially restricted (*N* - 2)-th derivative of action.** Consider at last the case, when the frequency $\omega_{1mu}$ corresponds to a segment of broken line $20\lg \mu u_1(\omega)$ with inclination $-(N-K-2)\cdot 20$ dB/dec, i.e. $\omega_{1mu} \le \sqrt{g_M^{(N)} / g_M^{(N-2)}}$. Then the magnitude $\tau_{\text{int}}$ is first of all determined by the necessity of the inequality (6.1) execution at $i = N-2$. Integrating (6.65) obtain the Fourier series for the function $g_{mu}^{(N-2)}(t)$

$$g_{mu}^{(N-2)}(t) = -\frac{4 g_M^{(N)}}{\pi \lambda^2} \sum_{l=1}^{\infty} \frac{1}{(2l-1)^3} \sin \frac{(2l-1)\lambda \tau_{int}}{2} \cos[(2l-1)\lambda t],$$

whence

$$\max g_{mu}^{(N-2)}(t) = \frac{4 g_M^{(N)}}{\pi \lambda^2} \sum_{l=1}^{\infty} \frac{1}{(2l-1)^3} \sin \frac{(2l-1)\lambda \tau_{int}}{2} = \frac{g_M^{(N)} \tau_{int}}{4} \left( \frac{\pi}{\lambda} - \frac{\tau_{int}}{2} \right). \qquad (6.69)$$

And for execution of condition $\max g_{mu}^{(N-2)}(t) = g_M^{(N-2)}$ the equality $\frac{g_M^{(N)} \tau_{\text{int}}}{4} \left( \frac{\pi}{\lambda} - \frac{\tau_{\text{int}}}{2} \right) = g_M^{(N-2)}$ must be fulfilled.

The solution of this quadratic equation in relationship to $\tau_{\text{int}}$ is:

$$\tau_{\text{int}} = \frac{\pi}{\lambda} \left( 1 - \sqrt{1 - \frac{8 \lambda^2 g_M^{(N-2)}}{\pi^2 g_M^{(N)}}} \right). \qquad (6.70)$$

Note, that $\lambda^2 g_M^{(N-2)} / g_M^{(N)} \le 1$.

As the frequency $\lambda$, which usually is a little smaller than frequency $\omega_{1mu}$, should lie near the resonance peak of curve $|H_{e(N-2)}(j\omega)|$ (see Fig. 6.9), and the frequencies $3\lambda$, $5\lambda$ ... lie in significant distance to the right from this peak, it is possible to consider the following inequalities fulfilled

$$|H_{e(N-2)}(j\lambda)| \ge |H_{e(N-2)}[j(2l-1)\lambda]|, l > 1 .$$

Then with the use of (6.66) for the maximum error the estimation

$$e_M \le \left|H_{e(N-2)}(j\lambda)\right| \max_t g_{mu}^{(N-2)}(t) = \left|H_{e(N-2)}(j\lambda)\right| \frac{g_M^{(N)}\tau_{\text{int}}}{4}\left(\frac{\pi}{\lambda} - \frac{\tau_{\text{int}}}{2}\right). \quad (6.71)$$

can be obtained.

Comparing expressions (6.71) and (6.58) at $I = N-2$, according to (6.70) and the condition $\left|H_{e(N-2)}(j\omega_{1mu})\right| \ge \left|H_{e(N-2)}(j\lambda)\right|$, provides the inequality

$$\begin{aligned} \frac{e_M}{e_{*1M}} &\le \frac{g_M^{(N)}\tau_{\text{int}}}{4g_M^{(N-2)}}\left(\frac{\pi}{\lambda} - \frac{\tau_{\text{int}}}{2}\right) = \\ &= \frac{\pi^2 g_M^{(N)}}{8\lambda^2 g_M^{(N-2)}}\left(1 - \sqrt{1 - \frac{8\lambda^2 g_M^{(N-2)}}{\pi^2 g_M^{(N)}}}\right)\left(1 + \sqrt{1 - \frac{8\lambda^2 g_M^{(N-2)}}{\pi^2 g_M^{(N)}}}\right) = 1, \end{aligned} \quad (6.72)$$

i.e. $e_M \le e_{*1M}$. It testifies that in the observed case shown in Fig. 6.6*b* or 6.6*a* the action is not the most unfavorable, because the processing of the harmonic action can have the greater error. A similar outcome can be made as the result of investigation of the action shown in Fig. 6.6*c*. Note that a few other calculations also giving the relation (6.72), are described in the book [8] on page 138.

**Outcomes.** Thus, it is proved, that the value of entered coefficient $\Re$ cannot exceed 1.48, and in the majority of practically interesting cases does not exceed even 1.27. This is the greatest relative underestimation of estimates for the maximum error obtained by the harmonic action or, especially, on the action from class $\Xi$. The relative underestimation of the estimate decreases at the strengthening of restrictions on higher derivatives of actions, and also at a decreasing order *I* of the curve $\left|H_{eN}(j\omega)\right|$ section. Under certain conditions such underestimation can be reduced to zero. In particular, it happens almost all the time in a case when the system order *n* and the higher order from essentially restricted derivatives of actions *N*, are in the ratio $N \ge n+2$.

The carried out investigation allows using rather simple frequency methods to determine the maximum error estimation. This method is also convenient to construct the machine algorithms for parametric optimization of a system by criterion $e_M \to \min$, and also at the synthesis of system with a guaranteed accuracy level on the basis of forbidden areas in plane of gain plot (see the chapter 7). The described frequency method successfully completes the research techniques of maximum error in time domains, in particular — the perturbations accumulation method, which has a greater accuracy and allows for the refining of the maximum error on the final stage of system synthesis and, if it is necessary, to correct the value of coefficient $\Re$ included in formula (6.59).

The calculation example, which illustrates the operation theory of methods explained here for maximum error estimation in detail, is given below.

## §6.7. Design example

With use of the methods explained in §6.2 — 6.6, calculate the maximum dynamic error in a second order system with first order astatism and with the transfer function of open loop

$$W(s) = \frac{K_1(1+b_1 s)}{s(1+T_1 s)}.$$

Such transfer functions can have, for example, the tracking radar range finder, where $K_1$ is the gain coefficient by velocity, $T_1$ and $b_1$ are the time constants for proportional-integrating amplifier, and $T_1 > b_1$. The distance to the target at radar tracking is the reference action in this case.

Let the maximum values of the first and second derivative of reference action $g_M^{(1)} = 300$ m/sec, $g_M^{(2)} = 25$ m/sec$^2$ be known, i.e. $K = 1$, $N = 2$.

Values of system parameters for the first variant are: $K_1 = 12$ sec$^{-1}$, $T_1 = 20$ sec, $b_1 = 1$ sec or $b_1 = 1.9$ sec for the second one. Giving two possible values of time constant $b_1$ on can compare by accuracy the versions of systems with different stability margins.

According to the formulas (6.2) and (6.4) for the transfer function $H_{eK}(s)$ and for step response $h_{eK}(t)$, describing the law of error variation at the step modification of the reference action rate, the following expressions are obtained

$$H_{e1}(s) = \frac{1}{[1+W(s)]s} = \frac{1+T_1 s}{K_1 + (K_1 b_1 + 1)s + T_1 s^2} = \frac{(1+T_1 s)T_1^{-1}}{q^2 + 2\xi q s + s^2},$$

$$h_{e2}(t) = L^{-1}\left\{\frac{H_{e1}(s)}{s}\right\} = K_1^{-1}\left[1 - \exp(-\gamma t)\left(\cos(\lambda_* t) + \frac{\gamma - K_1}{\lambda_*}\sin(\lambda_* t)\right)\right].$$

Here

$$q = \sqrt{K_1/T_1} = 0.775 \text{ sec}^{-1},\ \xi = (K_1 b_1 + 1)/\left(2\sqrt{K_1 T_1}\right),\ \gamma = q\xi,\ \lambda_* = q\sqrt{1-\xi^2}.$$

The substitution of numerical values equals in the first variant $\xi = 0.420$, $\gamma = 0.325$ sec$^{-1}$, $\lambda_* = 0.703$ sec$^{-1}$, and in the second variant $\xi = 0.736$, $\gamma = 0.570$ sec$^{-1}$, $\lambda_* = 0.525$ sec$^{-1}$. The rather large value of damping coefficient $\xi$ in the second variant testifies for the good system stability margin.

In order to determine the maximum error, frequency methods are first used. Having accepted the limiting condition, that the reference action is harmonic, consider the gain plot

$$|H_{e1}(j\omega)| = \sqrt{\frac{1+T_1^2\omega^2}{K_1^2 + \left[(K_1 b_1 + 1)^2 - 2K_1 T_1\right]\omega^2 + T_1^2\omega^4}}$$

and construct the graph for functions $-20\lg\left|H_{e1}(j\omega)\right|$. A solid line in Fig. 6.11 shows this. The graph of function $20\lg\mu u_1(\omega)$ for which, according to formula (6.53), it is possible to write

$$\mu u_1(\omega)=\mu\min\left\{g_M^{(1)},\frac{g_M^{(2)}}{\omega}\right\}=\begin{cases}\mu g_M^{(1)} & \text{at } \omega\le g_M^{(2)}/g_M^{(1)},\\ \dfrac{\mu g_M^{(2)}}{\omega} & \text{at } \omega> g_M^{(2)}/g_M^{(1)}.\end{cases}$$

is shown in Fig. 6.11 by dotted lines.

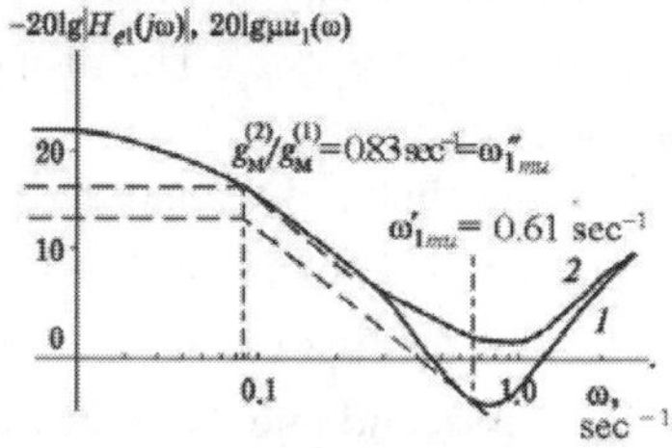


Fig. 6.11

By vertical transitions of a dotted line that corresponds to the variation of coefficient $\mu$, it is necessary to ensure the tangency of dotted line and solid lines in Fig. 6.11. The abscissa of tangency points provides the frequency for the most unfavorable harmonic action. In the first variant $\omega'_{1mu}=0.61\ \text{sec}^{-1}$ is obtained, and in second one $\omega''_{1mu}=0.083\ \text{sec}^{-1}$. The amplitudes for the first derivative of action is: $u'_1=g_M^{(2)}/\omega'_{1mu}=25/0.61=41.0$ m/sec, $u''_1=g_M^{(1)}=300$ m/sec.

According to formula (6.58) at $I=2$

$$e_{g*1M}=\frac{\left|H_{e1}(j\omega_{1mu})\right|g_M^{(2)}}{\omega_{1mu}},$$

whence for the first and second variants of the system, the equations

$$e'_{g*1M}=\frac{\left|H_{e1}(j\cdot 0.61)\right|\cdot 25}{0.61}=54.9\ \text{m},\ e''_{g*1M}=\frac{\left|H_{e1}(j\cdot 0.083)\right|\cdot 25}{0.083}=48.5\ \text{m}.$$

are obtained.

In the first version, the function $g^{(1)}(t)$ can have the stationary component $g_M^{(1)}-u'_1$, except for a harmonic component with amplitude $u'_1<g_M^{(1)}$. According to this stationary component, the refined maximum error is:

$$e_{g*M}=e_{g*1M}+\left(g_M^{(1)}-u'_1\right)h_{e2}(\infty)=54.9+(300-41.0)/12=76.5\ \text{m}.$$

The same outcome can be developed when searching for the maximum error for reference actions that belong to the class $\Xi$. The optimal values of the coefficient for approximating polynomial $C_1(\omega)=c_1+c_2\omega$, suiting condition $C_1(\omega)\geq |H_{e1}(j\omega)|$, $\omega\in[0,\infty)$, is $c_1' = 0.0833$ sec, $c_2' = 2.06$ sec$^2$ for the first one and $c_1' = 0.0833$sec, $c_2' = 0.940$ sec$^2$ for the second version of the system. It is easy to make sure that they ensure the tangency of graphs for functions $C_1(\omega)$ and $|H_{e1}(j\omega)|$ at points $\omega=0$, and $\omega=\omega_{1mu}$, i.e. actually at points of canonical representation. Therefore it is possible to use formula (6.52), according to which $e_{gM}\geq c_1 g_M^{(1)}+c_2 g_M^{(2)}$ or

$$e'_{gM}\geq 0.0833\cdot 300+2.06\cdot 25=76.5 \text{ m},$$

$$e''_{gM}\geq 0.0833\cdot 300+0.940\cdot 25=48.5 \text{ m}.$$

Further, find an approximated estimation for the maximum error from above by formula (6.32). The accumulation coefficient $k_p=k_{ss}$ for an oscillatory unit with transfer function

$$H_p(s)=\frac{1}{1+2\xi s/q+s^2/q^2}$$

can be determined from the graph in Fig. 6.4. For the first version the following is obtained: $k_p' = 1.61$, for second one: $k_p'' = 1.06$.

According to transfer function

$$H_s(s)=(1+T_1 s)K_1^{-1}$$

with the use of formula (6.36) for the maximum value of function $s(t)=\left[g^{(1)}(t)+T_1 g^{(2)}(t)\right]K_1^{-1}$ the following is obtained:

$$s_M=\left(g_M^{(1)}+T_1 g_M^{(2)}\right)K_1^{-1}=(300+20\cdot 25)/12=66.7 \text{ m}.$$

Then formula (6.32) gives the required upper-bound estimate $e_M'\leq K_p' s_M=1.61\cdot 66.7=107$ m for the first and $e_M''\leq K_p'' s_M=$ $=1.06\cdot 66.7=71.3$ m for the second system versions.

Evaluate last, the exact value for the maximum error by considering the most unfavorable reference action, whose second derivative is shown in Fig. 6.6*a*, and $\lambda=\lambda_*$. Consider for this purpose the function

$$\Delta h_{e2}(t)=\int_0^t\left[w_{e2}(\tau)-w_{e2}(\infty)\right]d\tau=\frac{T_1\exp(-\gamma t)}{K_1^2}\times$$

$$\times\left[(2\gamma-K_1)\cos(\lambda t)+\frac{\gamma^2-\lambda^2-\gamma K_1}{\lambda}\sin(\lambda t)\right]+\frac{(K_1-2\gamma)T_1}{K_1^2}.$$

included in expression (6.27). Its graph is shown in Fig. 6.12 (curves 1 and 2 for the first and second variants accordingly).

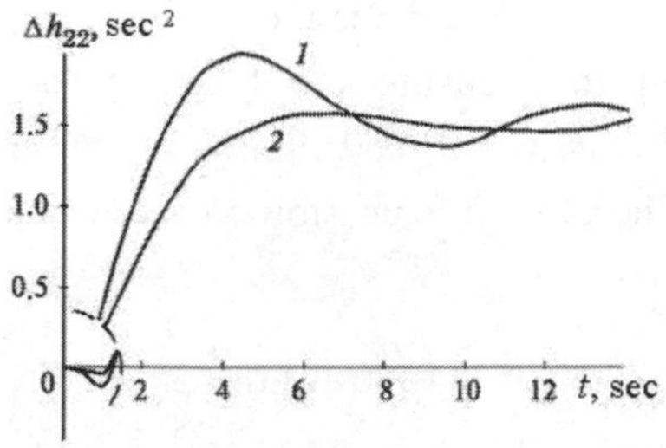


Fig. 6.12

Having equated the derivative of this function to zero, it is possible to determine its extreme points

$$t_i = \frac{1}{\lambda}\left[\operatorname{arctg}\frac{\lambda}{K_1-\gamma}+(i-1)\pi\right],\ i=1,2,3,\ldots$$

The values of extremums are

$$\Delta h_{e2}(t_i) = (-1)^i \frac{\sqrt{q^2T_1^2-2\gamma T_1+1}}{K_1 q}\times$$

$$\times\exp\left\{-\frac{\gamma}{\lambda}\left[\operatorname{arctg}\frac{\lambda}{K_1-\gamma}+(i-1)\pi\right]\right\}+\frac{(K_1-2\gamma)T_1}{K_1^2}.$$

Thus according to formula (6.28) it is possible to evaluate the value of an integral

$$\int_0^\infty |w_{e2}(\tau)-w_{e2}(\infty)|d\tau = 2[-\Delta h_{e2}(t_1)+\Delta h_{e2}(t_2)-\Delta h_{e2}(t_3)+\ldots]-\Delta h_{e2}(\infty)=$$

$$=\frac{2\sqrt{q^2T_1^2-2\gamma T_1+1}}{K_1 q[1-\exp(-\pi\gamma/\lambda)]}\exp\left(-\frac{\gamma}{\lambda}\operatorname{arctg}\frac{\lambda}{K_1-\gamma}\right)-\frac{(K_1-2\gamma)T_1}{K_1^2}.$$

The last expression is obtained as a result of summarizing the decreasing geometrical progression that has the denominator $\exp(-\pi\gamma/\lambda)$ and the first member $\pm g_M^{(n)}T_*/2$.

Further, use formulas (6.27) and (6.29), which provide inequalities:

$$e_{gM} \le \frac{g_M^{(1)}}{K_1}+g_M^{(2)}\int_0^\infty |w_{e2}(\tau)-w_{e2}(\infty)|\,d\tau,$$

$$e_{gM} \geq \frac{g_M^{(1)} - g_M^{(2)} t_1}{K_1} + g_M^{(2)} \int_0^{\infty} \left| w_{e2}(\tau) - w_{e2}(\infty) \right| d\tau.$$

After the substitution of numerical values, find the required value for maximum error $88.6 \text{ m} \leq e'_M \leq 88.7 \text{ m}$ or $e'_M = 89 \text{ m}$ for the first version, $65.3 \text{ m} \leq e''_M \leq 65.4 \text{ m}$ or $e''_M = 65 \text{ m}$ for the second version of a system.

At last find the relative underestimation of an estimate $e_{*M}$, obtained by the frequency method, in comparison with the precise value for maximum error $e_M$. For the first version of a system it is 89/76.5 = 1.16, for the second version: 65/48.5 = 1.34. Thus, inequality (6.61) is fulfilled, and the stronger inequality (6.62) is also fulfilled for the first version.

It is necessary to note, that the numerical data in an example are taken, in order to create rather unfavorable conditions for an estimation of maximum error by the frequency method. In the majority of practical cases it is possible to consider that inequality (6.69) is fulfilled with the margin.

The given example confirms the conclusion made in §6.5 that the frequency methods allow us to obtain rather simply the estimation for maximum error with relative inaccuracy, which docs not exceed the particularly small value.

*Questions*

1. Why is the maximal error analysis important?
2. Is the maximal error the static characteristic of control accuracy?
3. What is the technology of application of the perturbation accumulation method at maximal error analysis?
4. What is the accumulation coefficient? How can it be determined and what does it depend on?
5. At what condition does the error of a dynamic system reach involving the action with the constant maximal value of restricted derivative?
6. At what condition does the maximal error of a dynamic system reach its maximum with the variation of a higher restricted derivative by meander law?
7. Characterize the possible variants of action types, which cause the maximal value of error.
8. Explain the mode of approximate estimation from above for the maximal error, based on a separate consideration of numerator and denumerator for the transfer function of a closed loop system and for error.
9. What are the advantages and disadvantages of the frequency method for maximal error analysis?
10. At what restriction on a class of considered actions can the analysis of maximal error by the frequency method be produced without the account of phase relations between the spectral components of action?
11. Why is the analysis of maximal error by the frequency method the subject for problems of moments in a mathematical sense?
12. How can the frequency of the most unfavorable harmonic action be determined, which causes the maximal error of a system?

13. What is the difference between the most unfavorable harmonic action in the sense of maximal error creation and the “equivalent” harmonic action?
14. How can the maximal possible underestimation for maximal error estimation be determined with the application of the frequency method? What does it depend on and what values can it obtain?
15. What do the frequency methods of maximal error and r.-m.-s. error analysis have in common? What is the difference between them?

# Chapter 7

## Synthesis of control systems with preset ensured operation accuracy

### §7.1 Principles of construction: the forbidden areas for Bode diagrams of open loop systems

**Preliminary notes.** Considering the accuracy of control systems as the major quality characteristic of its operation, it is natural to first require the acceptable value for selected precision factor when realizing the dynamic synthesis of a system. An average quadrate or the dispersion of error can be used as such a factor only in a case when the spectral densities of input actions are given. The smaller a priori information causes the necessity to use the rougher accuracy characteristics. As it is shown in the chapters 5 and 6, rather small amounts of a priori information is required for calculation of the upper-bound estimate of error dispersion or maximum error. It provides the foundation to set the problem of robust dynamic system synthesis at partially unknown characteristics of input actions. In the first case the problem is: the synthesis on dispersions of action and its derivatives or on other generalized moments of action spectral density, in the second case — on maximum absolute values of action and its derivatives.

In both cases the problem of synthesis is often expedient to be set in a way, which ensures the given acceptable value of precision factors at a sufficient stability margin, speed of response and minimal complexity of systems. For this purpose it is necessary to allocate the class of systems, for which the precision factor lies within the acceptable limits, and then to narrow this class of systems gradually, sequentially superimposing the additional requirements on stability margin, speed of response, ease of realization and other qualitative factors.

Such procedure of synthesis is widely applied in engineering practice. For selection of a class of systems, which satisfy the requirement on accuracy, the forbidden areas in the Bode diagram plane of open loop system are usually used.

**Requirements on the accuracy conditions to the frequency transfer function of closed loop systems for error**. The selection of a system transfer function to ensure the restriction of dynamic error dispersion can be produced more simply as contrasted to a case of maximum error restriction, because it is necessary to take the phase relations between spectral components of actions into account. Therefore the fundamental ideas to construct the forbidden areas by accuracy will be explained at first for a case of preset dispersions of action derivatives and admissible dispersion of dynamic error.

It is necessary to remember, that the dynamic error can be presented as an outcome of a reference action passing through the filter with a frequency transfer function:

$$H_e(j\omega) = [1 + W(j\omega)]^{-1} = H(j\omega)/W(j\omega), \tag{7.1}$$

where $H(j\omega) = W(j\omega)/[1 + W(j\omega)]$ and $W(j\omega)$ are the frequency transfer functions for a closed and open systems accordingly.

Therefore the restriction of any precision factor describing the dynamic error is always connected to the superposition of those or other conditions on a kind of frequency transfer function of closed-loop system for an error $H_e(j\omega)$. The requirements for the frequency transfer function of the open loop system, which are more convenient to use at synthesis, are secondary and can be formulated in view of relation (7.1). Similarly the requirements for restriction of precision factors describing an error from interference can only be secondary.

As it was shown in §5.3, if the condition

$$|H_e(j\omega)|^2 \le C_{2N}(\omega) = c_K \omega^{2K} + c_{K+1}\omega^{2(K+1)} + \ldots + c_N \omega^{2N}, \tag{7.2}$$

is fulfilled, where $\{c_i\}_K^N$ are some real coefficients, then for dynamic error dispersion the following relation is valid:

$$D_{eg} = \frac{1}{\pi}\int_0^\infty |H_e(j\omega)|^2 S_g(\omega)d\omega \le \frac{1}{\pi}\int_0^\infty \sum_{i=K}^N c_i\omega^{2i} S_g(\omega)d\omega = \sum_{i=K}^N c_i D_i \,.$$

Therefore, it does not exceed the value

$$\overline{D}_{eg} = c_K D_K + c_{K+1} D_{K+1} + \ldots + c_N D_N \,. \tag{7.3}$$

If the problem was at the analysis of accuracy: to select the factors $c_i$ for a given transfer function of a system in order to ensure the realization of (7.2), there is now an inverse problem. For some set of factors $\{c_i\}_K^N$, at which the equality is fulfilled

$$c_K D_K + c_{K+1} D_{K+1} + \ldots + c_N D_N = D_{eg}^0 \tag{7.4}$$

it is required to allocate the class of transfer functions, in which condition (7.2) is satisfied. Here $D_{eg}^0$ is the given admissible dispersion of dynamic error.

For example, if only the dispersion for the first derivative of reference action is known, expression (7.4) becomes $c_1 D_1 = D_{eg}^0$, whence the unique nonzero factor $c_1 = D_{eg}^0 / D_1$ in the right hand side of (7.2) can be found. As the result the following is obtained: $|H_e(j\omega)|^2 \le c_1\omega^2$ or

$$\left|H_e(j\omega)\right| \le \sqrt{D_{eg}^0 / D_1}\,\omega . \tag{7.5}$$

The inequality (7.5) determines the requirement to the module of frequency transfer function of closed-loop systems for an error, whose realization ensures the deriving of system accuracy no worse than the given one. On the basis of this inequality and the relations in (7.1) the requirements to the frequency transfer function of an open loop system should be formulated in the considered example.

However, it necessary to first show , that the inequality (7.5) can be developed by another method, which is not connected to the use of relation (7.2). For this purpose it is necessary to consider only one spectral component of reference action on some frequency $\omega$ with the dispersion $\rho_0$ and with spectral density

$$S_g(\omega') = \pi\rho_0\left[\delta(\omega' - \omega) + \delta(\omega' + \omega)\right]. \tag{7.6}$$

The dispersion of the first derivative for this component $\rho_1 = \rho_0\omega^2$ can not exceed the value $D_1$, i.e. $\rho_0\omega^2 \le D_1$, whence $\rho_0 \le D_1/\omega^2$ . The error dispersion for processing of a considered component of actions is:

$$D_{eg} = \left|H_e(j\omega)\right|^2 \rho_0 \le \left|H_e(j\omega)\right|^2 D_1/\omega^2 . \tag{7.7}$$

As the requirement $D_{eg} \le D_{eg}^0$ takes place, from (7.7) thc following inequality is obeyed:

$$\left|H_e(j\omega)\right| \le \sqrt{D_{eg}/\rho_0} \le \sqrt{D_{eg}^0 / D_1}\,\omega . \tag{7.8}$$

In spite of the fact that the requirement (7.8) formally coincides with (7.5), it has some other sense. The matter is that (7.8) was obtained only as the necessary condition to cnsure the given accuracy. If (7.8) is not fulfilled, it inevitably causes the failure of accuracy requirements, which are broken at a harmonic action. However, the realization (7.8) does not give the foundation to claim, that the accuracy requircments are fulfilled at reference action with an arbitrary spectrum. Unlike this, the realization of inequality (7.5), whose derivation is based on inequality (7.2) and is irrelevant with any suppositions about the spectral content of action, is necessary and the sufficient condition ensures the given accuracy.

In §7.3 and 7.4 it will be shown, that in more complicated cases, including the representation of several dispersions $\{D_i\}_K^N$, the necessary conditions of achieving the required accuracy derived similarly to an inequality (7.8), generally do not coincide with sufficient conditions obtained similarly to (7.5), and have an independent value.

**Requirements on accuracy conditions to the frequency transfer function of open loop systems.** If condition $\left|H_e(j\omega)\right| \le \left|H_e^0(j\omega)\right|$ is obtained with reference to the frequency transfer function of a closed-loop system for an error, according to the requirements of accuracy, (in the example analyzed above: $\left|H_e^0(j\omega)\right| = \sqrt{D_{eg}^0 / D_1} \cdot \omega$), then according to (7.1) it is possible to write

$$\left|W(j\omega)\right| \geq \left|H(j\omega)\right| \left|H_e^0(j\omega)\right|^{-1}. \tag{7.9}$$

Relation (7.9) allows for imposing the following restrictions on a kind of frequency transfer function of an open loop system. Their realization does not allow the system accuracy to be worse than the required one.

On frequencies essentially smaller than the cutoff frequency, where $\left|H(j\omega)\right| \approx 1$, the gain plot of the open loop system should obey a criterion:

$$\left|W(j\omega)\right| \geq \left|H_e^0(j\omega)\right|^{-1}, \tag{7.10}$$

but the phase plot can be arbitrary.

On frequencies essentially greater than the cutoff frequency, where $H(j\omega) \approx W(j\omega)$ and $\left|H_e^0(j\omega)\right| \approx 1$, inequality (7.9) becomes an equality at any gain plot and phase plot of an open loop system. Therefore, the kind of frequency transfer function for an open loop system on high frequencies does not affect the system accuracy and can be arbitrary.

In the medium frequencies area the requirements to the frequency transfer function of an open loop system are most complicated. The reason is that the resonance peak of gain plot in the closed-loop system here is possible. Its value depends both on gain plot, and on phase plot of an open loop system. Therefore the restriction of only one of these two curves does not give the desirable outcome.

If the resonance frequency of closed-loop system $\omega_{res}$ was precisely known, then the following inequality must be obeyed in reference to the gain plot of open loop systems.

$$\left|W(j\omega_{res})\right| \geq M \left|H_e^0(j\omega_{res})\right|^{-1}, \tag{7.11}$$

Here

$$M = \left|H(j\omega)\right|_{\max} \left[H(j0)\right]^{-1} = \left|H(j\omega_{res})\right| \left[H(j0)\right]^{-1} \approx \left|H(j\omega_{res})\right| \tag{7.12}$$

is an oscillation index. In astatic systems $M = \left|H(j\omega_{res})\right|$. The value $M$ is usually set at synthesis. It determines the requirements on a stability margin of a closed-loop system. The methods to construct a system with preset stability margin and with the appropriate requirements for a frequency transfer function of open loop systems are well developed [6, 9, 72, 80].

However, the magnitude $\omega_{res}$ becomes known only after the selection of function $W(j\omega)$, therefore such a selection cannot be produced on the basis of a direct use of relation (7.11). It is possible only to check the membership of some function $W(j\omega)$ to the admissible class of these functions.

A requirement of realization of the inequality

$$|W(j\omega)| \geq M|H_e^0(j\omega)|^{-1} \tag{7.13}$$

in a whole area of medium frequencies is quite sufficient for deriving the required accuracy. However, at $M > 1$ it appears excessively strong on frequencies, which do not coincide with $\omega_{res}$. Therefore it is inexpedient to require categorically the strict realization of (7.13) at synthesis. Although such requirements would guarantee the realization of (7.9), they could cause the invalid expansion of a system passband.

Taking into account the above stated, it is meaningful to restrict the admissible class of transfer functions of open loop systems with weaker conditions as contrasted to (7.13)

$$|W(j\omega)| \geq |H_e^0(j\omega)|^{-1}, \tag{7.14}$$

and to require together the large stability margin for closed-loop systems to be ensured. The condition (7.14) should be fulfilled in a low frequency area, where it coincides with (7.10), and also in the medium frequencies area. This can be presented by the forbidden area in the Bode diagram plane of an open loop system $L(\omega) = 20\lg|W(j\omega)|$. This forbidden area is restricted by the curve

$$L^0(\omega) = 20\lg|H_e^0(j\omega)|^{-1} = -20\lg|H_e^0(j\omega)|. \tag{7.15}$$

Take into consideration that the right boundary of a medium frequency area can be allocated precisely only for any of particular kind of functions $W(j\omega)$. In this case inequality (7.14) should be fulfilled. The curve $L^0(\omega)$ must be conditionally broken off at level 0 dB when constructing forbidden area, i.e. in a point of its intersection with abscissa axis, though this curve could be prolonged a little bit below that of level 0 dB in many cases.

The forbidden area, obtained by a described mode, in a case when the accuracy requirements set the upper bound of dynamic error dispersion and the dispersion of the first derivative of reference action is known, i.e. the relation (7.5) is valid, as shown in Fig. 7.1.

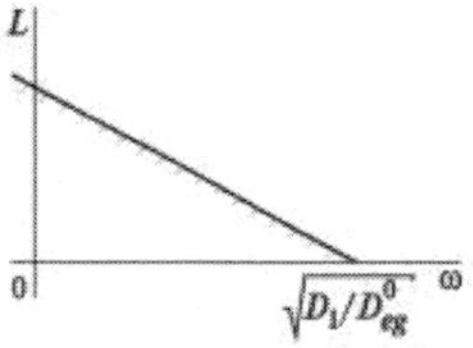


Fig. 7.1

Its boundary has the single negative inclination – 20 dB/dec. The requirements of accuracy are fulfilled, if the precise gain plot does not get inside the forbidden area and the stability margin of a closed-loop system appears large enough.

The requirements for the shape of gain plot for the open loop system in medium frequencies area, defining the stability margin, can be formulated more particularly at the representation even with a general view of function $W(j\omega)$. The examples of such investigation for transfer functions of a kind $W(j\omega)= = K_1[j\omega(1+j\omega T_1)]^{-1}$ (gain plot of a kind 1 — 2) and $W(j\omega)=K_2(1+j\omega\tau)(j\omega)^{-2}$ (gain plot of a kind 2 — 1) are given in the book [8] on pages 187 — 189. As the result it was possible to supplement the forbidden area shown in Fig. 7.1 with forbidden areas for salient points of an asymptotic Gain plot.

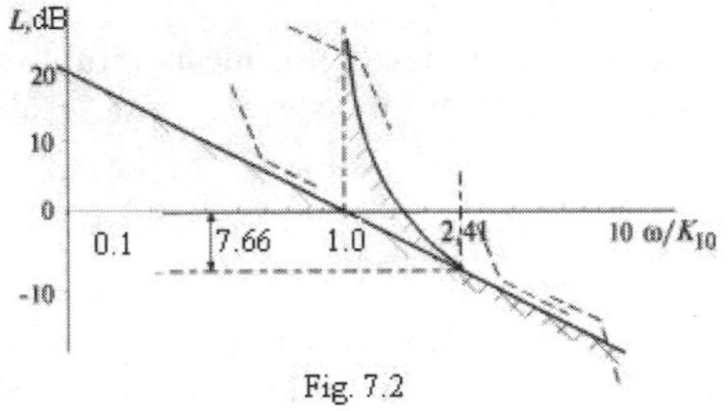


Fig. 7.2

They are shown in Fig. 7.2, where $K_{10}=\sqrt{D_1/D_{eg}^0}$.

## §7.2. Restriction of dynamic error dispersion

**Common expression for bounds of forbidden areas on Bode diagram plain.** Expressions (7.2) and (7.4) must be taken as a basis at the determination of forbidden areas on Bode diagram plane, appropriate to sufficient conditions to restrict the error dispersion for given admissible magnitude $D_{eg}^0$. Having exposed factorization, the polynomial $C_{2N}(\omega)$ on the right hand side of (7.2), is written as:

$$C_{2N}(\omega)=\Gamma_N(j\omega)\Gamma_N(-j\omega), \tag{7.16}$$

where

$$\Gamma_N(j\omega)=\gamma_K(j\omega)^K+\gamma_{K+1}(j\omega)^{K+1}+\ldots+\gamma_N(j\omega)^N,\ \{\gamma_i\}_K^N\in[0,\infty) \tag{7.17}$$

Then from inequality (7.2) the following is obtained:

$$|H_e(j\omega)|\le|\Gamma_N(j\omega)|. \tag{7.18}$$

The coefficients of polynomials $C_{2N}(\omega)$ and $\Gamma_N(j\omega)$ are connected by the relations

$$c_0=\gamma_0^2,\ c_1=\gamma_1^2-2\gamma_0\gamma_2,\ c_2=\gamma_2^2-2\gamma_1\gamma_3+2\gamma_0\gamma_4,\ldots,\ c_N=\gamma_N^2. \tag{7.19}$$

From (7.4) and (7.19) follows, that at $K=0$

$$\gamma_0^2D_0+(\gamma_1^2-2\gamma_0\gamma_2)D_1+(\gamma_2^2-2\gamma_1\gamma_3+2\gamma_0\gamma_4)D_2+\ldots+\gamma_N^2D_N=D_{eg}^0,$$

or, in other kind:

$$\gamma_0 = \frac{\sqrt{D_{eg}^0}}{\sqrt{D_0 + \left[\left(\gamma_0^2 - \gamma_0\gamma_2\right)D_1 + \left(\gamma_2^2 - 2\gamma_1\gamma_3 + 2\gamma_0\gamma_4\right)D_2 + \ldots + \gamma_N^2 D_N\right]\gamma_0^{-2}}}. \quad (7.20)$$

According to (7.20) and (7.17), (7.18) can be rewritten as

$$|H_e(j\omega) \quad | \le \gamma_0\left|1 + \gamma_1\gamma_0^{-1} j\omega + \gamma_2\gamma_0^{-1}(j\omega)^2 + \ldots + \gamma_N\gamma_0^{-1}(j\omega)^N\right| =$$

$$= \frac{\sqrt{D_{eg}^0}\left[1 + \gamma_1\gamma_0^{-1} j\omega + \gamma_2\gamma_0^{-1}(j\omega)^2 + \ldots + \gamma_N\gamma_0^{-1}(j\omega)^N\right]}{\sqrt{D_0 + \left[\left(\gamma_0^2 - \gamma_0\gamma_2\right)D_1 + \left(\gamma_2^2 - 2\gamma_1\gamma_3 + 2\gamma_0\gamma_4\right)D_2 + \ldots + \gamma_N^2 D_N\right]\gamma_0^{-2}}}$$

and, finally

$$|H_e(j\omega)| \le \left|\gamma_0 + \gamma_1 j\omega + \gamma_2(j\omega)^2 + \ldots + \gamma_N(j\omega)^N\right| / K_*, \quad (7.21)$$

where

$$K_* = \sqrt{\left[\gamma_0^2 D_0 + \left(\gamma_1^2 - 2\gamma_0\gamma_2\right)D_1 + \left(\gamma_2^2 - 2\gamma_1\gamma_3 + 2\gamma_0\gamma_4\right)D_2 + \ldots + \gamma_N^2 D_N\right] / D_{eg}^0}. \quad (7.22)$$

On the basis of (7.14) to gain plot of an open loop system, for a general case $K \ge 0$, when $\gamma_i = 0$ at $i = \overline{0, K-1}$, from (4.21) the following condition is obtained:

$$W(j\omega) \ge \frac{K_*}{\left|\gamma_K(j\omega)^K + \gamma_{K+1}(j\omega)^{K+1} + \ldots + \gamma_N(j\omega)^N\right|}. \quad (7.23)$$

If the requirements on stability margin for closed-loop systems (see §7.2) are fulfilled, then condition (7.23) is sufficient for the upper bound of dynamic error dispersion not exceeding value $D_{eg}^0$. It looks like the forbidden area for a precise Bode diagram of the open loop system. The shape of the forbidden area depends not only on values of dispersions $\{D_i\}_K^N$, but also on coefficients $\{\gamma_i\}_K^N$. These coefficients should be selected, so that the appropriate forbidden area must put the least hidden minimization of passband and to realization of other requirements, applied to a system.

**Case of known *N*-th derivative of action.** At $N = K$ the coefficient $K_*$ expressed by formula (7.22) can be finite only when all polynomial $\Gamma_N(j\omega)$ coefficients are equal to zero, except for $\gamma_N$. Therefore formula (7.23) according to (7.22) gives

$$|W(j\omega)| \ge \sqrt{D_N / D_{eg}^0} / \omega^N. \quad (7.24)$$

The boundary of appropriate forbidden areas on the Bode diagram plane is a straight line with inclination $-N \cdot 20$ dB/dec intersecting abscissa axis on the point $\omega = \sqrt[2N]{D_N / D_{eg}^0}$. The allowed Bode diagrams belong to systems with astatism not below the $N$-th order.

At $N = K = 0$, when only the dispersion of the action is known, the Bode diagram of the open loop system cannot be lower than level $10\lg\left(D_0 / D_{eg}^0\right)$ in all frequency areas, where the spectral components of actions are possible.

The case $N = K = 1$, when only the dispersion of the first derivative of action is known, was analyzed as an example in §7.2. The use of the Bode diagram with constant values of inclination – 20 dB/dec is possible in this case. At $N = K \geq 2$, use of fractureless Bode diagrams is impossible, because to ensure the required stability of closed-loop system in the medium frequencies area there should be an interval on the Bode diagram with the unit inclination – 20 dB/dec.

If $N = K = 2$, i.e. only the dispersion of the second derivative of action is known, which has the elementary shape of Bode diagram such as 2 — 1. In addition if an interval with inclination – 40 dB/dec passes directly on the boundary of a forbidden area, then the salient point should lie at least on 6 dB higher than the abscissa axis.

**Case of known ($N$ – 1)-th and $N$-th derivative of action.** From (7.23) and (7.22) at $\gamma_i = 0,\ i = \overline{0, N-2}$ the following is obtained:

$$|W(j\omega)| = \sqrt{\left(\gamma_{N-1}^2 D_{N-1} + \gamma_N^2 D_N\right) / D_{eg}^0}\,\left|\gamma_{N-1}(j\omega)^{N-1} + \gamma_N (j\omega)^N\right|^{-1}$$

or

$$|W(j\omega)| \geq \frac{\sqrt{\left(D_{N-1} + T_1^2 D_N\right) / D_{eg}}}{\omega^{N-1}\sqrt{1 + \omega^2 T_1^2}}, \qquad (7.25)$$

where $T_1 = \gamma_N / \gamma_{N-1}$ is some stationary time constant, for which it is possible to select any value from an interval $T_1 \in [0, \infty)$.

A set of forbidden areas for precise Bode diagram corresponds to condition (7.25). Each area is constructed at a defined value of time constant $T_1$. Boundaries of such forbidden areas at $N = 1$ (the dispersions $D_0$ and $D_1$ are known) and at $N = 2$ (the dispersions $D_1$ and $D_2$ are known) they are shown accordingly in Fig. 7.3*a* by solid lines.

The position of a precise Bode diagram for open loop systems outside any of these forbidden areas is the sufficient condition for deriving the required accuracy

$$\overline{D}_{eg} \leq D_{eg}^0 .$$

In practice, when choosing a Bode diagram for synthesized system, it is convenient not to construct the full set of forbidden areas but to use only one forbidden area with a more simple shape. However this is absolutely forbidden in the sense that it is completely entered into the limits of any forbidden area of the given set. Such an absolutely forbidden area is marked in Fig. 7.3 by a dotted line.

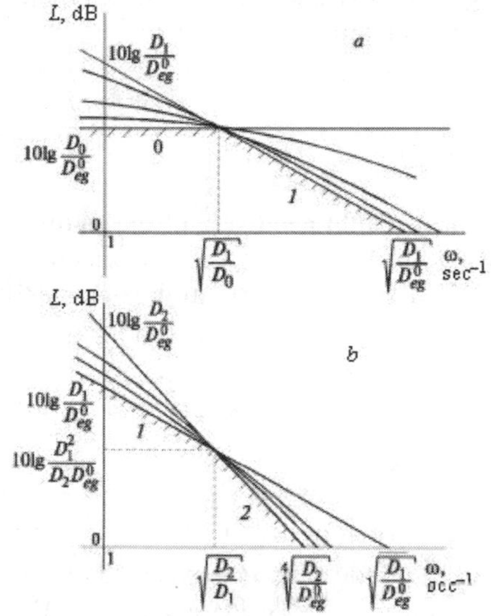

Fig. 7.3

The passing of the precise Bode diagram of the system outside the absolutely forbidden area is necessary, but not the sufficient condition for realization of accuracy requirements. It is easy to make sure of it by an example of frequency transfer function $W(j\omega)=k(1+j\omega T)\left[1+2\xi j\omega T+(j\omega T)^2\right]^{-1}$, where $k=\sqrt{D_0/D_{eg}^0}$, $T=\sqrt{D_0/D_1}, 0<\xi\le 0.707$. The bode diagram appropriate to this function (see graph in [8] in page 193) passes outside the absolutely forbidden area, but does not satisfy requirement (7.25) at any value of a time constant.

The analytical expression for the boundary of an absolutely forbidden area at arbitrary value $N$ can be obtained on the basis of reviewing the condition of processing the harmonic action with spectral density (7.7). Dispersions of $(N-1)$-th and $N$-th derivatives of this action are equal accordingly to $\rho_{N-1}=\rho_0\omega^{2N-2}$ and $\rho_N=\rho_0\omega^{2N}$. As these dispersions are restricted by magnitudes $D_{N-1}$ and $D_N$, the following relation of admissible value of action dispersion $\rho_0$ from frequency $\omega$ can be written as:

$$\rho_0\le\min\left\{\frac{D_{N-1}}{\omega^{2N-2}},\frac{D_N}{\omega^{2N}}\right\}=\begin{cases}D_{N-1}/\omega^{2N-2} & \text{at } \omega\le\sqrt{D_N/D_{N-1}},\\ D_N/\omega^{2N} & \text{at } \omega\ge\sqrt{D_N/D_{N-1}}.\end{cases}\tag{7.26}$$

The error dispersion from processing the considered action is: $D_{eg}=\left|H_e(j\omega)\right|^2\rho_0$ and it should not exceed the admissible value $D_{eg}^0$. From here the following is obtained: $\left|H_e(j\omega)\right|^2 \le D_{eg}^0/\rho_0$ or, passing to gain plot of an open loop system according to (7.14) and taking into account (7.26), the frequency transfer function $W(j\omega)$ is equal to:

$$\left|W(j\omega)\right| = \begin{cases} \sqrt{D_{N-1}/D_{eg}^0}\Big/\omega^{N-1} & \text{at } \omega \le \sqrt{D_N/D_{N-1}}, \\ \sqrt{D_N/D_{eg}^0}\Big/\omega^N & \text{at } \omega \ge \sqrt{D_N/D_{N-1}}. \end{cases} \tag{7.27}$$

The graphic image for the right hand side of inequality (7.27) in the Bode diagram plane is the boundary of an absolutely forbidden area.

**Case of known ($N-2$)-th, ($N-1$)-th and $N$-th derivatives of action.** At $\gamma_i = 0$, $i = \overline{0, N-3}$, from (7.23) and (7.22) the following is obtained:

$$\left|W(j\omega)\right| \ge \frac{\sqrt{\left[\gamma_{N-2}^2 D_{N-2} + \left(\gamma_{N-1}^2 - 2\gamma_{N-2}\gamma_N\right)D_{N-1} + \gamma_N^2 D_N\right]\Big/D_{eg}^0}}{\omega^{N-2}\sqrt{\gamma_{N-2}^2 + \left(\gamma_{N-1}^2 - 2\gamma_{N-2}\gamma_N\right)\omega^2 + \gamma_N^2\omega^4}}. \tag{7.28}$$

At each set of factors $\gamma_i$ values of requirement (7.28) can be drown by forbidden areas in the plane of the Bode diagram. Eventually the population of forbidden areas is obtained due to arbitrary selection of factors $\{\gamma_i\}_{N-2}^N \in [0, \infty)$. The passing of the precise Bode diagram outside even one forbidden area from this population is the sufficient condition to obtain the required accuracy. In the most typical case, when $N = 2$, (7.28) gives

$$\left|W(j\omega)\right| \ge \frac{\sqrt{\left[\gamma_0^2 D_0 + \left(\gamma_1^2 - 2\gamma_0\gamma_2\right)D_1 + \gamma_2^2 D_2\right]\Big/D_{eg}^0}}{\sqrt{\gamma_0^2 + \left(\gamma_1^2 - 2\gamma_0\gamma_2\right)\omega^2 + \gamma_2^2\omega^4}}. \tag{7.29}$$

At $\gamma_1 \ge 2\sqrt{\gamma_0\gamma_2}$ the boundaries of appropriate forbidden areas coincide with the Bode diagram of aperiodic second order units, and at $\gamma_1 < 2\sqrt{\gamma_0\gamma_2}$ — with the Bode diagram of oscillatory units. Some of these boundaries are shown in Fig. 7.4.

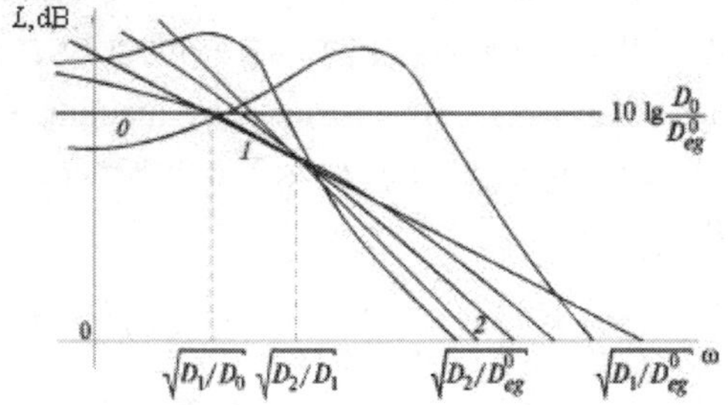


Fig. 7.4

Instead of populating forbidden areas, in practice it is convenient to construct one absolutely forbidden area. Passing of the precise Bode diagram outside this area is the necessary condition of obtaining the required accuracy. An analytical expression for its boundary at arbitrary value $N$ can be developed on the basis of analysis for the condition of handling the harmonic action with spectral density (7.6). The dispersions of $(N-2)$-th, $(N-1)$-th and $N$-th derivatives of such action are accordingly equal to $\rho_{N-2}=\rho_0\omega^{2N-4}$, $\rho_{N-1}=\rho_0\omega^{2N-2}$ and $\rho_N=\rho_0\omega^{2N}$. As these dispersions are restricted by the values $D_{N-2}, D_{N-1}$ and $D_N$, for arbitrary frequency $\omega\in[0,\infty)$ the following inequality should be fulfilled

$$\rho_0\le\min\left\{D_{N-2}/\omega^{2N-4}, D_{N-1}/\omega^{2N-2}, D_N/\omega^{2N}\right\}=$$
$$=\begin{cases} D_{N-2}/\omega^{2N-4} & \text{at} \quad 0<\omega\le\sqrt{D_{N-1}/D_{N-2}}, \\ D_{N-1}/\omega^{2N-2} & \text{at} \quad \sqrt{D_{N-1}/D_{N-2}}\le\omega\le\sqrt{D_N/D_{N-1}}, \\ D_N/\omega^{2N} & \text{at} \quad \omega\ge\sqrt{D_N/D_{N-1}}. \end{cases} \tag{7.30}$$

The action may turn out not to be a harmonic one and, except for spectral components with frequency $\omega$ and dispersion $\rho_0$, and may have other spectral components. In this case they will be characterized by dispersions $D_{N-2}-\rho_{N-2}=D_{N-2}-\rho_0\omega^{2N-4}$, $D_{N-1}-\rho_{N-1}=D_{N-1}-\rho_0\omega^{2N-2}$ and $D_N-\rho_1=D_N-\rho_0\omega^{2N}$, which should submit to inequality of a kind (3.3) as well as the dispersions $D_{N-2}$, $D_{N-1}$ and $D_N$. Otherwise the action with dispersions of derivatives $D_{N-2}$, $D_{N-1}$ and $D_N$ cannot be considered as appropriate to any physically feasible process. Taking it into account, by analogy with (3.3), write

$$\left(D_{N-1}-\rho_0\omega^{2N-2}\right)/\left(D_{N-2}-\rho_0\omega^{2N-4}\right)\le\left(D_N-\rho_0\omega^{2N}\right)/\left(D_{N-1}-\rho_0\omega^{2N-2}\right),$$

whence

$$\rho_0\le\frac{D_N D_{N-2}-D_{N-1}^2}{\omega^{2N-4}\left(D_N-2D_{N-1}\omega^2+D_{N-2}\omega^4\right)}. \tag{7.31}$$

Analyzing inequalities (7.30) and (7.31), it is possible to show, that (7.30) is stronger then (7.31) only at $\sqrt{D_{N-1}/D_{N-2}}<\omega<\sqrt{D_N/D_{N-1}}$. Therefore, after the integration of these two inequalities, the expression for the greatest possible dispersion of spectral component of action on frequency $\omega$ looks like:

$$\rho_0\le\begin{cases} \dfrac{D_N D_{N-2}-D_{N-1}^2}{\omega^{2N-4}\left(D_N-2D_{N-1}\omega^2+D_{N-2}\omega^4\right)} & \text{at } 0<\omega\le\sqrt{\dfrac{D_{N-1}}{D_{N-2}}} \text{ and } \omega>\sqrt{\dfrac{D_N}{D_{N-1}}}, \\ \dfrac{D_{N-1}}{\omega^{2N-2}} & \text{at } \sqrt{\dfrac{D_{N-1}}{D_{N-2}}}0\le\omega\le\sqrt{\dfrac{D_N}{D_{N-1}}}. \end{cases} \tag{7.32}$$

The dispersion for spectral components of system error on frequency $\omega$ is: $D_{eg}=\left|H_e(j\omega)\right|^2\rho_0$ and should not exceed the admissible value $D_{eg}^0$. Therefore,

$|H_e(j\omega)|^2 \le D_{eg}^0 / \rho_0$ is possible or, passing to gain plot of the open loop system according to relation (7.14) and taking into account (7.32),

$$|W(j\omega)| \ge \begin{cases} \dfrac{\sqrt{\left(D_N D_{N-2} - D_{N-1}^2\right)/D_{eg}^0}}{\omega^{N-2}\sqrt{D_N - 2D_{N-1}\omega^2 + D_{N-2}\omega^4}} \\ \text{at } 0 \le \omega \le \sqrt{\dfrac{D_{N-1}}{D_{N-2}}} \text{ and } \omega > \sqrt{\dfrac{D_N}{D_{N-1}}}, \\ \dfrac{\sqrt{D_{N-1}/D_{eg}^0}}{\omega^{N-1}} \text{ at } \sqrt{\dfrac{D_{N-1}}{D_{N-2}}} \le \omega \le \sqrt{\dfrac{D_N}{D_{N-1}}}. \end{cases} \tag{7.33}$$

As at failure of condition (7.33) the error dispersion of processing the considered action will exceed the admissible value $D_{eg}^0$ anyway, and condition (7.33) is necessary for obtaining the required accuracy. Taking the logarithm of the right and left hand sides (7.33) gives the analytical expression for the boundary of the absolutely forbidden area for the Bode diagram plane. However, the indicated condition is not sufficient for deriving the required accuracy, because even at its realization, the dispersion of error can exceed the value $D_{eg}^0$ at the expense of spectral components on other frequencies.

The absolutely forbidden area constructed according to inequality (7.33) at $N =$ = 2, is shown in Fig. 7.5.

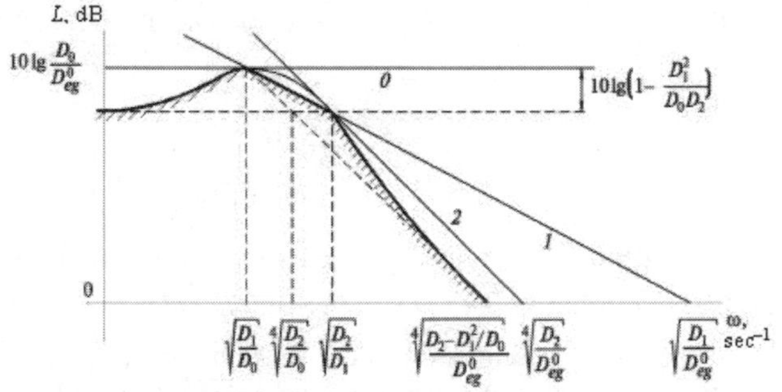


Fig. 7.5

The shape of an absolutely forbidden area can be simplified if all the coefficients in the denominator of the gain plot quadrate for synthesized system in the open loop state are supposed as nonnegative

$$|W(j\omega)|^2 =$$

$$= \left|1 + b_1 j\omega + \ldots + b_{n-1}(j\omega)^{n-1}\right|^2 \times,$$

$$\times \left|a_0 + a_1 j\omega + \ldots + a_n (j\omega)^n\right|^{-2}$$

i.e. the following inequality is fulfilled

$$a_0^2 \geq 0,\; a_1^2 - 2a_0 a_2 \geq 0, a_2^2 - 2a_1 a_3 + 2a_0 a_4 \geq 0, \ldots, a_n^2 \geq 0. \qquad (7.34)$$

Practically this means, that the open loop of a system should not contain the oscillatory units with a damping factor less than 0.707, because the conditions in (7.34) in a stable system can be broken only when the denominator contains the function $|W(j\omega)|^2$ of multiplicand, such as $\left|1 + 2\xi Tj\omega + (j\omega T)^2\right|^2 =$ $= 1 - 2T^2\left(1 - 2\xi^2\right)\omega^2 + T^4\omega^4$ at $\xi < 1/\sqrt{2} = 0.707$.

Then the possible values of factors $\gamma_i$ in expression (7.28) can be obeyed similar to (7.34) to inequalities

$$\gamma_0^2 \geq 0,\; \gamma_1^2 - 2\gamma_0\gamma_2 \geq 0,\; \gamma_2^2 - 2\gamma_1\gamma_3 + 2\gamma_0\gamma_4 \geq 0, \ldots, \gamma_n^2 \geq 0.$$

In the book [3] on pages 196-197 it is proved, that thus the necessary condition to ensure the required accuracy, contrary to (7.33), looks like:

$$|W(j\omega)| \geq \begin{cases} \sqrt{D_{N-2}/D_{eg}^0}\,/\omega^{N-2} & \text{at } 0 < \omega \leq \sqrt{D_{N-1}/D_{N-2}}, \\ \sqrt{D_{N-1}/D_{eg}^0}\,/\omega^{N-1} & \text{at } \sqrt{D_{N-1}/D_{N-2}} \leq \omega \leq \sqrt{D_N/D_{N-1}}. \\ \sqrt{D_N/D_{eg}^0}\,/\omega^N & \text{at } \omega \geq \sqrt{D_N/D_{N-1}}, \end{cases} \qquad (7.36)$$

It is directly implied from (7.30) without the use of (7.31).

The absolutely forbidden area constructed according to expression (7.36) at $N=$ $= 2$, is shown in Fig. 7.6.

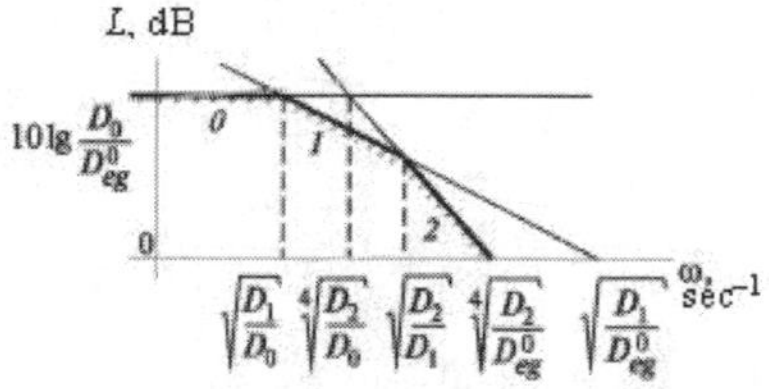


Fig. 7.6

**Case of known dispersions for more than three action derivatives.** When the dispersions of more than three derivatives of action are known, then the greatest number of possible variants for the shape of a forbidden area appropriate to condition (7.23), especially underlines the expediency to construct the absolutely forbidden area for the Bode diagram at the synthesis stage. Generally, absolutely forbidden areas can be constructed on the basis of a generalization of relations (7.32) and (7.33) as

$$|W(j\omega)| \geq \min\left\{\sqrt{\frac{\left(D_0 D_2 - D_1^2\right)/D_{eg}^0}{D_2 - 2D_1\omega^2 + D_0\omega^4}}, \frac{\sqrt{\left(D_1 D_3 - D_2^2\right)/D_{eg}^0}}{\omega\sqrt{D_3 - 2D_2\omega^2 + D_1\omega^4}}, \ldots\right.$$

$$\left.\ldots, \frac{\sqrt{\left(D_N D_{N-2} - D_{N-1}^2\right)/D_{eg}^0}}{\omega^{N-2}\sqrt{D_N - 2D_{N-1}\omega^2 + D_{N-2}\omega^4}}\right\}. \tag{7.37}$$

The construction of absolutely forbidden area at $K = 0, N = 3$ is shown as an example in Fig. 7.7, i.e. at known values for dispersions of reference action and three of its younger derivatives.

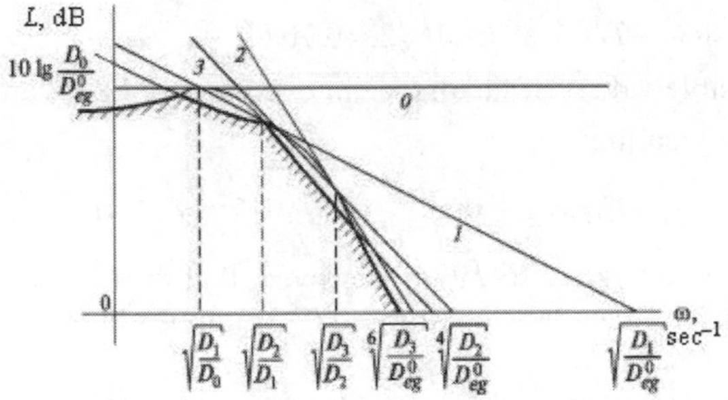


Fig. 7.7

The shape of the absolutely forbidden area can be simplified when the open loop of a synthesized system is accepted to have no weakly oscillatory units with a damping factor $\xi < 0.707$. Then the inequalities in (7.35) are valid for factors $\gamma_i$, included in expression (7.23). Prove, that thus the necessary condition to ensure the required accuracy looks like:

$$|W(j\omega)| \geq \begin{cases} \sqrt{D_0/D_{eg}^0} & \text{at} \quad 0 < \omega \leq \sqrt{D_1/D_0}, \\ \sqrt{D_1/D_{eg}^0}\,/\omega & \text{at} \quad \sqrt{D_1/D_0} \leq \omega \leq \sqrt{D_2/D_1}, \\ \ldots\ldots\ldots & \ldots \quad \ldots\ldots\ldots\ldots \\ \sqrt{D_{N-1}/D_{eg}^0}\,/\omega^{N-1} & \text{at} \quad \sqrt{D_{N-1}/D_{N-2}} \leq \omega \leq \sqrt{D_N/D_{N-1}}, \\ \sqrt{D_N/D_{eg}^0}\,/\omega^N & \text{at} \quad \omega \geq \sqrt{D_N/D_{N-1}}. \end{cases} \tag{7.38}$$

Analyze the difference of quadrates in the right hand sides of inequalities (7.23) and (7.38) at frequency values taken from an interval $\sqrt{D_i/D_{i-1}} \leq \omega \leq \sqrt{D_{i+1}/D_i}, 1 \leq i \leq N-1$. Having executed the reduction to a common denominator, this difference can be written as

$$\Delta_i = B/A, \tag{7.39}$$

where

$$A = \omega^{2i}\left|\gamma_0 + \gamma_1 j\omega + \gamma_2 (j\omega)^2 + \ldots + \gamma_N (j\omega)^N\right|^2 D_{eg}^0 > 0, \tag{7.40}$$

$$B=\omega^{2i}\left[\gamma_0^2 D_0+\left(\gamma_1^2-2\gamma_0\gamma_2\right)D_1+\ldots+\left(\gamma_i^2-2\gamma_{i-1}\gamma_{i+1}+2\gamma_{i-2}\gamma_{i+2}-\ldots\right)D_i+\ldots\right.$$
$$\left.\ldots+\gamma_N^2 D_N\right]-\left[\gamma_0^2+\left(\gamma_1^2-2\gamma_0\gamma_2\right)\omega^2+\ldots+\left(\gamma_i^2-2\gamma_{i-1}\gamma_{i+1}+\right.\right. \tag{7.41}$$
$$\left.\left.+2\gamma_{i-2}\gamma_{i+2}-\ldots\right)\omega^{2i}+\ldots+\gamma_N^2\omega^{2N}\right]D_i .$$

After the transformations expression (7.41) can be rewritten as:

$$B=\omega^{2i}D_i\left[\frac{\gamma_0^2 D_0}{D_i}\left(1-\frac{D_i}{\omega^{2i}D_0}\right)+\left(\gamma_1^2-2\gamma_0\gamma_2\right)\frac{D_1}{D_i}\left(1-\frac{D_i}{\omega^{2i-2}D_i}\right)+\ldots\right.$$
$$+\ldots+\left(\gamma_{i-1}^2-2\gamma_{i-2}\gamma_i+2\gamma_{i-3}\gamma_{i+1}-\ldots\right)\frac{D_{i-1}}{D_i}\left(1-\frac{D_i}{\omega^2 D_{i-1}}\right)+$$
$$+\left(\gamma_{i+1}^2-2\gamma_i\gamma_{i+2}+2\gamma_{i-1}\gamma_{i+3}-\ldots\right)\frac{D_{i+1}}{D_i}\left(1-\frac{\omega^2 D_i}{D_{i+1}}\right)+ \tag{7.42}$$
$$+\left(\gamma_{i+2}^2-2\gamma_{i+1}\gamma_{i+3}+2\gamma_i\gamma_{i+4}-\ldots\right)\frac{D_{i+2}}{D_i}\left(1-\frac{\omega^4 D_i}{D_{i+2}}\right)+\ldots$$
$$\left.\ldots+\frac{\gamma_N^2 D_N}{D_i}\left(1-\frac{\omega^{2(N-i)}D_i}{D_N}\right)\right].$$

As according to (4.5) the following inequalities are valid

$$\sqrt[i]{\frac{D_i}{D_0}}\le\ldots\le\sqrt{\frac{D_i}{D_{i-2}}}\le\frac{D_i}{D_{i-1}}\le\omega^2\le\frac{D_{i+1}}{D_i}\le\sqrt{\frac{D_{i+2}}{D_i}}\le\ldots\le\sqrt[N-i]{\frac{D_N}{D_i}}$$

and on the other hand, the relations in (7.35) are fulfilled, and each addend in the right hand side of formula (7.42) is nonnegative. From here it follows, that $B\ge 0$ and, according to (7.39) and (7.40), $\Delta_i\ge 0$ .

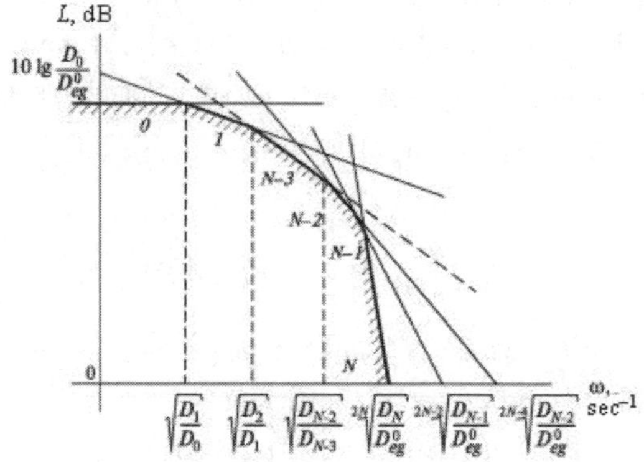


Fig. 7.8

The difference of quadrates in right members (7.23) and (7.38) is nonnegative in a whole frequency interval $\omega \in [0, \infty)$ because the given outcome is valid at arbitrary index $i$. Therefore at any value of coefficients $\gamma_i$, submitting to relation (7.35), the inequality (7.23) is stronger than the inequality (7.38). Thus, the realization of inequality (7.38) is the necessary condition for obtaining of required accuracy.

The absolutely forbidden area for the precise Bode diagram, constructed according to expression (7.38) at an arbitrary number of known dispersions $\{D_i\}_0^N$, is shown in Fig. 7.8.

## §7.3. Restriction of maximum dynamic error

**Sufficient conditions to restrict the maximum error**. When the maximum values for derivatives of reference actions are known, then the restriction of the maximum dynamic error consists of a selection of the systems class, for which the estimation of maximum error does not exceed some admissible value $e_M^0$. Actually the membership of this or that system to such class is determined by the type of its frequency transfer function for an error $H_e(j\omega)$. However, with the use of relation (7.14) it is possible to impose the restrictions on a gain plot of open loop system $|W(j\omega)|$ and to construct the appropriate forbidden areas in the Bode diagram plane.

As well as at restriction of error dispersion, it is possible to construct the forbidden areas on the basis of necessary or sufficient conditions to obtain the required accuracy. In the first case the lower estimation of the maximum error is being restricted, in the second one — the upper-bound estimate. The lower estimation is determined by formula (6.55) applying to the harmonic reference action, an upper-bound estimate — with reference to an arbitrary kind action by formula (6.59).

The forbidden areas appropriate to sufficient conditions for maximum error restrictions must be found first. According to (6.59) it is possible to claim, that the maximum error does not exceed the preset value $e_M^0$, if the module of frequency transfer function of the closed-loop system for an error obeys inequality

$$|H_e(j\omega)| \le c_K \omega_K + c_{K+1}\omega_{K+1} + \ldots + c_N \omega_N \tag{7.43}$$

at

$$\Re\left(c_K g_M^{(K)} + c_{K+1} g_M^{(K+1)} + \ldots + c_N g_M^{(N)}\right) = e_M^0 . \tag{7.44}$$

Having rewritten (7.43) as $|H_e(j\omega)| \le c_K \omega_K \left(1 + c_{K+1} c_K^{-1} \omega + \ldots + c_N c_K^{-1} \omega^{N-K}\right)$ and having substituted the values of coefficient

$$c_K = \Re^{-1} e_M^0 \Big/ \left(g_M^{(K)} + c_{K+1} c_K^{-1} g_M^{(K+1)} + \ldots + c_N c_K^{-1} g_M^{(N)}\right),$$

expressed from (7.44), it is possible to obtain the requirement to the gain plot of a closed-loop system for error, whose realization is enough to restrict the maximum error

$$|H_e(j\omega)| \le \frac{e_M^0 \omega^K \left(c_K + c_{K+1}\omega + \ldots + c_N \omega^{N-K}\right)}{\Re\left(c_K g_M^{(K)} + c_{K+1} g_M^{(K+1)} + \ldots + c_N g_M^{(N)}\right)}. \tag{7.45}$$

In order to explain the sense of expression (7.45), it is necessary to analyze one more way of its deriving. Let the $K$-th derivative of action belong to the class $\Xi$ and the total of $n$ harmonic functions with not be multiple frequencies $\omega_i$ and amplitudes $u_i$, $i = \overline{1, \nu}$, i.e. the equalities are valid:

$$g_M^{(K)} = \sum_{i=1}^{\nu} g_{iM}^{(K)} = \sum_{i=1}^{\nu} u_i \,,\; g_M^{(K+1)} = \sum_{i=1}^{\nu} \omega_i u_i \,,$$

$$g_M^{(K+2)} = \sum_{i=1}^{\nu} \omega_i^2 u_i, \ldots, g_M^{(N)} = \sum_{i=1}^{\nu} \omega_i^{N-K} u_i \,.$$

The maximum error from processing such an action is represented by the total of maximum errors from processing each of its harmonic components

$$e_{*M} = \sum_{i=1}^{\nu} e_{iM} = \sum_{i=1}^{\nu} |H_{eK}(j\omega_i)| u_i \,.$$

The requirements on accuracy can be guaranteed, if $e_{*M} \le e_M^0 / \Re$.

How can the admissible value of error from processing each of its harmonic components be assigned? How is it possible to arrange the resultant error $e_M^0 / \Re$ between all frequencies, on which the component of action can exist in a spectrum? For this purpose it is expedient to insert into accordance for each harmonic component of action $g_i(t)$ the dimensionless coefficient

$$\beta_i = \frac{c_K g_{iM}^{(K)} + c_{K+1} g_{iM}^{(K+1)} + \ldots + c_N g_{iM}^{(N)}}{c_K g_M^{(K)} + c_{K+1} g_M^{(K+1)} + \ldots + c_N g_M^{(N)}} = \frac{u_i \left(c_K + c_{K+1}\omega + \ldots + c_N \omega^{N-K}\right)}{c_K g_M^{(K)} + c_{K+1} g_M^{(K+1)} + \ldots + c_N g_M^{(N)}}. \tag{7.46}$$

This describes the contribution of such components into the formation of given values for maximum derivatives of actions. The total of coefficients $\beta_i$ for all components can not exceed one. Therefore, if the admissible resultant error is arranged between harmonic components proportionally to the values of coefficients $c$ according to the formula $e_{iM} = \beta_i e_M^0 / \Re$, then the resultant error does not exceed the given level $e_M^0 / \Re$. From here the requirement $|H_{eK}(j\omega_i)| u_i \le \beta_i e_M^0 / \Re$ can be obtained or, with the account of (7.46) and (6.4), it can be written as follows:

$$\left|H_e(j\omega_i)\right| = \omega_i^K \left|H_{eK}(j\omega_i)\right| \le \frac{e_M^0 \omega_i^K \left(c_K + c_{K+1}\omega_i + \ldots + c_N \omega_i^{N-K}\right)}{\Re\left(c_K g_M^{(K)} + c_{K+1} g_M^{(K+1)} + \ldots + c_N g_M^{(N)}\right)}. \tag{7.47}$$

This inequality coincides with (7.45) because at derivation of (7.47) there were no restrictions for the frequency $\omega_i$. Thus, the sense of requirement (7.45) is: that the admissible maximum error when processing an action as the sum of several harmonic components is arranged between these components proportionally to their contribution into the formation of given maximum values of action derivatives estimated by magnitude $\beta_i$.

When passing to a gain plot of an open loop system with use of relation (7.14), from (7.45), the following inequality is obtained:

$$\left|W(j\omega)\right| \ge \frac{\Re\left(c_K g_M^{(K)} + c_{K+1} g_M^{(K+1)} + \ldots + c_N g_M^{(N)}\right)}{e_M^0} \times$$
$$\times \frac{1}{\omega^K \left(c_K + c_{K+1}\omega + \ldots + c_N \omega^{N-K}\right)}. \tag{7.48}$$

Remember that according to the strict expression (7.9) it would be necessary to write an inequality

$$\left|W(j\omega)\right| \ge \frac{\Re\left(c_K g_M^{(K)} + c_{K+1} g_M^{(K+1)} + \ldots + c_N g_M^{(N)}\right)}{e_M^0} \times$$
$$\times \frac{\left|H(j\omega)\right|}{\omega^K \left(c_K + c_{K+1}\omega + \ldots + c_N \omega^{N-K}\right)}.$$

However, if the resonance peak in the Bode diagram of a closed-loop system $\left|H(j\omega)\right|$ is restricted, i.e. the closed-loop system has large enough stability margin, then it is possible to obtain inequality (7.48) instead.

It is essential, that the right hand side of (7.48) principally differs from the right member in a similar way, as compared to expression (7.23) from § 7.3, which is the function of $\omega^2$, but not of $\omega$. The inequality (7.48) is drawn in a gain plot plane by a forbidden area, in which the shape depends on the maximum values of derivatives of action $\left\{ g_M^{(i)} \right\}_K^N$ and on coefficients $\left\{ c_i \right\}_K^N \in [0, \infty)$. As the values of these factors can be selected arbitrarily, then there is the set of forbidden areas but not a single one. The passing of the precise Bode diagram outside even one of them is the sufficient condition to restrict the maximum error, if the requirements on stability margin of closed-loop systems are fulfilled.

**Necessary conditions for maximal error restriction.** Use formulas (6.53) — (6.55), obtained with reference to the condition of processing action, $K$-th derivative of which is the harmonic function with frequency $\omega_1$ and amplitude $u_1$. According to (6.54) the error thus makes $e_{*1} = \left|H_e(j\omega_1)\right| u_1 / \omega_1^K$ and does not exceed the value $e_M^0$, if the following condition is satisfied

$$\left|H_e(j\omega_1)\right| \le e_M^0 \omega_1^K / u_1 . \tag{7.49}$$

At $K = 0$, when $u_1$ is the amplitude of action, condition (6.54) becomes: $\left|H_e(j\omega_1)\right| \le e_M^0 / u_1$. According to formula (6.53), the amplitude $u_1$ cannot exceed the value $u_{1M} = \min\limits_i g_M^{(i)} / \omega_1^{i-K}$, $i = \overline{K, N}$, otherwise the accepted restrictions on maximum values of action derivatives will be broken. From here considering (6.54) and passing on the basis of (7.14) to the gain plot of the open loop system, the following is obtained

$$\left|W(j\omega)\right| \ge \frac{\min\limits_i g_M^{(i)} / \omega^i}{e_M^0} = \min\left\{ \frac{g_M^{(K)}}{e_M^0 \omega^K}, \frac{g_M^{(K+1)}}{e_M^0 \omega^{K+1}}, \ldots, \frac{g_M^{(N)}}{e_M^0 \omega^N} \right\}. \tag{7.50}$$

When violating condition (7.50) the error from the processing of the harmonic action exceeds the admissible value $e_M^0$, therefore this condition is necessary to derive the required accuracy. It looks like an absolutely forbidden area in the Bode diagram plane, whose boundary generally consists of straight line segments with inclinations $-K \cdot 20$ dB/dec, $-(K+1) \cdot 20$ dB/dec, ..., and where $-N \cdot 20$ dB/dec has the shape shown in Fig. 7.9.

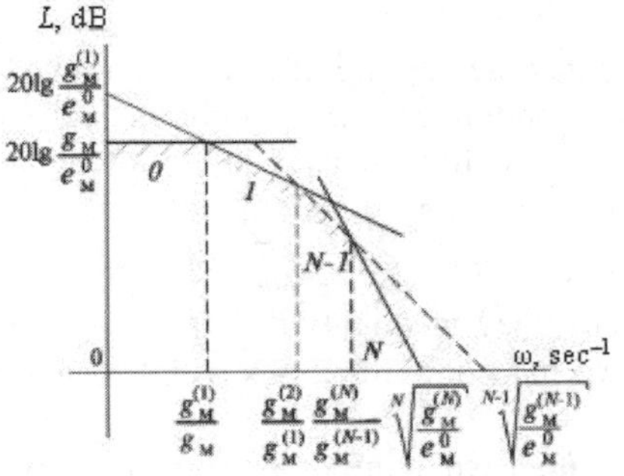


Fig. 7.9

The interval with inclination $-i \cdot 20$ dB/dec corresponds to an interval of frequencies $g_M^{(i)} / g_M^{(i-1)} \le \omega \le g_M^{(i+1)} / g_M^{(i)}$. If the relation between the maximum values of derivative is such, that $g_M^{(i)} / g_M^{(i-1)} \le 2 g_M^{(i+1)} / g_M^{(i)} \le 2 g_M^{(i)} / g_M^{(i-1)}$ (it does not contradict to (4.9)), then the boundary of the absolutely forbidden area has no interval with inclination $-i \cdot 20$ dB/dec and, on point $\omega = \sqrt{g_M^{(i+1)} / g_M^{(i-1)}}$ the inclination of this boundary changes at once on – 40 dB/dec.

It is natural, that the absolutely forbidden area does not fall outside the limits of any forbidden area appropriate to (7.48).

**Comparison of sufficient and necessary restriction conditions.** The singularities of constructing the forbidden and absolutely forbidden areas for Bode diagrams according to expressions (7.48) and (7.50) at particular values $K$ and $N$ are described in the book [8] on pages 204 — 210. The boundaries of such areas are shown as an example in Fig. 7.10 at $K = 0$, $N = 2$.

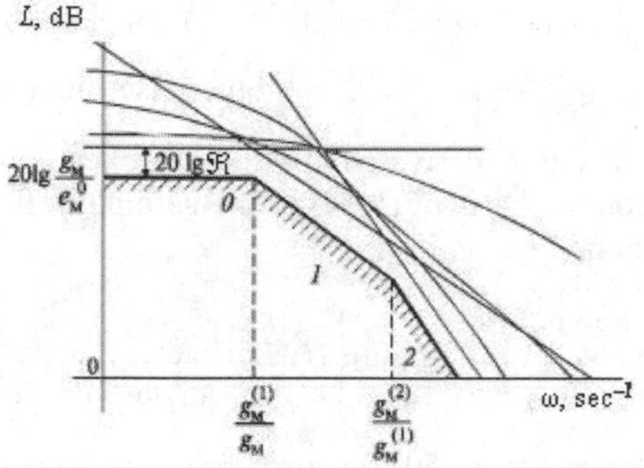


Fig. 7.10

It is visible, that unlike the described case in § 7.3 for restriction of error dispersion, here even the lowest in particular frequency interval boundary of forbidden area passes on 20lg $\Re$ [dB], which is higher than the boundary of the absolutely forbidden area. As it is shown in §6.6, the value $\Re$ in essence cannot exceed the value 1.48, and in the majority of practical cases — 1.27. Relative upward biases in the Bode diagram plane in 3.4 and 2.1 dB correspond to the numbers there.

In order to illustrate that the passing of Bode diagram of system outside the absolutely forbidden area is only necessary, but not the sufficient condition to derive the required accuracy, it is useful to analyze the following example.

Let the frequency transfer function of an open loop look like: $W(j\omega) = K_0/(1 + j\omega T_0)$, where $T_0 = g_M / g_M^{(1)}$, $K_0 = \sqrt{2}\, g_M / e_M^0$. Such frequency transfer functions obey inequality (7.50) at $K = 0$, $N = 1$. Let the reference action be the total of two harmonic functions with frequencies $\omega_1$ and $\omega_2$, amplitudes $g_{1M}$ and $g_{2M}$ and casual initial phases, and $\omega_1 = \dfrac{g_M^{(1)}}{10 g_M}$, $\omega_2 = \dfrac{10 g_M^{(1)}}{g_M}$, $g_{1M} = \dfrac{10}{11} g_M$, $g_{2M} = \dfrac{1}{11} g_M$.

The similar action has the restricted derivative $\left| g^{(1)}(t) \right| \le \omega_1 g_1 + \omega_2 g_2 = g_M^{(1)}$ and, therefore, belongs to the admissible class of actions. The maximum error of the system is:

$$e_M = \sum_{i=1}^{2} g_{iM} \left| H_e(j\omega_i) \right| = \sum_{i=1}^{2} g_{iM} \sqrt{\frac{1 + \omega_i T_0^2}{(K_0 + 1)^2 + \omega_i^2 T_0^2}} =$$

$$= \frac{10 g_M \sqrt{1 + 0.01}}{11\sqrt{(K_0 + 1)^2 + 0.01}} + \frac{g_M \sqrt{1 + 100}}{11\sqrt{(K_0 + 1)^2 + 100}},$$

or at $K_0 \geq 100$, $e_M = \dfrac{20 g_M}{11 K_0} = \dfrac{20 e_M^0}{11\sqrt{2}} = 1.29 e_M^0$. Thus, the realization of inequality (7.50) does not guarantee the required accuracy to be derived, because the retrieved error exceeds the admissible value 1.29 times. Realization of condition (7.48) can ensure that the required accuracy be obtained.

It is easy to prove with the other example, that the value of coefficient $\Re = 1.48$ in inequality (7.48), expressing the sufficient condition to derive the required accuracy, generally is not excessively overstated. For this purpose it is necessary to research the accuracy of a closed-loop system with a frequency transfer function of open loop

$$W(j\omega) = \frac{K_2(1 + j\omega\tau)}{(j\omega)^2}$$

when processing an action with the first derivative, restricted by an inequality $|g(t)| \leq g_M^{(1)}$. Compare two cases: in the first case the action can be arbitrary, and in second one it is necessarily harmonic.

With use of the formulas from §6.2 it is possible to develop the formulas for the true maximum error achievable on an action with the first derivative as a kind of meander:

$$e_M = \frac{2}{\sqrt{K_2}} \exp\left( -\frac{\tau\sqrt{K_2}}{\sqrt{4 - K_2\tau^2}} \operatorname{arctg} \frac{\sqrt{4 - K_2\tau^2}}{\tau\sqrt{K_2}} \right) \left[ 1 - \exp\left( \frac{\pi\tau\sqrt{K_2}}{\sqrt{4 - K_2\tau^2}} \right) \right]^{-1} g_M^{(1)}$$

at $\tau\sqrt{K_2} \leq 2$.

For an error, which is achieved on a harmonic action with the amplitude of first derivative $g_M^{(1)}$ and with the most unfavorable frequency, according to expression (6.55) it is possible to obtain

$$e_{*1M} = \max_{\omega} \left| \frac{1}{1 + W(j\omega)} \right| \frac{g_M^{(1)}}{\omega} = \left. \frac{\omega g_M^{(1)}}{\sqrt{(K_2 - \omega^2)^2 + (K_2\tau\omega)^2}} \right|_{\omega = \sqrt{K_2}} = \frac{g_M^{(1)}}{K_2\tau}.$$

Having researched the ratio of the determined values of error $e_M / e_{*1M}$ on extremums, find out that at $\tau\sqrt{K_2} = 2$, it reaches the maximum value $e_M / e_{*1M} = 4\exp(-1) = 1.472$, which characterizes the greatest underestimation of maximum error estimation in a case, when the only harmonic actions are taken into account in a considered example.

Generally the obtained value $e_M / e_{*1M} = 1.47$ practically coincides with the value $\Re = 1.48$, determined in §6.6 as an upper-bound estimation. However, the analyzed example is especially selected and it is not characteristic in practice. For a majority of practical cases the value $\Re = 4/\pi = 1.27$ (see §6.6) is usually quite enough.

**Physical treatment of necessary conditions for restriction at $N = 2$.** As it is visible from Fig. 7.10, the boundary of absolutely forbidden area in the Bode diagram plane consists of three straight sections with zero, single and double inclinations at known maximum values of action and two of its derivatives. Physically it is explained as follows [7].

If the frequency of harmonic action does not exceed the value $g_M^{(1)}/g_M$, then the restrictions by velocity and by acceleration of action variation will not be broken even at its greatest possible amplitude $g_M$. Therefore all frequencies from an interval $\omega \in \lfloor 0, g_M^{(1)}/g_M \rfloor$ can compare the same value of amplitude $g_M$ that causes the persistence of requirements to values for a gain plot of the system in this frequency interval.

If the frequency of harmonic action exceeds the value $g_M^{(1)}/g_M$, its amplitude cannot accept the greatest possible value $g_M$, because it would cause the intolerable high action rate. If the amplitude of the action rate is accepted to be equal to the maximum possible value of $g_M^{(1)}$, then the amplitude of action is $g_M^{(1)}/\omega$. The decrease of possible amplitudes of action with increasing frequency causes the appropriate drop down of the requirements to values of gain plot for the system, that gives the single negative inclination for the boundary of absolutely forbidden area in the Bode diagram.

Lastly, if the frequency of harmonic action exceeds the value $g_M^{(2)}/g_M^{(1)}$, then neither amplitude of action, nor amplitude of its rate can accept the maximum possible values, because it would cause the intolerably large acceleration of action variation. Having accepted the amplitude of acceleration equal to the maximum possible value $g_M^{(2)}$, the amplitude of action is determined as $g_M^{(2)}/\omega^2$. Therefore, with the rise of frequency, the requirement to the gain plot of a system decreases proportionally to the frequency quadrate, and the boundary of the absolutely forbidden area for the Bode diagram gains the double negative inclination.

## §7.4. Restriction of error from interference and total error measures

**Variants of problem statement.** If the interference and the reference action are applied to the system, it causes the additional control error component to appear. This should be taken into account to ensure the required accuracy. Obviously for this purpose it is necessary to expand the forbidden areas for the Bode diagram, constructed in § 7.3 and § 7.4 with the analysis for only a dynamic component of error.

It is possible to allocate two characteristic cases: in the first one the disturbing action is applied directly to the controlled plant, and in the second one — to the system input. It is easy to show [7, 9], that in the first case with a decrease of error from interference at the fixed transfer function of control plant, the variation of transfer function for a control unit should ensure the expansion of passband for the open loop system. This also causes the decrease of dynamic error from the processing of reference action. In this sense the influence of applied disturbing action to a controlled plant on system accuracy is equivalent to increasing intensity of reference action variation. For example, it is possible to find the equivalent values for derivatives of reference action that exceed some of the true values of such derivatives of reference action when the disturbance is absent. It causes the expansion of forbidden areas in the Bode diagram plane. However, in essence the procedure of constructing forbidden areas does not change. The forbidden areas still envelop only the bottom left of the Bode diagram.

If the interference is applied (or is transformed) to a system input, i.e. it is directly added to the reference action, at expansion of passband for closed loop systems, then the error from interference rises, and the dynamic error must decrease. Therefore when selecting a system transfer function there is the optimal solution in the sense of obtaining the best accuracy. If the spectral densities of external actions are known, such solution is easy to be found, for example, as the Wiener filter. Extremization of precision factors can also be produced at inexact a priory information on properties of actions (see chapter 8). Consider here the problem in determining the class of transfer functions ensuring the preset accuracy. This class of transfer functions can be compared to the allowed area in the Bode diagram plane of the open loop system. If the problem has a solution, then such allowed area should envelop the Bode diagram with this or that width, optimal by the best accuracy criterion. Unlike the first case, the allowed area will be restricted not only from below, but also from above, because the excessive expansion of a system passband can cause the violation of accuracy requirements at the expense of increasing error from interference.

The determination of boundaries for the described allowed area generally is connected to rather complicated calculations. Therefore, the expediency of its construction can be justified, apparently, only in a specific case, which is not characteristic for practice, when the spectral density of interference is known and is uniform within the limits of the supposed passband of a synthesized system. In other cases the synthesis can be produced by successive approximations to the acceptable solution with use of forbidden areas constructed only of dynamic component of error.

**Case of white noise interference**. Let the spectral density of interference be expressed by the formula $S_\upsilon(\omega) = S_\upsilon$. Then the dispersion of error and practical maximum error (see § 3.4) from interference are:

$$D_{ev} = S_\upsilon \Delta f_{eq}, e_{\upsilon p} = 3\sigma_{e\upsilon} = \sqrt{S_\upsilon \Delta f_{eq}}, \tag{7.51}$$

where $\Delta f_{eq} = \frac{1}{2\pi} \int_{-\infty}^{\infty} |H(j\omega)|^2 d\omega$ is the equivalent passband of a closed-loop system for the white noise.

It is shown in the book [8], that the value $\Delta f_{eq}$ with good accuracy can be estimated directly by the Bode diagram of open loop system. For this purpose it is necessary to find the base frequency $\omega_0$ and to take into account the inclination of asymptote $-l \cdot 20$ dB/dec, which intersects with the abscissa axis giving the indicated base frequency, and then using the elementary formula

$$\Delta f_{eq} \approx (\omega_0 l)/2 . \tag{7.52}$$

For typical low order transfer functions usually obtaining the result of optimization by the best accuracy criterion at the uniform spectral density of interference, formula (7.52) has no errors. The following transfer functions concern this kind:

$$W(s) = \frac{K_1}{s} \, (l = 1,\ \omega_0 = K_1), \tag{7.53}$$

and also $W(s) = K_1[s(1 + T_1 s)]^{-1}$, $T_1 \in [0, \infty)$;

$$W(s) = \frac{K_2(1 + \tau s)}{s^2} \left(l = 2,\ \omega_0 = K_2^{1/2}\right) \tag{7.54}$$

at $K_2\tau^2 = 1$ (if $K_2\tau^2 = 1.5$ the error is 2 %);

$$W(s) = \frac{K_3\left(1 + \tau_1 s + \tau_2^2 s^2\right)}{s^3} \left(l = 3,\ \omega_0 = K_3^{1/3}\right) \tag{7.55}$$

at $K_3\tau_1^3 = 1$, $\tau_2 = \sqrt{2}\tau_1$ (if $K_3\tau^3 = 8$, $\tau_2 = \tau_1/\sqrt{2}$ the error is 11 %).

When researching the inaccuracy of formula (7.52) with the presence of additional slowing and forcing stationary time constants in transfer functions, it is possible to show [8] that it does not exceed 5 — 12% for the majority of cases, which have a practical interest, when the stability margin of system is enough.

Formula (7.52) can be simplified by essentially the investigations of guaranteed control accuracy (and generally accuracy) and expansion of the number of problems of control systems synthesis, whose solution can be obtained analytically or graphically excluding the passage to numerical optimization methods.

**Construction of absolutely forbidden areas for the Bode diagram.** Write the formula (7.51) with use of (7.52) as

$$D_{e\upsilon} \approx S_\upsilon \omega_0 l/2, e_{\upsilon p} \approx 3\sqrt{S_\upsilon \omega_0 l/2} .$$

Accepting $D_{e\upsilon} = D_e^0, e_{\upsilon p} = e_M^0$, determine the maximum possible value of base frequency $\omega_{0\upsilon}$, at which excess of the accuracy requirement will be broken, even in the case of zero dynamic error: $\omega_{0\upsilon} = \dfrac{2D_e^0}{S_\upsilon l}$ or $\omega_{0\upsilon} = \dfrac{2\left(e_M^0\right)^2}{9 S_\upsilon l}$.

The inequality $\omega_0 \le \omega_{0\upsilon}$, of which the realization is the necessary condition to obtain the required accuracy, is possible to draw as the absolutely forbidden area in the Bode diagram plane of open loop systems. For constructing this absolutely forbidden area it is necessary to mark the points

$$\omega'_{0\upsilon} = \frac{2D_e^0}{S_\upsilon}\left(\text{or } \omega'_{0\upsilon} = \frac{2\left(e_M^0\right)^2}{9S_\upsilon}\right),\ \omega''_{0\upsilon} = \frac{\omega'_{0\upsilon}}{2},\ \omega'''_{0\upsilon} = \frac{\omega'_{0\upsilon}}{3} \tag{7.56}$$

on the abscissa axis and to draw straight lines with inclinations – 20 dB/dec, – 40 dB/dec and – 60 dB/dec accordingly through them. Such construction is shown in Fig. 7.11.

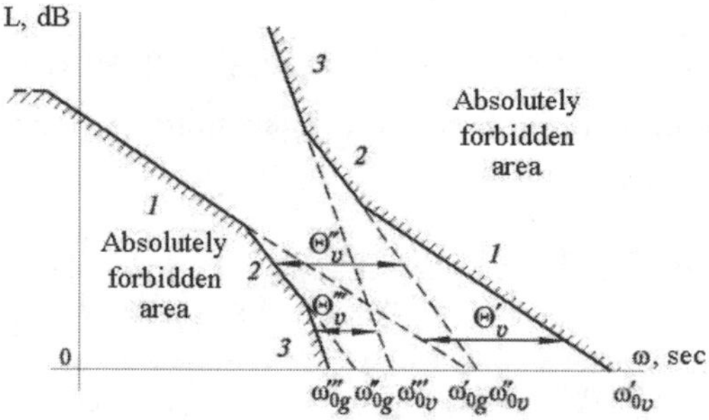


Fig. 7.11

The absolutely forbidden area on hand dynamic error is only constructed there (see § 7.3 and 7.4) for a case $K = 1$, $N = 3$.

Its boundary is formed also by straight line segments of single, double and triple inclination, whose position is determined by the base frequencies

$$\omega'_{0g} = \sqrt{D_1/D_e^0},\ \omega''_{0g} = \sqrt[4]{D_2/D_e^0},\ \omega''_{0g} = \sqrt[6]{D_3/D_e^0} \tag{7.57}$$

in the case of given dispersions or

$$\omega'_{0g} = g_M^{(1)}/e_M^0,\ \omega''_{0g} = \sqrt{g_M^{(2)}/e_M^0},\ \omega'''_{0g} = \sqrt[3]{g_M^{(3)}/e_M^0} \tag{7.58}$$

in the case of given maximum values.

It is visible from the figure, that the allowed area for the Bode diagram appears restricted both from below and from above.

If the condition $\omega_{0\upsilon} > \omega_{0g}$ for frequencies $\omega_{0\upsilon}$ and $\omega_{0g}$ with any identical superscript is broken, then it testifies to the impossibility to obtain the required accuracy with use of the Bode diagram with the appropriate inclination. But realization of this condition is not enough to derive the required accuracy. For example, if $\omega'_{0\upsilon} < \omega'_{0g}, \omega''_{0\upsilon} > \omega''_{0g}$, $\omega'''_{0\upsilon} > \omega'''_{0g}$, then obtaining the required accuracy in a system with the Bode diagram of single inclination is impossible. However, even at $\omega'_{0\upsilon} > \omega'_{0g}$ it is still impossible to claim that the selection of the Bode diagram of single inclination can ensure the required accuracy. For validity of such a statement the inequality $\omega'_{0\upsilon} > \omega'_{0g}$ should be fulfilled with the particular reserve.

**Graphical analysis of possibility to obtain the required accuracy**. Estimate the minimal possible ratio of frequencies $\omega_{0\upsilon}/\omega_{0g}$ with identical superscripts, sufficient for existence of the Bode diagram, which ensures the required accuracy. Expressions for dispersion and for the maximal value of total error can be written with the use of base frequency $\omega_0$, appropriate to asymptote of the Bode diagram with inclination $-l \cdot 20$ dB/dec:

$$D_e = \frac{D_l}{\omega_0^{2l}} + \frac{S_\upsilon \omega_0 l}{2}, \; e_M = \frac{g_M^{(l)}}{\omega_0^l} + 3\sqrt{\frac{S_\upsilon \omega_0 l}{2}}. \tag{7.59}$$

Having written the base frequency $\Theta = \omega_0 / \omega_{0g}$ normalized to value $\omega_{0g}$ and having used the indication $\Theta_\upsilon = \omega_{0\upsilon} / \omega_{0g}$, according to (7.56) – (7.58) the expression (7.59) looks like:

$$D_e = \frac{D_e}{(\omega_{0g}\Theta)^{2l}} + \frac{S_\upsilon \omega_{0\upsilon} \Theta l}{2\Theta_\upsilon} = \left( \frac{1}{\Theta^{2l}} + \frac{\Theta}{\Theta_\upsilon} \right) D_e^0,$$

$$e_M = \frac{g_M^{(l)}}{(\omega_{0g}\Theta)^l} + 3 \cdot \sqrt{\frac{S_\upsilon \omega_{0\upsilon} \Theta l}{2\Theta_\upsilon}} = \left( \frac{1}{\Theta^l} + \sqrt{\frac{\Theta}{\Theta_\upsilon}} \right) e_M^0.$$

From here for deriving the required accuracy, when $D_e \le D_e^0$ or $e_M \le e_M^0$, the value $\Theta$ should be selected to satisfy condition

$$1/\Theta^{2l} + \Theta/\Theta_\upsilon \le 1 \tag{7.60}$$

with the restriction of error dispersion and

$$1/\Theta^l + \sqrt{\Theta/\Theta_\upsilon} \le 1 \tag{7.61}$$

with the restriction of maximum error.

The left hand sides of (7.60) and (7.61) accordingly reach their minimums

$$\frac{1}{\Theta_{01}^{2l}} + \frac{\Theta_{01}}{\Theta_\upsilon} = \frac{1+2l}{(2l\Theta_\upsilon)^{2l/(2l+1)}}, \; \frac{1}{\Theta_{01}^l} + \sqrt{\frac{\Theta_{02}}{\Theta_\upsilon}} = \frac{1+2l}{\left(2l\sqrt{\Theta_\upsilon}\right)^{2l/(2l+1)}}.$$

on points $\Theta = \Theta_{01} = (2l\Theta_v)^{1/(2l+1)}$ and $\Theta = \Theta_{02} = \left(2l\sqrt{\Theta_\upsilon}\right)^{2/(2l+1)}$.

The realization of inequalities (4.60) and (4.61) is possible only in a case, when these minimums are less than one. This gives the requirements

$$\Theta_\upsilon \ge (1+2l)^{1+1/2l} / (2l) \tag{7.62}$$

with the restriction of error dispersion and

$$\Theta_\upsilon \ge (1+2l)^{2+1/l} / \left(4l^2\right) \tag{7.63}$$

with the restriction of maximum error.

With the violation of conditions (7.62), (7.63) it becomes impossible to derive the required accuracy in a system with the Bode diagram, having an asymptote with inclination $-l \cdot 20$ dB/dec. The minimal acceptable values of magnitude $\Theta_\upsilon$, calculated according to (7.62) and (7.63), are reduced in Tab. 7.1. The optimal values of normalized base frequencies $\Theta_{01}$ and $\Theta_{02}$ are given there also.

Table 7.1 Values of normalized base frequency, ensuring the derivation of preset accuracy

| Inclination of Bode diagram asymptote | Restriction of error dispersion | | Restrictioern of maximal error | |
|---|---|---|---|---|
| | $\Theta_\upsilon$ | $\Theta_{01}$ | $\Theta_\upsilon$ | $\Theta_{02}$ |
| 1 | 2.60 | 1.73 | 6.75 | 3.00 |
| 2 | 1.87 | 1.50 | 3.49 | 2.24 |
| 3 | 1.61 | 1.38 | 2.60 | 1.91 |

In an extreme case, when the inequality (7.62) or (7.63) is converted into the equality, reaching the required accuracy is possible only with the use of the Bode diagram with the base frequency $\omega_0 = \Theta_{01}\omega_{0g}$ or $\omega_0 = \Theta_{02}\omega_{0g}$. At an increasing value $\Theta_\upsilon$, the interval of accuracies for base frequencies, admissible on accuracy conditions, is expanding and its low bound approaches the value $\omega_{0g}$.

If at the synthesis stage it is required to minimize the system passband when obtaining a given accuracy, then the determination of minimum possible values for base frequency $\omega_0 = \Theta\omega_{0g}$ is of interest. If the value $\Theta_\upsilon$ essentially exceeds the extreme values reduced in Tab. 7.1, the minimal possible normalized base frequency should be close to 1. Accepting $\Theta \approx 1$ in (7.60), the following is obtained: $1/\Theta^{2l} + 1/\Theta_\upsilon \le 1$ or

$$\Theta \ge \left[\Theta_\upsilon/(\Theta_\upsilon - 1)\right]^{1/2l}. \tag{7.64}$$

Similarly from (7.61) for a case of maximum error restriction the following is obtained:

$$\Theta \ge \left[\sqrt{\Theta_\upsilon}/\left(\sqrt{\Theta_\upsilon} - 1\right)\right]^{1/l}. \tag{7.65}$$

At $\Theta_\upsilon > 10^2$ the value $\Theta = 1$ or $\omega_0 = \omega_{0g}$ is practically admissible i.e. the synthesis can be carried out excluding the account of disturbing action.

**Analytical analysis of possibility to obtain the required accuracy.** The possibility to reach the required accuracy with the account of dynamic error and error from interference can be estimated without the constructions shown in Fig. 7.11. For this purpose, it is necessary to test the realization of inequalities (7.62), (7.63), meaning that

$$\Theta_\upsilon = \omega_{0\upsilon}/\omega_{0g} = 2\left(D_e^0\right)^{1+1/2l}\left(lD_l^{1/2l}S_\upsilon\right)^{-1}$$

with the restriction of error dispersion and

$$\Theta_\upsilon = \omega_{0\upsilon}/\omega_{0g} = 2\left(e_M^0\right)^{2+1/l}\left[\left(9lg_M^{(l)}\right)^{1/l} S_\upsilon\right]^{-1}$$

with the restriction of maximum error. Then the following requirements are developed:

$$D_e^0 \geq \left[\frac{(1+2l)^{2l+1}}{4^{2l}} D_l S_\upsilon^{2l}\right]^{\frac{1}{2l+1}}, \; e_M^0 \geq \left[\frac{9^l (1+2l)^{2l+1}}{8^l l^l} g_M^{(l)} S_\upsilon^l\right]^{\frac{1}{2l+1}},$$

or at $l = 1$

$$D_e^0 \geq 1.19\left(D_1 S_\upsilon^2\right)^{1/3}, \; e_M^0 \geq 3.12\left(g_M^{(1)} S_\upsilon\right)^{1/3}, \tag{7.66}$$

at $l = 2$

$$D_e^0 \geq 1.65\left(D_2 S_\upsilon^4\right)^{1/5}, \; e_M^0 \geq 3.97\left(g_M^{(2)} S_\upsilon^2\right)^{1/5}, \tag{7.67}$$

at $l = 3$

$$D_e^0 \geq 2.13\left(D_3 S_\upsilon^6\right)^{1/7}, \; e_M^0 \geq 4.60\left(g_M^{(3)} S_\upsilon^3\right)^{1/7}. \tag{7.68}$$

If inequalities (7.66) – (7.68) are not fulfilled, then the system synthesis problem with the given precision factor has no solution. If these inequalities are fulfilled with the greatest reserve, then the synthesis can be produced practically without taking into account the disturbance.

*Questions*

1. Why is it convenient to select the class of systems ensuring the given accuracy with the construction of forbidden areas in the plane of Bode diagram of an open loop system?
2. What is the difference between the procedures of determining the sufficient and necessary conditions for deriving the required accuracy?
3. Why are the forbidden areas in the Bode diagram plane constructed only above the abscissa axis?
4. What shape do the forbidden areas in the Bode diagram have, constructed on the basis of sufficient conditions for restriction of the dynamic error dispersion?
5. What shape do the forbidden areas in the Bode diagram have, constructed on the basis of sufficient conditions for restriction of the maximal dynamic error?
6. What shape do the forbidden areas in the Bode diagram have, constructed on the base of necessary conditions for restriction of the total error?
7. Does the passing of the precise Bode diagram of open loop systems outside the absolutely forbidden area ensure the realization of sufficient conditions for the restriction of the dynamic error?
8. Describe the physical treatment of necessary conditions for the restriction of the maximal dynamic error at known maximal values of action and its two derivatives.

9. What is the difference between the procedures for restriction of the dynamic error measure and the procedure for restriction of the measure of error from interference?
10. How can the precise measure of error from input interference such as white noise be evaluated?
11. What is the equivalent passband of closed-loop systems for white noise?
12. Explain the procedure of approximated estimation for the passband of a closed-loop system on the Bode diagram of open loop systems. In what cases does it ensure good accuracy of estimation?
13. What is the shape of the forbidden area in the Bode diagram, constructed with the condition of restriction of the measure of error from interference?
14. What is the shape of the allowed "corridor" for the Bode diagram with the restriction of total control error measure?
15. What is the minimal width of the allowed "corridor" in the Bode diagram that ensures deriving of required accuracy, which is still theoretically possible?

# Chapter 8

## Optimization of control systems by criterion of highest ensured accuracy

### §8.1. Analytical and numeric methods

**Singularities of problem definition.** Similar to the problem of Wiener filter synthesis, which minimizes a total error dispersion at known spectral densities of reference action and interference, it is possible to apply the system synthesis problem, in which the upper-bound estimate of total error dispersion is minimal at known generalized or power moments of reference action and interference spectral densities. The system optimization by minimum upper-bound estimate criterion of total maximal error is also possible, if the available a priori information allows for the calculation of such an estimate for any system admissible variant. In the both cases it is possible to speak about an optimal synthesis of control systems by criterion of the highest accuracy or about minimax robust system synthesis [8].

Precision factor estimation methods with the lack of a priori information, described in chapters 5 and 6, can be applied to the analysis of defined order systems with the transfer function such as in (5.1). Therefore, the objective function of optimization depends on particular parameters $\{a_i\}_K^N$, $\{b_j\}_1^{n-1}$, whose optimal values are to be found. It makes the optimization problem parametrical. However, if the initial order of transfer function (5.1) is great enough (for example, it is not less then the order $N$ of the highest restricted derivative of reference action), for which the objective function of optimization is written, the optimization of system structure will be carried out practically. This is due to the fact that the optimal values of some parameters can be equal to zero. For example, assume $n = 3$ and consider the initial transfer function

$$W(s) = \frac{1 + b_1 s + b_2 s^2}{a_0 + a_1 s + a_2 s^2 + a_3 s^3}.$$

After the optimization of six absolute terms $a_0, a_1, a_2, a_3, b_1, b_2$, it is possible to find the transfer function of a system that posses the best accuracies among all first, second and third order systems, both static and astatic. If $a_3 = 0$, the system order becomes equal to two, at $a_0 = 0$ the system becomes astatic etc.

The objective optimization function like in (1.8) should take into account all error components and, therefore, should contain so many addends, from the amount of mutually uncorrelated actions applied to a system. The minimal number of such addends is two.

The optimization complexity is: not all the addends can be expressed analytically — some of them can be described only in a algorithmic way that does not give the required value immediately but only the algorithm of its determination. For example, if the upper-bound estimate of dynamic error dispersion is determined by an approximative method, the necessary coefficients values $\{c_i\}_K^N$ can be found numerically according to the procedure described in § 5.3. Therefore, the call to this procedure is necessary on each step of searching the objective function extremum for system optimization. The optimal values of systems parameters appropriate to an indicated extremum cannot be expressed analytically.

The examples of robust systems numerical optimization are given, for example, in [8].

**Ability of analytical solving.** The problem of optimization by the best accuracy criterion has an analytical solution and can be reduced to the optimization of transfer function parameters of those such as in (7.53) - (7.55), appropriate to the $N$-th order system with the $N$-th order astatism at a known dispersion or maximum value of only the $N$-th derivative of reference action and in the level of spectral density reduced to a system input of disturbing action $S_\upsilon$. It is possible to be convinced by it, making the control numerical solutions for the problem of parametric optimization for more complicated transfer functions. The outcomes practically coincide with indicated analytical solutions at the expense of zeroing the "excess" parameters. The physical treatment of this effect is connected to transfer functions like (7.53) - (7.55) that give the minimal passband of closed loop systems at the defined gain coefficient and realization of specific requirements on stability margin.

**Case of restricted first derivative of action.** If only the dispersion of the first derivative of reference action is known, then according to (5.22) and (7.53) the objective function looks like:

$$\overline{D}_e(K_1) = \overline{D}_{eg}(K_1) + D_{e\upsilon}(K_1) = D_1/K_1^2 + S_\upsilon K_1/2 \to \min .$$

Its research on extremum gives the optimal value for gain coefficient

$$K_1^0 = 2^{2/3}(D_1/S_\upsilon)^{1/3} = 1.59(D_1/S_\upsilon)^{1/3} \tag{8.1}$$

and minimum value of upper-bound estimate for error dispersion appropriate to it:

$$\overline{D}_{e\min} = \overline{D}_e = 3 \cdot 2^{-4/3} D_1^{1/3} S_\upsilon^{2/3} = 1.19 D_1^{1/3} S_\upsilon^{2/3} . \tag{8.2}$$

If the maximal value of only the first derivative of reference action is known, then according to §6.2 similarly the following is obtained:

$$e_M(K_1) = g_M^{(1)}/K_1 + 3\sqrt{K_1 S_\upsilon/2} \to \min,$$
$$K_1^0 = 2\cdot 3^{-2/3}\left(g_M^{(1)}\right)^{2/3} S_\upsilon^{-1/3} = 0.962\left(g_M^{(1)}\right)^{2/3} S_\upsilon^{-1/3}. \tag{8.3}$$

$$e_{M\min} = e_M\left(K_1^0\right) = 3^{5/3}\cdot 2^{-1}\left(g_M^{(1)} S_\upsilon\right)^{1/3} = 3.12\left(g_M^{(1)} S_\upsilon\right)^{1/3}. \tag{8.4}$$

**Case of a restricted second derivative of action.** If only the dispersion of a second derivative of reference action is known, then for transfer function (7.54) parameters, having determined the approximating polynomial coefficient $c_2$ from the condition $c_2 = \max_{\omega\in[0,\infty)} \left|H_e(j\omega)\right|^2/\omega^4$ and having used the exact formula for $\Delta f_{ex} = \left(1 + K_2\tau^2\right)/(2\tau)$ at $K_2\tau^2 \le 2$, it is possible to accept an objective optimization function

$$\overline{D}_e(K_2,\tau) = \frac{D_2}{K_2^3\tau^2\left(1 - K_2\tau^2/4\right)} + S_\upsilon \frac{1 + K_2\tau^2}{2\tau} \to \min.$$

When researching this function on extremum the following is obtained:

$$K_2^0 = 8\cdot 3^{-1/5}\cdot 5^{-4/5}\left(D_2/S_\upsilon\right)^{2/5} = 1.77\left(D_2/S_\upsilon\right)^{2/5}, \tag{8.5}$$

$$\tau_0 = \sqrt{3/\left(2K_2^0\right)} = 1.22/\sqrt{K_2^0} = 0.92\left(S_\upsilon/D_2\right)^{1/5}, \tag{8.6}$$

$$\overline{D}_{e\min} = \overline{D}_e\left(K_2^0,\tau^0\right) = (5/4)(5/3)^{3/5} D_2^{1/5} S_\upsilon^{4/5} = 1.70 D_2^{1/5} S_\upsilon^{4/5}. \tag{8.7}$$

The optimal closed-loop system has a weakly oscillating step response with an overshoot $\sigma = 24.3\%$. The gain plot of a closed-loop system is characterized by the oscillations index $M = 1.33$. If it is necessary to ensure some other value of oscillations index, then through the relation [7]

$$K_2\tau^2 = 2\left(M^2 - M\sqrt{M^2-1}\right)/\left(M^2-1\right),$$

which superimposes limitation on optimized parameters and slightly reduces the potential accuracy.

If only the maximum value of the second derivative of reference action is known, then according to (7.54) by a similar way the following is valid:

$$e_M(K_2,\tau) = \frac{2\,\Re\, g_M^{(2)}}{K_2^{3/2}\tau\sqrt{4 - K_2\tau^2}} + \frac{3\sqrt{1 + K_2\tau^2}\, S_\upsilon^{1/2}}{\sqrt{2\tau}} \to \min,$$

$$K_2^0 = (16\Re/15)^{4/5}(8/3)^{1/5}\left(g_M^{(2)}\right)^{4/5} S_\upsilon^{-2/5} = 1.28\left(\Re g_M^{(2)}\right)^{4/5} S_\upsilon^{-2/5}, \tag{8.8}$$

$$\tau^0 = \sqrt{3/\left(2K_2^0\right)} = 1.22/\sqrt{K_2^0} = 1.08 S_\upsilon^{1/5}\left(\Re\, g_M^{(2)}\right)^{-2/5}, \tag{8.9}$$

$$e_{M\min} = 3^{1/2} 5^{13/10} 2^{-9/5}\left(\Re g_M^{(2)} S_\upsilon^2\right)^{1/5} = 4.03\left(\Re g_M^{(2)} S_\upsilon^2\right)^{1/5}, \tag{8.10}$$

At $\Re = 1.27$ the following is given:

$$K_2^0 = 1.55\left(g_M^{(2)}\right)^{4/5} / S_\upsilon^{4/5},\ \tau^0 = 0.98\, S_\upsilon^{1/5} / \left(g_M^{(2)}\right)^{2/5},\ e_{M\min} = 4.23\left(g_M^{(2)} S_\upsilon^2\right)^{1/5}.$$

Note that both from (8.6), and from (8.8) follows $K_2^0\left(\tau^0\right)^2 = 3/2$.

**Case of a restricted third derivative of action.** If only the dispersion of a third derivative of reference action is known, then it is necessary to optimize the parameters of transfer function (7.55). When searching for the approximating polynomial coefficient $c_3 = \max\limits_{\omega\in[0,\infty)} \left|H_e(j\omega)\right|^2 / \omega^6$ included in the first component of optimization objective functions, it is expedient to use the formulas, developed in [8] on page 93. Write an objective function having also used the exact expression for passband $\Delta f_{eq}$,

$$\overline{D}_e\left(K_3,\tau_1,\tau_2\right) = \frac{D_3}{K_3^2 P_M\left(K_3,\tau_1,\tau_2\right)} + S_\upsilon \frac{K_3\tau_1\tau_2^4 + \tau_1^2 + \tau_2^2}{2\left(\tau_1\tau_2^2 - 1/K_3\right)} \to \min,$$

where the coefficient $P_M$ is expressed through the parameters $K_3, \tau_1$ and $\tau_2$ by the bulky formula from [8]. When researching this function on extremum the following is obtained:

$$K_3^0 = \left[18 D_3/\left(5 S_\upsilon\right)\right]^{3/7} = 1.73\left(D_3/S_\upsilon\right)^{3/7}, \tag{8.11}$$

$$\tau_1^0 = \left(8/K_3^0\right)^{1/3} = 1.67\left(S_\upsilon/D_3\right)^{1/7}, \tag{8.12}$$

$$\tau_2^0 = \tau_1^0/\sqrt{2} = \sqrt{2}/\left(K_3^0\right)^{1/3} = 1.18\left(S_\upsilon/D_3\right)^{1/7}, \tag{8.13}$$

$$\overline{D}_{e\min} = \overline{D}_e\left(K_3^0,\tau_1^0,\tau_2^0\right) - 35\left(D_3/5\right)^{1/7}\left(S_\upsilon/18\right)^{6/7} = 2.33 D_3^{1/7} S_\upsilon^{6/7}. \tag{8.14}$$

At such values of parameters the closed-loop system oscillations index is $M = 2.0$. The system with the transfer function such as $W(s) = K_3(1+\tau s)^2/s^3$ and optimal parameters values founded in [8] on page 223 has a greater stability margin ($M = 1.71$). At additional restriction $\tau_2 = \tau_1/2$, this transfer function gives the worse value of precision factor $\overline{D}_e\left(K_3^0,\tau^0\right) = 2.54 D_3^{1/7} S_\upsilon^{6/7}$, because it is the special case of (7.55). The loss in r.-m.-.s error against (8.14) is 4.4 %.

If only the maximum value of the third derivative of reference action is known, then for optimal values of transfer function (7.55) parameters, similarly as a result of objective function research on extremum

$$e_M\left(K_3,\tau_1,\tau_2\right) = \frac{\Re g_M^{(3)}}{K_3\sqrt{P_M\left(K_3,\tau_1,\tau_2\right)}} + 3\sqrt{\frac{K_3\tau_1\tau_2^4 + \tau_1^2 - \tau_2^2}{2\left(\tau_1\tau_2^2 - 1/K_3\right)} S_\upsilon} \to \min,$$

the following is developed:

$$K_3^0 = \left(6\Re g_M^{(3)} / \sqrt{15 S_\upsilon}\right)^{3/7} = 1.46\left(\Re g_M^{(3)} / \sqrt{S_\upsilon}\right)^{6/7}, \tag{8.15}$$

$$\tau_1^0 = \left(8/K_3^0\right)^{1/3} = 1.76\left(\Re g_M^{(3)} / \sqrt{S_\upsilon}\right)^{-2/7}, \tag{8.16}$$

$$\tau_2^0 = \tau_1^0 / \sqrt{2} = 1.25\left(\Re g_M^{(3)} / \sqrt{S_\upsilon}\right)^{-2/7}, \tag{8.17}$$

$$e_{M\min} = 7(15/36)^{3/7}\left(\Re g_M^{(3)} S_\upsilon^3\right)^{1/7} = 4.81\left(\Re g_M^{(3)} S_\upsilon^3\right)^{1/7}, \tag{8.18}$$

At $\Re = 1.27$ the following is given

$$K_3^0 = 1.79\left(g_M^{(3)} / \sqrt{S_\upsilon}\right)^{6/7}, \quad \tau_1^0 = 1.65\left(g_M^{(3)} / \sqrt{S_\upsilon}\right)^{-2/7},$$

$$\tau_2^0 = 1.16\left(g_M^{(3)} / \sqrt{S_\upsilon}\right)^{-2/7}, \quad e_{M\min} = 4.98\left(g_M^{(3)} S_\upsilon^3\right)^{1/7}.$$

**Choice of structure for an optimized system.** The use of transfer functions (4.53) - (4.55) sometimes can be justified even with the restriction of two or greater numbers of action derivatives. It is expedient in the cases when only one of the restricted derivatives essentially affects the potential accuracy. The expressions (8.1) - (8.18) allow for the finding of conditions, at which one of the indicated first, second or third order transfer functions can be considered as practically optimal.

From (8.4) and (8.10) follows, that the second order system ensures the smaller maximum error, than the first order system with realization of inequality.

$$\left(g_M^{(2)}\right)^3 \left(g_M^{(1)}\right)^{-5} S_\upsilon \le 0.0104 \tag{8.19}$$

If (8.19) is not fulfilled, and its left-hand side greatly exceeds the right one, then the first order system with transfer function (7.53) ensures the accuracy, which practically does not differ from the potential one. If it is fulfilled with large reserve, and the third derivative of action is not restricted, then the transfer function (7.54) gives the practically potential accuracy.

From (8.10) and (8.18) follows, that the third order system ensures the smaller maximum error to be present, than the second order system, with the realization of inequality

$$\left(g_M^{(3)}\right)^5 \left(g_M^{(2)}\right)^{-7} S_\upsilon < 3.30 \cdot 10^{-3}. \tag{8.20}$$

If (8.20) is fulfilled with a large reserve, then transfer function (7.55) is practically optimal. If it is not fulfilled, the knowledge of value $g_M^{(3)}$ practically does not increase the potential accuracy in comparison with the case, when the value $g_M^{(2)}$ (or $g_M^{(1)}$ and $g_M^{(2)}$) is known only, and it is possible to consider the transfer function (7.54) as optimal.

Research, similar in sense, with reference to the known dispersions of action derivatives shows accordingly

$$D_2^3 D_1^{-5} S_\upsilon^2 \leq 4.75 \cdot 10^{-3} \tag{8.21}$$

and

$$D_3^5 D_2^{-7} S_\upsilon^2 \leq 3.57 \cdot 10^{-8}. \tag{8.22}$$

instead of formulas (8.19) and (8.20).

## §8.2. Graphic-analytical method

**Sense of method.** An approximated graph-analytic method of control system optimization on precision criterion well satisfies simplicity and universality requirements. This is based on construction and gradual narrowing of an allowed "corridor" for the Bode diagram of an open loop system. As a result the optimal location of the Bode diagram, appropriate to the allowed "corridor" shape is determined. The procedure for construction of the allowed "corridor" restricted by the forbidden areas from the below left and from the above right, is considered in § 7.1-7.4. Its possible shape is shown in Fig. 7.11.

The preset restrictions on action derivative dispersions $\{D_i\}_K^N$ or on maximum values of derivatives $\{g_M^{(i)}\}_K^N$ are used when constructing an allowed "corridor". The level of uniform spectral density of entering additive interference $S_\upsilon$ is also used there. Besides, some initial approximate (overestimated) measure of control error – dispersion $D_e^0$ or maximum value $e_M^0$ must also be necessarily given. This value can be rated heuristically, according to minimal control accuracy requirements, or analytically according to the restriction on any derivative of action by formulas (8.2), (8.4) and similar to them.

If a minimal constructed "corridor" on the horizontal width between two forbidden areas of the Bode diagram exceeds approximately one quarter or one third of a decade, it should be narrowed to the indicated value by raising the boundary of the left forbidden area on ν and displacing the right forbidden area boundary to the left on $\nu/10$. The value ν, at which the advisable "corridor" width is reached, is easily selected by sequential approximations. The described contraction of the allowed "corridor" corresponds to the substitution of the precision factor initial value $D_e^0$ (or $e_M^0$) by some smaller value $D_e^1 = D_e^0/10^{\nu/10}$ (or $e_M^1 = e_M^0/10^{\nu/20}$), which characterizes the control potential accuracy better.

Note that the indicated advisable "corridor" width of one quarter of a decade is admissible for practice rounding of exact values, shown in Tab. 7.1 and the component, accordingly, $\lg 2.60$, $\lg 1.87$ and $\lg 1.61$ dec at inclinations of segments boundaries in – 20, – 40 and – 60 dB/dec.

A synthesis final stage is the selection of the Bode diagram, passing in the middle part of the obtained allowed "corridor", corresponding to the system stability margin requirements and having the simplest form.

**The case when the action is the sum of addends with the numeric characteristics of derivatives.** The above-described graph-analytical method with some additions can be effective also with the optimization of a system, in which the reference action is the sum of the casual independent centered components

$$g(t)=\sum_{j=1}^{l} g_j(t),\, l>1\,,$$

for each of them the dispersions (or maximum values) of some, probably, different derivatives $M\left\{\left[g_j^{(i)}(t)\right]^2\right\}\le D_{ij}$ (or $\left|g_j^{(i)}(t)\right|\le g_{jM}^{(i)}$), $K_j\le i\le N_j$ are restricted. An additive interference on system input is still white noise with a known level of spectral density $S_\upsilon$.

The system optimization by criterion $\overline{D}_e\to\min$ (or $e_M\to\min$) is done as follows.

The initial approximate "overestimated" value of control error measure $D_e^0$ (or $e_M^0$) should be determined at the first stage. It can be done according to the numerical characteristics of action derivative components of only that order, for which such characteristics are available for all components. If these derivatives have the order *k*, i.e. $\max\limits_j K_j\le k\le\min N_j$, then when reviewing a *k*-th order system with *k*-th order astatism and optimal parameters, the formulas (8.1), (8.4) and similar formulas allow for the expression of an error measure, analytically through the values $D_k=\sum_{j=1}^{l} D_{k_j}$ (or $g_M^{(k)}=\sum_{j=1}^{l} g_{jM}^{(k)}$) and $S_\upsilon$.

Then the absolutely forbidden areas in the Bode diagram should be constructed, according to the requirement $\overline{D}_e\le D_e^0$ (or $e_M\le e_M^0$) at the separate registration of characteristics for each of action components and magnitude $S_\upsilon$ for the interference. In outcome, *l* partially overlapped forbidden areas in the bottom left and in the top right of the Bode diagram plane are shaped.

Afterwards *l* forbidden areas at the bottom left part of the Bode diagram plane in one resultant area are being united. Its boundary ordinates determine the magnitudes preset by the graphs in logarithmic scale according to summing rules. The boundary of an obtained resultant forbidden area is located not lower then the boundaries of the joined forbidden areas and, unlike them, can have curvilinear segments. Thus, the possibility to observe only two forbidden areas is ensured. The width of the allowed "corridor" between them characterizes the possibility of the control accuracy rise in relation to a level appropriate to the value $D_e^0$ (or $e_M^0$).

The further procedure of optimal Bode diagram determination does not differ in essence from the procedure described above with reference to a case of action representation, which is not separated on components.

*Example.* $l=3$, $K_1=2$, $N_1=3$, $K_2=1$, $N_2=2$, $K_3=N_3=2$,

$D_{21} = 3.6 \cdot 10^{-6}$ m$^2$/sec$^4$, $D_{31} = 1.7 \cdot 10^{-12}$ m$^2$/sec$^6$, $D_{22} = 9.10 \cdot 10^{-4}$ m$^2$/sec$^4$, $D_{12} = 1.3 \cdot 10^{-4}$ m$^2$/sec$^2$, $D_{23} = 4.0 \cdot 10^{-8}$ m$^2$/sec$^4$, $S_\upsilon = 0.055$ m$^2$/Hz.

Optimization of the Bode diagram by criterion $\overline{D}_e \to \min$ is required.

When the second derivative for each of three action components is restricted, then $k = 2$ and $D_2 = \sum_{j=1}^{3} D_{2j} = 9.14 \cdot 10^{-4}$ m$^2$/sec$^4$. The evaluation under formula (8.7) gives $D_e^0 = 1.70 D_2^{1/5} S_\upsilon^{4/5} = 0.0414$ m$^2$.

The constructed boundaries of four absolutely forbidden areas in the Bode diagram are drawn by dashed lines in Fig. 8.1, where

$$\omega'_{0\upsilon} = 2D_e^0/S_\upsilon, \omega''_{0\upsilon} = \omega'_{0\upsilon}/2, \omega'''_{0\upsilon} = \omega'_{0\upsilon}/3,\ \ \omega'_{0g} = \left(D_{12}/D_e^0\right)^{1/2},$$

$$\omega''_{0g} = \left(D_{23}/D_e^0\right)^{1/4},\ \ \omega'''_{0g} = \left(D_{31}/D_e^0\right)^{1/6},\ \ \omega_{cg1} = \left(D_{31}/D_{21}\right)^{1/2}$$

The magnitude $D_{22}$ has no fracture of boundary on frequency $\omega_{cg2} = \left(D_{22}/D_{12}\right)^{1/2}$, because $\omega_{cg2} > \omega'_{0g}$. The boundary of the resultant absolutely forbidden area, integrating three forbidden areas in the bottom left of Fig. 8.1, is drawn by the bold line.

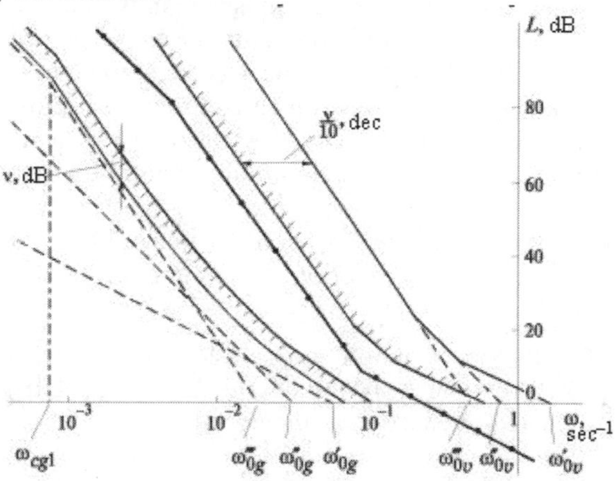


Fig. 8.1

As the width of the obtained allowed "corridor" exceeds the quarter decade, it can be narrowed down. Select the contraction parameter $\nu = 4.5$. Boundaries of the narrowed "corridor" are marked by the leader in Fig. 8.1. The left one is raised on 4.5 dB, and the right one is shifted to the left on 0.45 dec. The checking precision factor value $D_e^1 = D_e^0/10^{0.45} = 0.0165$ m$^2$ corresponds to these new boundaries.

The Bode diagram selection variant, passing in the middle part of the allowed "corridor", is marked by line with points in Fig. 8.1.

## §8.3. Estimation of loss in potential accuracy when lacking a priori information

**Loss due to the lack of information about action properties.** The minimum possible value of the upper-bound estimate for the total error dispersion $\overline{D}_{e\min}$ obtained at a system optimization, characterizes the potential guaranteed control accuracy at an unknown spectral density of reference action. The actual error dispersion in minimax robust systems cannot exceed the value $\overline{D}_{e\min}$ even at the most unfavorable shape of action spectrum from their set, within the limits of which the accepted restrictions on power moment values (2.20) are fulfilled. Moreover, actual error dispersion generally can remain smaller than the value $\overline{D}_{e\min}$ even at the most unfavorable spectrum of action, because the upper-bound estimate, and not the upper bound of error dispersion, is used as a precision factor.

It is natural, that the potential control accuracy at known action spectral densities characterized by an exact value of error dispersion and achievable in Wiener filter, is higher than the potential accuracy with the lack of a priori information. However, the error dispersion, restricted by an inequality $D_e \le \overline{D}_{e\min}$, is guaranteed at any admissible action spectral density shape in the minimax robust system synthesized with the lack of a priori information. In the Wiener filter the deviation of the spectral density actual shape from the accepted model can cause the essential accuracy degradation. This is the advantage of robust systems.

The estimation of the greatest possible loss of robust systems in the potential accuracy against Wiener filter synthesized at full a priori information is of interest. For such a solution to this problem it is possible to use the expressions for the lower estimation of dynamic error dispersion, obtained in §5.5, and to find the result from system optimization by criterion of minimum lower estimation of the total error dispersion $\underline{D}_e = \underline{D}_{eg} + D_{ev} \to \min$.

The total error dispersion in an optimal system synthesized at known a spectral density of reference action, must take an intermediate position between the values $\underline{D}_{e\min}$ and $\overline{D}_{e\min}$. Therefore the dimensionless coefficient

$$\eta = \sqrt{\overline{D}_{e\min} / \underline{D}_{e\min}} \tag{8.23}$$

characterizes the maximum possible ratio of r.-m.-s. errors in optimal systems, which are synthesized at known dispersions $\{D_i\}_K^N$ and at completely known function $S_g(\omega)$. If the initial order $n$ of transfer function (2.1), for which the optimization objective function is defined, is great enough, or at least $n \ge N$, the coefficient $\eta$ shows the maximal loss in accuracy of synthesized robust system against the Wiener filter.

Outcomes for the lower estimation $\underline{D}_{eg}$ explained in §5.5 allow for the generating of an objective function of system parameter optimization by criterion $\underline{D}_e \to \min$ at voluntary orders *n* and *N*. In common cases the optimization is carried out by numerical methods.

Similar research is carried out in [8], where the following estimations of value $\eta$ for the two most typical cases are obtained.

If the order of higher action derivative with the restricted dispersion is equal to two and, accordingly, the minimax robust system has the same order $n = 2$, then

$$\eta = \eta'' = \left( \frac{D_2 D_0}{D_1^2} \right)^{1/10} . \tag{8.24}$$

If the order of action derivatives is higher with the restricted dispersion being equal to three and $n = 3$, then

$$\eta = \eta''' = \left( \frac{D_3 D_1}{D_2^2} \right)^{1/14} . \tag{8.25}$$

Expressions (8.24) and (8.25) clearly have a physical sense, because the ratio $D_{i+1} D_{i-1} / D_i^2$, $i = 1, 2, \dots$, as it was shown in §3.1, characterize the width of the action spectrum. At $D_{i+1} D_{i-1} / D_i^2 = 1$ the action is possible to be assimilated to the harmonic function, which has only one spectral line in a frequency $\omega_1 = \sqrt{D_{i+1} / D_i}$ . In this case the full a priori information is actually known and the robust synthesis problem becomes the classical problem of optimal linear filtration. The loss of robust systems in potential accuracy against the Wiener filter should be absent. This is confirmed by formulas (8.24) and (8.25), giving $\eta = 1$.

If $D_{i+1} D_{i-1} / D_i^2 > 1$, then the spectrum of reference action expands and cannot uniquely be recovered by finite number of dispersions $\{D_i\}_K^N$ . Loss of potential accuracy in robust systems increases with the expansion of the action spectrum. As it is shown in [8], the inequality $D_{i+1} D_{i-1} / D_i^2 < 10 - 50$ is usually fulfilled in practice. At $D_{i+1} D_{i-1} / D_i^2 < 10$ formulas (8.24) and (8.25) are, accordingly, $\eta < 1.26$ and $\eta < 1.18$, at $D_{i+1} D_{i-1} / D_i^2 < 50$ - $\eta < 1.48$ and $\eta < 1.32$. Thus, it is possible to claim, that the r.-m.-s. errors in minimax robust systems and in the Wiener filter differ, as a rule, by less than in 1.5 times at three known power moments of action spectral density.

With an increasing volume of a priori information on properties of actions, the loss of potential accuracy in robust systems decreases. This can be confirmed by numerical research on a computer.

**Effect of parametric perturbations**. Sometimes it is necessary to optimize the control system with the account of instability of some of its parameters, whose values can casually vary during the system operation or have an essential scatter on realizations. It is necessary to consider this approach when probability densities of unstable parameters are unknown, but finite areas of possible variation of these parameters, whose boundaries are the known functions of their nominal values, are preset. Such an approach is well entered within the frameworks of the common concept of robust systems synthesis. This approach can be explained with reference to a case, when only one of the system parameters $\alpha$ is essentially unstable.

Let $I(\alpha)$ be a unimodal function, accepted as the characteristic of system operation quality (similar functions in § 8.1 look like $\overline{D}_e(\alpha,\ldots)$ or $e_M(\alpha,\ldots)$. If the distribution law $\varphi(\alpha,\alpha_N)$ were known for parameter $\alpha$ as a slow variable within the neighborhood of nominal values $\alpha_N$, then at optimization of value $\alpha_N$ it would be possible to use the criterion (minimum of average risk)

$$R(\alpha_N)=\int_{-\infty}^{\infty} I(\alpha)\,\varphi(\alpha,\alpha_N)\,d\alpha \to \min .$$

The problem with researching "the optimum of nominal" is usually applied in this manner.

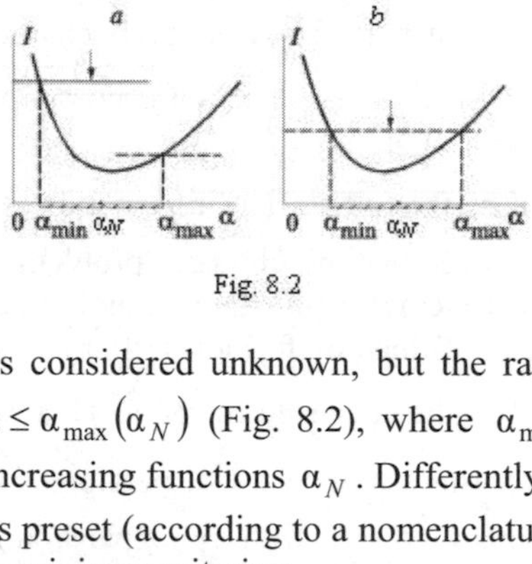


Fig. 8.2

The law $\varphi(\alpha,\alpha_N)$ is considered unknown, but the range of parameter $\alpha$ is given as $\alpha_{\min}(\alpha_N)\le\alpha\le\alpha_{\max}(\alpha_N)$ (Fig. 8.2), where $\alpha_{\min}(\alpha_N)$ and $\alpha_{\max}(\alpha_N)$ are the monotonically increasing functions $\alpha_N$. Differently, the band model of parametric perturbations is preset (according to a nomenclature, given in §1.5). Then it is expedient to use the minimax criterion

$$\max\{I[\alpha_{\min}(\alpha_N)], I[\alpha_{\max}(\alpha_N)]\} \to \min . \tag{8.26}$$

with the selection of $\alpha_N$.

Expression (8.26) determines an objective function of parametric optimization. However, it cannot be used immediately for an analytical determination of optimal value $\alpha_N=\alpha_N^0$. It is easily proven, that a sufficient condition for realization (8.26) is the validity of an equality

$$I[\alpha_{\min}(\alpha_N)]=I[\alpha_{\max}(\alpha_N)] \tag{8.27}$$

This justifies the possibility of searching for value $\alpha_N^0$, satisfying (8.26), by a analytical or numerical solution of equation (8.27).

**Choice of nominal values for unstable parameters**. Apply (8.27) to the particular problems of robust systems parametric optimization considered in § 8.1. Assume the systems gain coefficient $K_n$ ($n = 1, 2$ or 3 in transfer functions (7.53) - (7.55) accordingly) has slow uncontrollable fluctuations in the range $K_n \in [K_{nN}/\chi, K_{nN}\chi]$, $\chi \geq 1$, i.e. it can vary on $\pm 20\lg\chi$ [dB] in relationship to the nominal value $K_{n\mathrm{N}}$.

At $n=1$ according to quality characteristic $I(K_1) = \overline{D}_e(K_1)$ described in §8.1, (8.27) can be rewritten as

$$D_1(\chi/K_{1N})^2 + S_\upsilon K_{1N}/(2\chi) = D_1/(K_{1N}\chi)^2 + S_\upsilon K_{1N}\chi/2 .$$

The solution of this equation in relationship to $K_{1N}$ is:

$$K_{1N}^0 = \left[2\left(\chi + \chi^{-1}\right)\right]^{1/3} (D_1/S_\upsilon)^{1/3}, \tag{8.28}$$

Therefore an upper-bound estimate of error dispersion in an optimal system can be found as:

$$\overline{D}_{e\min} = \overline{D}_e\left(K_{1N}^0\right) = \frac{1+\chi^2\left(1+\chi^2\right)}{2^{2/3}\left(\chi+\chi^{-1}\right)^{2/3}\chi^2} D_1^{1/3} S_\upsilon^{2/3} . \tag{8.29}$$

Similar research for the maximum error is:

$$K_{1N}^0 = 2^{1/3} 3^{-2/3}\left(\chi^{1/2} + \chi^{-1/2}\right)^{2/3}\left(g_M^{(1)}\right)^{2/3} S_\upsilon^{-1/3}, \tag{8.30}$$

$$e_{M\min} = e_M\left(K_{1N}^0\chi\right) = \frac{3^{2/3}}{2^{1/3}} \frac{1+\chi}{\left(1+\chi\right)^{2/3}\chi^{2/3}} \left(g_M^{(1)} S_\upsilon\right)^{1/3} . \tag{8.31}$$

It is natural that at $\chi \to 1$ formulas (8.28) - (8.31) become accordingly (8.1) - (8.4).

Note that the optimal nominal value of a system gain coefficient $K_{1N}^0$ at $\chi > 1$ exceeds the value $K_{10}$ that adds a minimum to the function $I(K_1)$ and is optimal with the lack of parametric fluctuations. The reason for this is the function $I(\alpha) = I(K_1)$ to the left from the point $K_1 = K_{10}$ increases more abruptly, than the right (see Fig. 8.2). Therefore the underestimated values of the system gain coefficient are more undesirable, than overestimated ones. That is taken into account by selection $K_{1N}^0 > K_{10}$. It is characteristic also for other optimization examples considered below.

At $n = 2$ according to quality characteristic $I(K_2) = \overline{D}_e(K_2, \tau)$ described in §8.1, (8.27) looks like:

$$\frac{4D_2\chi^3}{K_{2N}^2 d(4-d/\chi)} + S_\upsilon\sqrt{K_{2N}}\,\frac{1+d/\chi}{2\sqrt{d}} = \frac{4D_2}{K_{2N}^2 d\chi^3(4-d\chi)} + S_\upsilon\sqrt{K_{2N}}\,\frac{1+d\chi}{2\sqrt{d}}, \tag{8.32}$$

where $d = K_{2N}\tau^2$, $\chi < 4/d$. The solution of this equation in relationship to $K_{2N}$ allows obtaining the following:

$$K_{2N}^0 = (\aleph D_2/S_\upsilon)^{2/5}, \tag{8.33}$$

where

$$\aleph = 8\left[4\chi(\chi^4 + \chi^2 + 1) - d(\chi^4 + 1)(\chi^2 + 1)\right] / \left[\chi d^{3/2}(4\chi - d)(4 - d\chi)\right].$$

Then the upper-bound estimate of error dispersion is

$$\overline{D}_e(K_2^0, \chi, \tau) = \left(\frac{4}{d\chi^3(4 - d\chi)} + \frac{\aleph(1 + d\chi)}{2\sqrt{d}}\right)\aleph^{-4/5} D_2^{1/5} S_\upsilon^{4/5}. \tag{8.34}$$

Investigating (8.34), it is possible to also find the correlation of optimal values $d$ by $\chi$. This correlation is rather weak. This allows, following (8.9), the acceptance of $d^0 = 3/2$, i.e. $\tau^0 = \sqrt{3/(2K_{2N}^0)}$.

In a case $\chi \to 1$, (8.5) and (8.7) are accordingly obtained from (8.33) and (8.34).

If $n = 3$ then the research procedure remains the same, but the calculations become more complicated.

Expression (8.27) can be used with parametric optimization of a system synthesized with completely known properties of actions and, therefore, not being robust in relation to them. For example, if $n = 2$, as a reference action, compared to the considered above case, has a constant known second derivative $g^{(2)}(t) = a$, then the dynamic error in a steady state is $e_g = a/K_2$. Thus expression (8.27) can be written in more simply , as contrasted to (8.32)

$$\frac{a^2\chi^2}{K_{2N}^2} + S_\upsilon\sqrt{K_{2N}}\,\frac{1 + d/\chi}{2\sqrt{d}} = \frac{a^2}{K_{2N}^2\chi^2} + S_\upsilon\sqrt{K_{2N}}\,\frac{1 + d\chi}{2\sqrt{d}}.$$

The solution of this equation by $K_{2N}$ gives

$$K_{2N}^0 = \left(\frac{2(\chi + 1/\chi)}{\sqrt{d}}\frac{a^2}{S_\upsilon}\right)^{2/5}.$$

Here the magnitude $d = K_{2N}^0\tau^2$, determining the system stability margin, as well as earlier, is selected near a value $d = 1.5$.

**Analysis of loss research outcomes.** The obtained formulas above illustrate well the circumstance that indeterminacy in system properties, as well as the indeterminacy in action properties, decreases the accessible control quality. However, such decrease can be minimized in robust systems synthesized in view of indicated indeterminacy. This can be reached, in particular, by special selection of nominal values of parameters by formulas such as (8.28), (8.30), and (8.33). With the selection of control system structure it is necessary to take into account that with the increase of system order, the criticality of its precision factor to parameter instability increases. It can be confirmed, for example, with the comparative analysis of formulas (8.29) and (8.34).

It is necessary to pay attention to some quantitative outcomes describing the relative decrease of potential accuracy when increasing the possible "amplitude" of parametric fluctuations, characterized in formulas (8.29), (8.31), and (8.34) by magnitude $\chi$. The potential system accuracy with constant, completely known optimal parameters is accepted as the "starting point".

At $\chi = \sqrt{2}$, when the maximum value of gain coefficient $K_{nN}\chi$ may exceed the minimum one $K_{nN}/\chi$ by two, the loss of robust system optimal by criterion $\overline{D}_e \to \min$ in r.-m.-s. error equals 5.9 % at $n=1$ and 7.8 % at $n=2$.

At $\chi = 2$, when the maximum value of the gain coefficient may exceed the minimum one by four times, the loss in r.-m.-s. error increases up to 22.8 % at $n=1$ and 28.1 % at $n=2$. Note for comparison, that in a system synthesized without the account of parameters instability and, therefore, which is not robust in relation to this instability (but robust with reference to action properties), the appropriate loss equals 29.1 % at $n=1$ and 33.8 % at $n=2$.

The precision factor in robust systems, optimal by criterion $e_M \to \min$, is less critical to parameters instability, than in a case of criterion $\overline{D}_e \to \min$. The loss of the first order robust system in a maximum error equals only 3.2 % at $\chi = \sqrt{2}$ and 12.0 % at $\chi = 2$.

*Questions*

1. What is the objective function of robust system optimization by precision criterion? How many addends does it have?
2. Why is only the numerical optimization of robust system possible in common cases?
3. What are the cases in which the problem of robust system optimization by precision criterion has an analytical solution? What do the transfer functions look like in this case? What does their parameter choice depend on?
4. Based on theoretical analysis explain the procedure of selection of the optimal robust structure systems among the first, second and third order systems.
5. How is the graph-analytic method of optimization of the system Bode diagram realized, based on precision criterion?
6. What singularities does construction of the forbidden area of the Bode diagram have with an action as the sum of components with different numerical performances?
7. Explain the procedure and outcome estimation from maximal possible loss of potential accuracy in robust systems compared to the Wiener filter.
8. How can the optimization problem of control system nominal values be compatible with parametric perturbations? How can the range of fluctuated system parameter be set?
9. In what direction does the optimal nominal value of unstable system parameters vary with the increase of parametric fluctuation depth?
10. How does the increasing depth of parametric fluctuations influence the guaranteed value of system accuracy measurement?

# Chapter 9

## Digital robust systems

### §9.1. Determination of characteristics for values of actions discrete in time domain

**Lattice casual processes.** Analyzing impulse systems and particularly digital ones, the characteristics of discrete, in time domain, values of action, obtained by the passing of initial continuous action through the sampler, are of interest. The lattice process $g[k]=g(t)|_{t=kT}$ is the mathematical model of such an impulse signal, where $T$ is the sampling period, and $k = 0, 1, 2$ is the discrete time (see §2.1).

Lattice processes passing through the system is analyzed on based on the difference equations describing it. The Z-conversion is used for their solving.

In the presence of casual actions, analysis of impulse systems is based on determination of the statistical characteristics of output casual lattice processes with known characteristics of the input lattice process. In turn, the statistical characteristics of input processes could be obtained either during direct processing of experimental data or from statistical characteristics of initial continuous action. The latter method is more preferable. Hence it is necessary to analyze how it is used to determine such characteristics of stationary lattice casual process as a correlation function, spectral density, dispersion and maximal values of various orders differences.

**Correlation function and spectral density.** If the correlation function of initial continuous process $R_g(\tau)$ is known, then the correlation function of lattice process $R_g[m]=\overline{g[k]g[k+m]}$ is obtained as a result of discretization by formula

$$R_g[m]=R_g(\tau)|_{\tau=mT},$$

The spectral density can be found as a two-sided z-conversion of correlation function

$$S_{g\perp}(z)=\sum_{m=-\infty}^{\infty}R_g[m]\ z^{-m}=F(z)+F(z^{-1})-R_g[0], \qquad (9.1)$$

where $F(z)=Z\{R_g[m]\}=\sum_{m=0}^{\infty}R_g[m]\ z^{-m}$.

Passing to the circular frequency $\omega$, and taking into account the parity of correlation function, from (9.1) the following is developed:

$$S_{g\perp}\left(e^{j\omega T}\right)=\sum_{m=-\infty}^{\infty} R_g[m]\; e^{-j\omega mT}=2\,\mathrm{Re}\,F\left(e^{j\omega T}\right)-R_g[0]. \tag{9.2}$$

When passing to pseudo-frequency $\lambda=\frac{2}{T}\mathrm{tg}\frac{\omega T}{2}$, the latter expression becomes:

$$S_{g\perp}^{*}(\lambda)=S_{g\perp}\left(\frac{1+j\lambda T/2}{1-j\lambda T/2}\right)=2\,\mathrm{Re}\,F\left(\frac{1+j\lambda T/2}{1-j\lambda T/2}\right)-R_g[0]. \tag{9.3}$$

The inversion formulas corresponding to (9.2) and (9.3) look like the following:

$$R_g[m]=\frac{T}{2\pi}\int_{-\pi/T}^{\pi/T} S_{g\perp}\left(e^{j\omega T}\right)e^{j\omega mT}d\omega=\frac{T}{\pi}\int_{0}^{\pi/T} S_{g\perp}\left(e^{j\omega T}\right)\cos\omega mTd\omega\,, \tag{9.4}$$

$$R_g[m]=\frac{T}{2\pi}\int_{-\infty}^{\infty}\left[\frac{1+j\lambda T/2}{1-j\lambda T/2}\right]^{m}\frac{S_{g\perp}^{*}(\lambda)}{\left|1+j\lambda T/2\right|^{2}}d\lambda=$$
$$\frac{T}{2\pi}\int_{0}^{\infty}\frac{S_{g\perp}^{*}(\lambda)\cos[2m\,\mathrm{arctg}(\lambda T/2)]}{\left|1+j\lambda T/2\right|^{2}}d\lambda. \tag{9.5}$$

If $m=0$, then expressions (9.4) and (9.5) give the following value of dispersion for the lattice casual process

$$R_g[0]=D_0=\frac{T}{\pi}\int_{0}^{\pi/T} S_{g\perp}\left(e^{j\omega T}\right)d\omega=\frac{T}{2}\int_{0}^{\infty}\frac{S_{g\perp}^{*}(\lambda)}{\left|1+j\lambda T/2\right|^{2}}d\lambda. \tag{9.6}$$

The spectral density of casual lattice process (9.2) can be found not only through its correlation function, but also directly from the spectral density of the initial continuous process $S_g(\omega)$ by the following formula

$$S_{g\perp}\left(e^{j\omega T}\right)=T^{-1}\sum_{r=-\infty}^{\infty} S_g(\omega+2\pi r/T). \tag{9.7}$$

In this formula is convenient to analyze the regularities of shape variations in the process spectral density, during its passage through the sampler.

Using (9.4) and (9.7) for dispersion of lattice action it is possible to write:

$$D_0=\frac{T}{\pi}\int_{0}^{\pi/T}\frac{1}{T}\sum_{r=-\infty}^{\infty} S_g\left(\omega+\frac{2\pi r}{T}\right)d\omega=\frac{1}{\pi}\int_{0}^{\infty} S_g(\omega)d\omega=R_g(0).$$

This proves again the coincidence of dispersions of lattice continuous and initial continuous processes. Their expectation values also coincide.

**Dispersion of differences.** It is necessary to find out in what way the dispersions of differences for lattice processes $g[k]$ are connected to dispersions of derivatives for initial continuous casual process $g(t)$. Here the frequency method is used. The expression for dispersions of first difference then looks like:

$$D_{\nabla 1} = \overline{(g[k]-g[k-1])^2} = \overline{[g(t)-g(t-T)]^2} = 2[R_g(0)-R_g(T)] =$$
$$= \frac{2}{\pi}\int_0^{\infty}(1-\cos\omega T)S_g(\omega)d\omega = \frac{1}{\pi}\int_0^{\infty}4\sin^2\frac{\omega T}{2}S_g(\omega)d\omega.$$

Using similar evaluations for dispersion of $j$-th difference it is possible to write:

$$D_{\nabla j} = \frac{1}{\pi}\int_0^{\infty}\left(2\sin\frac{\omega T}{2}\right)^{2j} S_g(\omega)\,d\omega,\ j = 0,1,\ldots \tag{9.8}$$

In turn, the dispersion of derivatives for the initial continuous action is expressed by formula (3.20). With the comparison of expressions (9.8) and (3.20) the following inequality can be found

$$D_{\nabla j} \le \varepsilon_{i,j} D_i, \tag{9.9}$$

where

$$\varepsilon_{i,j} = \max_{\omega}\frac{(2\sin\omega T/2)^{2j}}{\omega^{2i}} = \left.\frac{(2\sin\omega T/2)^{2j}}{\omega^{2i}}\right|_{\omega=\omega_M}. \tag{9.10}$$

The equalities on the left and right hand sides in (9.9) is achieved under action $g(t)$, which has only one spectral component with the frequency $\omega_M$.

At $j = i$ formula (9.10) is:

$$\varepsilon_{i,j} = \max_{\omega}\left(\frac{2\sin\omega T/2}{\omega}\right)^{2i} = \lim_{\omega\to 0}\left(\frac{2\sin\omega T/2}{\omega}\right)^{2i} = T^{2i}$$

or, according to (9.9),

$$D_{\nabla j} \le T^{2i} D_i. \tag{9.11}$$

If $j > i$ then the different values of coefficients $\varepsilon_{i,j}$ correspond to different pairs of indexes $i, j$.

For example when $i = 1, j = 2$, the frequency $\omega_M$ is determined by the transcendental equation $\operatorname{tg}\omega_M T/2 = \omega_M T$. This is a result of the analysis of (9.10) on extremum. After solving it is possible to find $\omega_M = 2.33/T$. Substituting it into (9.10), the value $\varepsilon_{1,2} = 2.10 \cdot T^2$.

The other coefficients $\varepsilon_{i,j}$ can similarly be found. Some of their values are given in Tab.9.1.

Table 9.1 Values of coefficients $\varepsilon_{i,j}$ in inequality (9.9)

| $i \backslash j$ | 1 | 2 | 3 | 4 | 5 |
|---|---|---|---|---|---|
| 1 | $T^2$ | $2.10\ T^2$ | $7.59\ T^2$ | $29.0\ T^2$ | $113\ T^2$ |
| 2 | – | $T^4$ | $1.42\ T^4$ | $4.41\ T^4$ | $15.5\ T^4$ |
| 3 | – | – | $T^6$ | $1.08\ T^6$ | $1.70\ T^6$ |

Note, if $j >> i$ then the analysis of expression (9.10) can obtain $\omega_M \approx \pi/T$ and

$$\varepsilon_{i,j} \approx 2^{2j} T^{2i} / \pi^{2i} \tag{9.12}$$

Formula (9.12) ensures the satisfactory accuracy at $j - i \geq 5$.

Thus, the restriction of the $i$-th derivative of initial continuous action ensures the restriction of $i$-th and all of the following differences of discrete readings of this action. By analyzing this problem more carefully, it is possible to show that in the case when two or more derivatives of continuous action are restricted, usually only the highest of them essentially affects restriction of differences.

**Maximum values of differences.** The differences of discrete readings of actions usually cannot assume infinite large values and must be restricted as:

$$\left|\nabla^j g[k]\right| \leq \nabla_M^j, \quad j = 1,\ 2,\ \ldots$$

The estimation of maximum values of differences $\nabla_M^j$ can be carried out on the basis of conditions (3.22), applied to maximum values of derivatives of initial continuous action $g(t)$. This can easily be done when only one $i$-th derivative $\left|g^{(i)}(t)\right| \leq g_M^{(i)}$ is restricted. Actually the $i$-th derivative of continuous action has a shape of meander with a period $2T$ and spasmodically varies between levels $\pm g_M^{(i)}$. Then action $g(t)$ itself varies with the same period according to the piecewise-parabolic law with semi-range $g_M^{(i)} T^i / i!$, and the lattice process $g[k]$ follow the following law

$$g[k] = \pm(-1)^k \frac{g_M^{(i)} T^i}{i!}. \tag{9.13}$$

Substituting (9.13) into the formula for $i$-th inverted difference [7]

$$\nabla^i g[k] = \nabla^{i-1} g[k] - \nabla^{i-1} g[k-1] = \ldots = \sum_{v=0}^{i} (-1)^v C_i^v g[k-v] \tag{9.14}$$

it is possible to find

$$\nabla^i g[k] = \pm \sum_{v=0}^{i} C_i^v \frac{g_M^{(i)} T^i}{i!} = \pm g_M^{(i)} T^i \sum_{v=0}^{i} \frac{1}{v!(i-v)!} = \pm g_M^{(i)} T^i$$

or

$$\nabla_M^i = \pm g_M^{(i)} T^i . \tag{9.15}$$

The value $\nabla^i g[k] = \pm g_M^{(i)} T^i$ could also be found when the $i$-th derivative of continuous action $g^{(i)}(t) = g_M^{(i)}$ has a constant value. Then $g[k] = g[0] + \frac{g_M^{(i)} (kT)^i}{i!}$, when substituted into (9.14) equals

$$\nabla^i g[k] = g_M^{(i)} T^i \sum_{v=0}^{i} (-1)^v C_i^v \frac{(i-v)!}{i!} = g_M^{(i)} T^i \sum_{v=0}^{i-1} \frac{(-1)^v (i-v)^i}{(i-v)! vv} = g_M^{(i)} T^i .$$

Note, that pseudocasual signal (9.13) and the continuous action generating it have δ- type distribution laws. Therefore their maximum values coincide with r.-m.-s. values. So, it is no wonder that relation (9.15) corresponds to (9.11).

It is essential, that under the finite maximum value of the $i$-th difference, the differences of higher orders cannot be infinite. This follows from relation (9.14). According to this relation,

$$\left|\nabla^{i+1} g[k]\right| \le \left|\nabla^i g[k]\right| + \left|\nabla^i g[k-1]\right| \le 2 \max\left|\nabla^i g[k]\right| = 2\nabla_M^i$$

or for general cases,

$$\left|\nabla^i g[k]\right| \le \sum_{v=0}^{j-i} C_{j-i}^v \left|\nabla^i g[k-v]\right| \le \sum_{v=0}^{j-i} C_{j-i}^v \nabla_M^i = 2^{j-i} \nabla_M^i, j > i.$$

Hence, according to (9.15) the following inequalities can be written:

$$\nabla_M^j < 2\nabla_M^{j-1} \le 4^{j-2} \nabla_M^i \le \ldots \le 2^{j-i} \nabla_M^i = 2^{j-i} g_M^{(i)} T^i . \tag{9.16}$$

They become equalities when the lattice processes are similar to (9.13).

If two or more derivatives of the initial continuous action are restricted, then the variations of lattice process by law (9.13) may become impossible. In this case the restrictions applied to differences should be strengthened. This operation is connected to the determination of lattice processes, having the maximum values of differences, and needing laborious calculations in order to be accomplished. An example of them is shown in the book [8].

In practice, the problem solving of determination of the maximum values of differences becomes essentially easier, because at rather small sampling periods the formula (9.15) becomes correct, independent from the number of derivatives of initial continuous action.

## §9.2. Synthesis of closed loop digital robust systems

**Variants of system structure.** Ways of giving the robust property to a digital system in relation to external actions of indeterminacy as well as with research methods for a guaranteed accuracy of its operation differ depending on system structure. As a rule, the digital system block diagram consists of digital and continuous dynamic units, definitely joint together. If the continuous units correspond to a controlled plant and to a sensor, and the digital part of a system implements the required control law, then the system block diagram with zero order extrapolator and single main feed-back looks like that shown in Fig. 9.1.

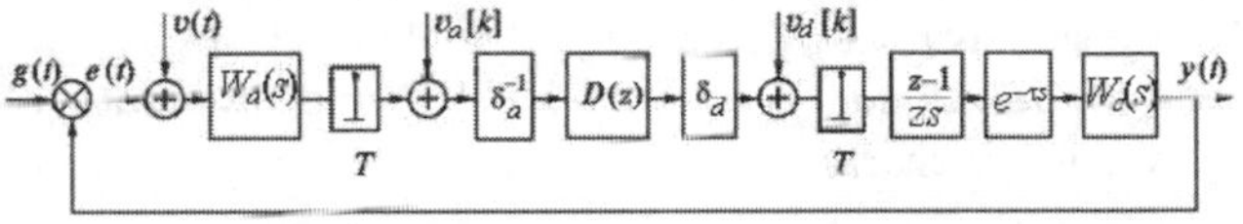


Fig. 9.1

Here $W_d(s)$ and $W_c(s)$ are the transfer functions of continuous units, $D(z)$ is the discrete transfer function of digital part, and $\delta_a^{-1}$ and $\delta_d$ are the transfer constants of linearized analog-to-digital and digital-to-analog converters. The other labels are standard. The dead time unit that considers the delay of giving a signal by digital part on time $\tau$ in each discretization phase $T$ ( $\tau < T$ ) is conditionally referred to as a continuous part of a system.

The kind of transfer function for an entering continuous dynamic unit $W_d(s)$ is basic. If this unit is considered inertialess, i.e. $W_d(s) = k_d$, then the system is digital-analog. Otherwise the system is analog-digital, more correctly, analog-digital-analog.

Unlike digital-analog systems, where the continuous units are only present on output, the analog-digital systems cannot be considered generalized impulse filters with input and output signals as lattice time functions. The output signal of analog-digital systems depends on values of the input signal at any moment of continuous time, but not just in clock points $t = kT$. Therefore, such a system cannot be adequately described by the difference equation or by the discrete transfer function, in which the research methods in time or in frequency domain create the basis for linear impulse classical theory and linearized digital-analog systems.

Note, that the majority of digital control systems are actually analog-digital systems, because the actions, applied to them, and control error usually correspond to continuous time functions of its physical nature. Even if the sensor used in a system, has a digital output, the measurement does not last instantaneous and is connected to accumulation of information on the quantity measured during some time interval, or preliminary filtering with the purpose of noise suppression and other dynamic procedures. The time delay of the smoothing device for the measured signal before the analog-to-digital conversion, serving for narrowing the spectrum of this signal subjected to quantization in time is often added to its own time delay with the use of an analog sensor. Thus, the continuous unit is often situated before the digital part of a system.

The automatic systems, whose sensors by principle of operation or by virtue of particular kinematical relations react not to input action itself, but to some of its dynamic transformation — the derivative, integral, etc. and are analog-digital in essence. For example, the inertial gauge measures the object acceleration along the sensitivity axis, i. e. the linear displacement second derivative. The Doppler radio engineering gauge, the aerial velocity gauge, and the Doppler log measure the components of the linear displacement rate. The two-power gyroscopes measure the angular velocities, and laser gyroscopes with digital output measure the increment of tilt angles during the clock time, etc.

At the same time, if the input dynamic unit can be considered inertialess within the first approximation, then it has been necessarily done in order to simplify the analysis and the system is researched as digital-analog.

**Research of digital-analog systems**. The dynamic properties of open loop digital-analog system are characterized by its discrete transfer function

$$W(z) = D(z)\frac{\delta_d}{\delta_a} W_0(z),$$

where

$$W_0(z) = \frac{z-1}{z} Z\left\{\frac{W_0(s)}{s}\right\} \tag{9.17}$$

is the discrete transfer function of a reduced continuous part.

The appropriate frequency transfer function of an open loop system looks like: $W\left(e^{j\omega T}\right) = W(z)\big|_{z=e^{j\omega T}}$.

Passing to absolute pseudo-frequency $\lambda = \frac{2}{T}\operatorname{tg}\frac{\omega T}{2}$, it is possible to obtain the frequency transfer function as:

$$W^*(j\lambda) = W(z)\Big|_{z=\frac{1+j\lambda T/2}{1-j\lambda T/2}}.$$

Unlike the function $W\left(e^{j\omega T}\right)$ it contains the fractional rational kind and is not periodic.

The use of the frequency transfer function of an open loop system $W^*(j\lambda)$ allows for research of the digital robust system on the basis of frequency methods, developed according to continuous systems.

Consider, for example, the problem of the dynamic error dispersion restriction

$$D_{eg} = \frac{T}{\pi} \int_0^{\infty} \frac{\left| H_e^*(j\lambda) \right|^2 S_g^*(\lambda)}{1+\lambda^2 T^2/4} d\lambda \tag{9.18}$$

at unknown spectral density of reference action $S_g^*(\lambda)$. Here $H_g^*(j\lambda) = \left[1 + W^*(j\lambda)\right]^{-1} = H^*(j\lambda)/W^*(j\lambda)$ is the frequency transfer function of a closed-loop system for an error, and $H^*(j\lambda)$ is the frequency transfer function of a closed-loop system.

The dispersions of differences of the orders from $K$ up to $N$, $N \geq K \geq 0$, of the lattice centered casual process $g[k] = g(t)\big|_{t=kT}$, $k = 0, 1, \ldots$ are considered known. They can be obtained either by direct research of lattice process $g[k]$ properties or by dispersions derivatives of its continuous enveloping (see §9.1).

Differences in dispersions of reference action are considered as the generalized moments of its spectral density

$$D_{\nabla i} = \frac{T}{\pi} \int_0^{\infty} \left( \frac{\lambda^2 T^2}{1+\lambda^2 T^2/4} \right)^i \frac{S_g^*(\lambda)}{1+\lambda^2 T^2/4} d\lambda, \; i = \overline{K,N} \,. \tag{9.19}$$

Let the values of coefficients $\{c_i\}_K^N$ be found at some frequency transfer function of a closed-loop system for an error $H_e^*(j\lambda)$. They ensure the realization of inequality

$$\sum_{i=K}^{N} c_i \left( \frac{\lambda^2 T^2}{1+\lambda^2 T^2/4} \right)^i \geq \left| H_e^*(j\lambda) \right|^2, \tag{9.20}$$

at least in pseudo-frequencies area, where the spectral components of reference action are possible.

Then the following formula is valid for an upper-bound estimate of dynamic error dispersion:

$$\overline{D}_{eg} = c_K D_{\nabla K} + c_{K+1} D_{\nabla(K+1)} + \ldots + c_N D_{\nabla N} \,. \tag{9.21}$$

From (9.18) — (9.21) it is possible to derive the requirement to a system frequency transfer function, at which realization of the dynamic error dispersion does not exceed the given admissible value $D_e^0$, i. e. $\overline{D}_{eg} \leq D_e^0$ :

$$\left|H_e^*(j\lambda)\right|^2 \le \frac{D_e^0}{\sum_{i=K}^{N} c_i D_{\nabla i}} \left(\frac{\lambda^2 T^2}{1+\lambda^2 T^2/4}\right)^K \times$$
$$\times \left[c_K + c_{K+1}\frac{\lambda^2 T^2}{1+\lambda^2 T^2/4} + \ldots + c_N \left(\frac{\lambda^2 T^2}{1+\lambda^2 T^2/4}\right)^{N-K}\right]. \tag{9.22}$$

By factorizing the right hand side of inequality (9.22) and passing to the gain plot of an open loop system by the formula $\left|W^*(j\lambda)\right| = \left|H^*(j\lambda)/H_e^*(j\lambda)\right|$, this inequality can be rewritten as following:

$$\frac{\left|W^*(j\lambda)\right|}{\left|H^*(j\lambda)\right|} \ge$$
$$\ge \frac{\sqrt{\left[\gamma_K^2 D_{\nabla K} + \left(\gamma_{K+1}^2 - 2\gamma_K \gamma_{K+2}\right) D_{\nabla(K+1)} + \ldots + \gamma_K^2 D_{\nabla K}\right] / D_e^0}}{\left(\frac{\lambda T}{\sqrt{1+\lambda^2 T^2/4}}\right)^K \left|\gamma_K + \gamma_{K+1}\frac{j\lambda T}{\sqrt{1+\lambda^2 T^2/4}} + \ldots + \gamma_N \left(\frac{j\lambda T}{\sqrt{1+\lambda^2 T^2/4}}\right)^{N-K}\right|}, \tag{9.23}$$

where the real coefficients $\{\gamma_i\}_K^N$ can accept any non-negative values.

The relation (9.23) allows for the formulation of the requirements of the gain plot of an open loop system and therefore to a stability margin of a closed-loop system characterized by an oscillations coefficient $M = \max_{\lambda}\left|H^*(j\lambda)\right|$ (see § 7.1). Thus it is possible to construct the forbidden and absolutely forbidden areas for the Bode diagram of an open loop system expressing the sufficient or only necessary conditions for the obtainment of given accuracy.

If the discretization period $T$ is small enough and at $\lambda = 2/T$ the right hand part in inequality (9.23) is less than one, then the forbidden areas for the Bode diagram, appropriate to this inequality, coincide by their shape with the forbidden areas constructed in § 7.2 for Bode diagrams of continuous systems. The expansion of the discretization period causes the non-linear deformation of boundaries for forbidden areas along the abscissa axis in the direction of forbidden areas expansion.

After the selection of an acceptable Bode diagram, passing outside the forbidden area, it is necessary to write the frequency transfer function $W^*(j\lambda)$, appropriate to it, and to find the discrete transfer function of an open loop system by the formula

$$W(z) = W^*(j\lambda)\Big|_{j\lambda = \frac{2}{T}\frac{z-1}{z+1}}.$$

It is necessary to note, that the conditions for restriction of dynamic error dispersion, similar in a sense to (9.23), can be obtained even with the use of regular circular frequency $\omega$, if the expression written according to (9.7) and periodicity property of function $H_e\left(e^{j\omega T}\right)$ is used as the basic expression instead of (9.18):

$$D_{eg} = \frac{T}{\pi}\int_0^{\pi/T}\left|H_e\left(e^{j\omega T}\right)\right|^2 S_{g\perp}\left(e^{j\omega T}\right)d\omega =$$
$$= \frac{T}{\pi}\int_0^{\pi/T}\left|H_e\left(e^{j\omega T}\right)\right|^2 T^{-1}\sum_{r=-\infty}^{\infty} S_g(\omega + 2\pi r/T)d\omega = \frac{1}{\pi}\int_0^{\infty}\left|H_e\left(e^{j\omega T}\right)\right|^2 S_g(\omega)d\omega. \quad (9.24)$$

It is possible to use the dispersions of derivatives, instead of differences of action, having exchanged (9.20) and (9.21) by expressions

$$\sum_{i=K}^{N} c_i\omega^{2i} \geq \left|H_e\left(e^{j\omega T}\right)\right|^2, \; \overline{D}_{eg} = \sum_{i=K}^{N} c_i D_i .$$

when determining the upper-bound estimate for the value $D_{eg}$ because the spectral density of initial continuous action $S_g(\omega)$ is considered here. However, the transcendental kind of function $\left|H_e\left(e^{j\omega T}\right)\right|$ makes it impossible to realize the system synthesis, constructing the simple form forbidden areas in the Bode diagram plane.

The amount of quantization noises by level in digital systems causes the necessity to expand the forbidden areas for the Bode diagram for the given accuracy conditions, because the admissible contribution of the dynamic component in a total error decreases. Analysis methods for control the error component at the expense of quantization by level in the analog-to-digital and digital-to-analog converters are described in [7, 60]. Underline only, that it is necessary to take into consideration, even a very small time lag in a digital part $\tau$ with such an investigation (see Fig. 9.1). The reason is that the value of quantization noise $v_a[k]$ or $v_d[k]$ can pass through the principal feedback circuit of a system and return to a start point at $\tau > 0$ no earlier than at the following moment of discrete time $k+1$. But quantization noise is the discrete white noise and two of its serial values are statistically and mutually irrelevant. Therefore the principal feedback in a digital system can increase the r.-m.-s. error from quantization noise or from another broadband action. The indicated effect is not revealed by the analysis if $\tau = 0$.

**Research of digital-analog systems.** The error dispersion of an analog-digital system is expressed through the spectral density of action by a more complicated formula than in (9.18) or (9.24). It is connected so that the spectral component of action at arbitrary frequency $\omega$ gives the spectral components of error on frequencies $\omega + 2\pi r/T$, $r = 0, \pm 1, \pm 2, \ldots$ owing to the effect of spectrum "reproduction" at quantization in time domain, and the power of each of them depends on a periodic frequency characteristic of a digital part of a system and on a nonperiodic frequency characteristic of an cntering continuous unit [60].

The indicated circumstance hampers the direct use of the explained above frequency research methods of error dispersion on dispersions of differences or on derivatives of action. However, it is possible to point two modes of overcoming these difficulties by approximated investigation.

The first way consists of approximated replacement of an analog-digital system by a digital-analog one, in which the discrete unit with close dynamic properties is included instead of an entering continuous unit with a transfer function $W_d(s)$. The frequency transfer function of such a unit can be found, for example, under the formula $W_d^*(j\lambda) = W_d(s)\big|_{s=j\lambda}$. The investigation of obtained equivalent digital-analog system is then further fulfilled.

The second way is produced when a continuous unit with dynamic properties that are close to it approximately replaces the digital part of a system, and the accuracy analysis of the obtained equivalent continuous system is carried out. Solving synthesis problems fulfills the inverse operations. At first the desirable frequency transfer function of an open loop robust system $W_{des}(j\omega)$ should be determined and the frequency transfer function of the continuous prototype for a digital part of the system is determined by the formula

$$W_{cp}(j\omega) = \frac{W_{des}(j\omega)}{W_d(j\omega)W_c(j\omega)}.$$

Further the discrete transfer function for the digital part of a system $D(z)$ must be selected to ensure the proximity of digital system dynamic properties and their continuous prototype. The methods of such problem solving for the discrete approximation of the continuous filter — prototype are well developed [40]. In particular, it is possible to use the elementary method of bilinear transformation, which gives the formula

$$D(z) = \frac{\delta_a}{\delta_d} W_{cp}\left(\frac{2}{T}\frac{z-1}{z+1}\right). \tag{9.25}$$

It is necessary to note, that the digital filter obtained according to (9.25) has the same order, as the order of initial continuous filter, i.e. a minimally possible one. Unlike the more complicated methods of discrete approximation, it ensures the advantage not only in simplification of digital calculator operation algorithms, but also in smaller criticality of filter dynamic properties to errors of its coefficients adjustment, connected to a finite width of a digit grid for the digital calculator. With a lower order of digital filters and at larger discretization period, such criticality is lowered. The last circumstance is especially essential at microprocessor realization of the digital part of a system.

## §9.3. Synthesis of digital robust functional elements of dynamic systems

**Robust prediction and smoothing filters.** The problems of stationary linear filtering or the prediction of lattice casual process $g[k]$, observable in a mixture with an additive interference $v[k]$, are easily solved analytically with the representation of spectral densities of both of these processes or by the representation of signal $g[k]$ as a linear combination of known functions with casual quasiconstant coefficients [7]. However, the deviation of actual properties of actions from the obtained models can cause the depreciation of a produced synthesis. Apply the synthesis problem of robust digital filter to the incomplete a priory information.

Let the lattice function $g[k]$ correspond to the discrete readings of initial continuous process $g(t)$, i.e. $g[k]=g(t)|_{t=kT}$, where $T$ is the discretization period, $k=0,1,\ldots$. It is necessary to consider two versions of giving the process $g(t)$ dynamic properties representation and, therefore, the process $g[k]$. In the first version the dispersions of several derivatives of process are considered known, i.e. the known even power moments of its spectral density are:

$$D_i = \frac{1}{\pi}\int_0^\infty \omega^{2i} S_g(\omega)\, d\omega,\ i=\overline{K,N},\ N \ge K \ge 0 .$$

In the second version the known magnitudes are the maximum values of these derivatives $\left|g^{(i)}(t)\right| \le g_M^{(i)}$. The interference $v[k]$ is considered as discrete white noise, for which in the first version the dispersion $D_v$, and in second one — the maximum value $|v[k]| \le v_M$ are preset.

Let the synthesis be produced in the class of nonrecursive pulsed filters of order $m$, i.e. the observable process $u[k]=g[k]+v[k]$ is supposed to be handled as:

$$y[k]=g[k+l]=b_0 u[k]+b_1 u[k-1]+\ldots+b_m u[k-m], \qquad (9.26)$$

where $y[k]$ is the filter current output value, and $l$ is an interval of prediction (at $l$ > 0). At $l=0$ the problem of filtering is set, at $l<0$ — the problem of smoothing (interpolation). The magnitude $m$ characterizes the admissible complicity of the discrete filter.

It is required to select the coefficients $\{b_i\}_0^m$ by an optimum way. The minimization of dispersion of an upper-bound estimate (in the first version) or maximum (in the second version) total error

$$e[k]=g[k+l]-g[k+l]=e_g[k]+e_v[k]. \qquad (9.27)$$

is the optimal criterion. Here the dynamic error $e_g[k]$ and error $e_v[k]$ from interference according to (9.26) are determined as

$$e_g[k]=g[k+l]-\sum_{i=0}^{m} b_i g[k-i]e_v[k]=-\sum_{i=0}^{m} b_i v[k-i].$$

The dispersion and maximum value of error from an uncorrelated interference is expressed by the convenient formulas

$$D_{ev}=\left(b_0^2+b_1^2+\ldots+b_m^2\right)D_v,\ e_{vM}=\left(|b_0|+|b_1|+\ldots+|b_m|\right)v_M. \tag{9.28}$$

For dynamic error investigation it is necessary to write the discrete transfer function for an error appropriate to algorithm (9.26)

$$H_e(z)=z^l-b_0-b_1z^{-1}-\ldots-b_mz^{-m}, \tag{9.29}$$

correlating the z-images of error and the input signal by the relation $E_g(z)=H_e(z)G(z)$.

Passing into the frequency domain, the frequency transfer function $H_e\left(e^{j\omega T}\right)=H_e(z)\big|_{z=e^{j\omega T}}$ is obtained from (9.29). The quadrate of its module after the long transformations can be presented as:

$$\begin{aligned}\left|H_e\left(e^{j\omega T}\right)\right|^2&=1+\sum_{i=0}^{m} b_i^2-2\sum_{i=0}^{m} b_i\cos[(l+i)\omega T]+\\&+2\sum_{v=1}^{m}\sum_{i=v}^{m} b_{i-v}b_i\cos[v\omega T].\end{aligned} \tag{9.30}$$

It is possible to use expression (9.30), as well as expression (9.29), with the determination of error characteristics considered only in clock points, i.e. as lattice casual process $e[k]=e(t)\big|_{t=kT}$.

**Research of error dispersion.** Let factors of some polynomial $C_{2N}(\omega)=\sum_{i=K}^{N} c_i\omega^{2i}$ be selected, so that the inequality $C_{2N}(\omega)\geq\left|H_e\left(e^{j\omega T}\right)\right|^2$ is fulfilled at $|\omega|\leq\pi/T$. Then according to (9.7) and known spectral density power moments of input signal for dynamic error dispersion, the following formula is valid:

$$\begin{aligned}D_{eg}&=\frac{T}{\pi}\int_0^{\pi/T}\left|H_e\left(e^{j\omega T}\right)\right|^2 S_{g\perp}\left(e^{j\omega T}\right)d\omega=\frac{1}{\pi}\int_0^{\infty}\left|H_e\left(e^{j\omega T}\right)\right|^2 S_g(\omega)d\omega\leq\\&\leq\frac{1}{\pi}\int_0^{\infty}C_{2N}(\omega)S_g(\omega)d\omega=\sum_{i=K}^{N} c_iD_i=\overline{D}_{eg}.\end{aligned} \tag{9.31}$$

If the coefficients $\{c_i\}_K^N$ obey to criterion $\sum_{i=K}^{N} c_i D_i \to \min$, then the upper-bound estimates $\overline{D}_{eg}$ expressed by (9.31) are the strongest.

In a characteristic case, when the discretization period is much less than the action correlation interval $g(t)$, the optimal values of coefficients $\{c_i\}_K^N$ practically coincide with the coefficients of function $\left|H_e\left(e^{j\omega T}\right)\right|^2$ with expansion in the McLoren series by powers ω. Then the following is obtained:

$$\begin{gathered} c_0 = 1 + \sum_{j=0}^{m} b_j\left(b_j - 2\right) + 2\sum_{v=1}^{m}\sum_{j=v}^{m} b_{j-v} b_j, \\ c_i = (-1)^{(i+1)} \frac{2T^{2i}}{(2i)!}\left(\sum_{j=0}^{m}(l+j)^2 b_j - \sum_{v=1}^{m}\sum_{j=v}^{m} v^{2i} b_{j-v} b_j\right), i = \overline{K,N}. \end{gathered} \tag{9.32}$$

The error dispersion will have the finite value only when the coefficients $c_i$ with indexes $i < K$ are equal to zero. Formula (9.32) for the coefficient $c_N$ is valid only at an odd *N*, because the realization of condition $c_N \geq 0$ is necessary. Generally, for determination of optimal values of coefficients $\{c_i\}_K^N$ it is necessary to use the numerical analysis algorithms described in § 5.3.

The expressions (9.28), (9.31) and (9.32) allow writing the object function of factors optimization $\{b_j\}_0^m$ as:

$$\overline{D}_e\left(\{b_j\}_0^m\right) = \overline{D}_{eg}\left(\{b_j\}_0^m, \{D_i\}_K^N\right) + D_{ev}\left(\{b_j\}_0^m, D_v\right) \to \min \tag{9.33}$$

in an explicit kind and to find the solution for the considered problem of robust digital filter synthesis by analytical or numerical methods.

In the elementary case of prediction on one clock period at $l = 1,\ m = 1,\ N = K = 1,\quad c_0 = (1 - b_0 - b_1)^2 = 0$ expression (9.33) becomes $(2 - b_0)^2 T^2 D_1 + \left(2b_0^2 - 2b_0 + 1\right)D_v \to \min$, whence

$$b_{0\text{opt}} = \left(2T^2 D_1 + D_v\right)/\left(T^2 D_1 + 2D_v\right), b_{1\text{opt}} = 1 - b_{0\text{opt}}. \tag{9.34}$$

At $D_v \to 0$ from (9.34) the following is obtained: $b_{0\text{opt}} = 2,\ b_{1\text{opt}} = -1$, at $D_v \to \infty$ – $b_{0\text{opt}} = b_{1\text{opt}} = 1/2$, which corresponds to the physical sense of the problem.

**Maximum error investigation.** At first it is necessary to determine the expression for a dynamic error component in a function of reciprocal differences $\nabla g[k] = g[k] - g[k-1],\ \nabla^2 g[k] = \nabla g[k] - \nabla g[k-1], \ldots$

Z-images of reciprocal $i$-th order difference and the input signal are connected by a discrete transfer function $H_{\nabla i}(z)=\left(1-z^{-1}\right)^i$. When $i=1$ it is possible to express $z^{-1}=1-H_{\nabla i}(z)$ and after its substitution into (9.29) it is possible to write:

$$\begin{aligned} H_e(z)=\{1-b_0[1-H_{\nabla 1}(z)]^l-b_1[1-H_{\nabla 1}(z)]^{l+1}-\ldots \\ \ldots-b_m[1-H_{\nabla 1}(z)]^{l+m}\}z^l. \end{aligned} \tag{9.35}$$

Taking into account, that $[H_{\nabla 1}(z)]^i=H_{\nabla i}(z)$, it is necessary to put the difference equations

$$e_g[k-l]=\beta_0 g[k]+\beta_1\nabla g[k]+\beta_2\nabla^2 g[k]+\ldots+\beta_{l+m}\nabla^{l+m}g[k], \tag{9.36}$$

where

$$\beta_0=1-\sum_{j=0}^{m}b_j, \beta_i=\frac{(-1)^{i+1}}{i!}\sum_{j=\max\{0,i-l\}}^{m}\frac{(l+j)!b_j}{(l+j-i)!}, i=\overline{1,l+m}, \tag{9.37}$$

into accordance to transfer functions (9.35).

As it is shown in §9.1, the differences of lattice process $g[k]=g(t)|_{t=kT}$ are also restricted by the inequalities in $\left|\nabla^i g[k]\right|\le\nabla^i_M$, $i=K, K+1,\ldots$, where $\nabla^i_M\le g^{(i)}_M T^i$ at restricted maximum values of continuous process derivative $g(t)$. Therefore, having assumed $\beta_i=0$ for $i<K$, the upper-bound estimate of maximum dynamic error $e_{gM}$ is obtained from (9.36) as

$$e_{gM}\le\sum_{i=K}^{l+m}|\beta_i|\nabla^i_M=\bar{e}_{gM}. \tag{9.38}$$

At $l+m-K=0$ and $l+m-K=1$, when the sum in the right hand side of an inequality (9.38) contains not more than two members, this inequality turns into an equality, i.e. the maximum error estimation is precise. At $l+m-K=2$ the precise upper-bound estimate according to formula (6.36) then looks like:

$$\bar{e}_{gM}=|\beta_M|\frac{2\nabla^K_M\nabla^{K+2}_M-\left(\nabla^{K+1}_M\right)^2}{2\nabla^{K+2}_M}+|\beta_{K+1}|\nabla^{K+1}_M+|\beta_{K+2}|\nabla^{K+2}_M, \tag{9.39}$$

if $\nabla^{K+2}_M/\nabla^{K+1}_M\ge\beta_K/\beta_{K+1}$.

With use of formulas (9.28), (9.37) - (9.39) the object function of factors $\{b_j\}_0^m$ optimization

$$\bar{e}_M\left(\{b_j\}_0^m\right)=\bar{e}_{gM}\left(\{b_j\}_0^m,\{\nabla^i_M\}_K^{l+m}\right)+e_{\upsilon M}\left(\{b_j\}_0^m,\upsilon_M\right)\to\min \tag{9.40}$$

can be written in an explicit way and the solution of the synthesis problem can be found by the numerical methods.

In the case of prediction for one clock period with known maximum values of differences of the first and second orders, when $l = 1$, $m = 1$, $K = 1$, $N = 2$, $\beta_0 = 1 - b_0 - b_1 = 0$, expression (9.40) becomes

$$1 + |b_1|\nabla_M + |b_1|\nabla_M^2 + (|b_0| + |b_1|)\, \upsilon_M \to \min ,$$

whence the following is obtained: $b_{0\text{opt}} = 2$, $b_{1\text{opt}} = -1$ at $\upsilon_M \le (\nabla_M - \nabla_M^2)/2$ and $b_{0\text{opt}} = 1$, $b_{1\text{opt}} = 0$ at $\upsilon_M \ge (\nabla_M - \nabla_M^2)/2$. Note that at an unknown maximum second order difference, based on (9.16) it is possible to assume $\nabla_M^2 = 2\nabla_M$, it is also obvious that $b_{0\text{opt}} = 1$, $b_{1\text{opt}} = 0$.

If the maximum values of the first, second and third order differences are known, at $l = 1$, $m = 2$, $K = 1.3$, $\beta_0 = 1 - b_0 - b_1 - b_2 = 0$ then expression (9.40) becomes

$$\begin{aligned} &|1 + b_1 + 2b_2|\left[2\nabla_M c - (\nabla_M^2)^2\right](2\nabla_M^3)^{-1} + |b_1 + 3b_2|\nabla_M^2 + \\ &+ |b_2|\nabla_M^3 + (|b_0| + |b_1| + |b_2|)\upsilon_M \to \min, \end{aligned}$$

Whence it is obtained:

$$b_{0\text{opt}} = 3,\ b_{1\text{opt}} = -3,\ b_{2\text{opt}} = 1,\quad \bar{e}_M\left(\{b_{j\text{opt}}\}_0^2\right) = \nabla_M^3 + 10\upsilon_M$$

at $\upsilon_M \le \min\left\{\dfrac{\nabla_M - \nabla_M^3}{9}, \dfrac{3\nabla_M^2 - \nabla_M^3}{16}, \dfrac{\nabla_M^2 - \nabla_M^3}{7}\right\}$;

$$b_{0\text{opt}} = 2,\ b_{1\text{opt}} = -1,\ b_{2\text{opt}} = 0,\quad \bar{e}_M\left(\{b_{j\text{opt}}\}_0^2\right) = \nabla_M^2 + 3\upsilon_M$$

at $\dfrac{\nabla_M^2 - \nabla_M^3}{7} \le \upsilon_M \le \min\left\{\dfrac{\nabla_M - \nabla_M^2}{2}, \dfrac{\nabla_M^2 + \nabla_M^3}{2}\right\}$;

$$b_{0\text{opt}} = 1.5,\ b_{1\text{opt}} = 0,\ b_{2\text{opt}} = -0.5,\ \bar{e}_M\left(\{b_{j\text{opt}}\}_0^2\right) = 1.5\nabla_M^2 + 0.5\nabla_M^3 + 2\upsilon_M$$

at $(\nabla_M^2 + \nabla_M^3)/2 \le v_M \le \nabla_M - 1.5\nabla_M^2 - 0.5\nabla_M^3$;

$$b_{0\text{opt}} = 1, b_{1\text{opt}} = b_{2\text{opt}} = 0,\quad \bar{e}_M\left(\{b_{j\text{opt}}\}_0^2\right) = \nabla_M + \upsilon_M$$

at $\upsilon_M \ge \max\left\{\dfrac{\nabla_M - \nabla_M^3}{9}, \dfrac{\nabla_M - \nabla_M^2}{2}, \nabla_M - 1.5\nabla_M^2 - 0.5\nabla_M^3\right\}$.

With the synthesis by criterion of maximal error minimization the frequency methods of its estimations described in § 6.5 can also be used. Basically the class of considered filters can be expanded at the expense of connection of the recursive filters to it. However, it may cause sophisticated evaluations.

**Robust differentiating filters.** The digital device, producing the estimate $\dot{g}_{ss}[k]$ of the initial continuous process $g(t)$ derivative based on discrete readings of this process $g[k]=g(t)|_{t=kT}$, observable in a mixture with interference $\upsilon[k]$, is called the digital differentiating filter. Similar devices are included in the structure of digital systems of combined control on error and on reference action (see §2.2), motion control systems, navigation systems and many others [7, 22, 60, 76]. The interference $\upsilon[k]$ is usually stipulated by noise of measurement and quantization by a level noise with useful signal transformation to a digital form.

The complexity of digital differentiating filters synthesis even at the full a priori information of properties of processes $g[k]$ and $\upsilon[k]$ is connected to the derivation operation for lattice time functions, which are not defined. Therefore it is impossible to enter "an ideal conversing operator", as it was done, for example, by the synthesis of predicting or smoothing digital Wiener filters [7, 8, 60].

It should be considered with linear robust filters investigation, that the dispersions $\{D_i\}_K^N$ of several derivatives of initial continuous action $g(t)$ are known. The interference $\upsilon[k]$ is assumed as the discrete white noise with the dispersion $D_v$. Stationary lattice casual processes $g[k]$ and $\upsilon[k]$ are considered centered and mutually not correlated.

Let the synthesis be produced in the class of nonrecursive digital filters of the order $m$, realizing the algorithm of a kind

$$\dot{g}_{ss}[k]=b_0u[k]+b_1u[k-1]+\ldots+b_mu[k-m], \tag{9.41}$$

where $u[k]=g[k]+\upsilon[k]$ is the observable process, $\{b_i\}_0^m \in(-\infty,\infty)$. It is necessary to consider the problem of determining the optimum values of coefficients $\{b_i\}_0^m$ by the criterion of minimum of upper-bound estimate $\overline{D}_e$ dispersions for a total error of derivation

$$e[k]=\dot{g}(t)|_{t=kT}-\dot{g}_{ss}[k]=e_g[k]+e_\upsilon[k].$$

Here $e[k]=\dot{g}(t)|_{t=kT}-\sum_{i=0}^{m}b_ig[k-i]$ is the methodical error, which is defined by an irregularity of the accepted algorithm (9.41) and dynamics of differentiable process variation, $e_\upsilon[k]=-\sum_{i=0}^{m}b_i\upsilon[k]$ is an error from interference with the dispersion

$$D_{e\upsilon}=\left(b_0^2+b^2\quad{}_1+\ldots+b_m^2\right)D_\upsilon. \tag{9.42}$$

The upper-bound estimate of methodical error dispersion $\overline{D}_e$ can also be expressed through the coefficients $\{b_i\}_0^m$ and values $\{D_i\}_K^N$ (see below). The object optimization function can then be written as

$$\overline{D}_e\left(\{b_i\}_0^m\right)=\overline{D}_{eg}\left(\{b_i\}_0^m,\{D_i\}_K^N\right)+D_{e\upsilon}\left(\{b_i\}_0^m,D_\upsilon\right)\to \min \tag{9.43}$$

in an explicit kind as function of $m+1$ variables and their optimum values $\{b_{i\text{opt}}\}_0^m$ as a result of numerical or analytical investigation of (9.43) on an extremum can be found.

The number of independent variables in (9.43) can be reduced with the superposition of the additional requirements to the filter dynamic properties. For example, if zero methodical error at stationary and at linearly varying functions $g[k]$ is required, then for the coefficients of algorithm (9.41) it is easy to write the conditions

$$b_0+b_1+\ldots+b_m=0, \tag{9.44}$$

$$b_1+2b_2+\ldots+mb_m=-1/T, \tag{9.45}$$

that are sufficient for the realization of the indicated requirements.

**Methodical error of digital differentiation**. For methodical error investigation it is necessary to enter the discrete transfer function

$$H(z)=b_0+b_1z^{-1}+\ldots+b_mz^{-m}=0 \tag{9.46}$$

which is appropriate to algorithm (9.41), and the lattice pulse response $w[k]=Z^{-1}\{H(z)\}$.

Using the convolution formula, the expressions for methodical error values at the moment of discrete time $k$ and $k+m$ can be written as:

$$e_g[k]=\dot{g}(t)\big|_{t=kT}\ \sum_{i=0}^{k}w[i]g[k-i],$$

$$e_g[k+m]=\dot{g}(t)\big|_{t=kT}-\sum_{j=0}^{k+m}w[j]g[k+m-j].$$

Having multiplied the left and right hand sides of these equalities, changing the order of toting and having averaged out discrete time $k$, the following expressions are developed for the correlation function of methodical error:

$$\begin{aligned}R_{eg}[m]=\overline{e_g[k]e_g[k+m]}=R_{\dot{g}}[m]-\sum_{i=0}^{\infty}w[i]R_{\dot{g}g}[m+i]-\\-\sum_{j=0}^{\infty}w[j]R_{g\dot{g}}[m+j]+\sum_{i=0}^{\infty}w[i]\sum_{j=0}^{\infty}w[i]R_g[m-j+i],\end{aligned} \tag{9.47}$$

where $R_{\dot{g}}[m]$ is the correlation function of process $\dot{g}[k] = g(t)|_{t=kT}$, $R_g[m]$ is the correlation function of process $g[k]$, and $R_{\dot{g}g}[m]$ and $R_{g\dot{g}}[m]$ are mutual correlation functions of the indicated lattice casual processes.

The bilateral z-transformation of correlation function (9.47) gives the spectral density of lattice process $e_g[k]$

$$
\begin{aligned}
S_{eg\perp}(z) = \sum_{m=-\infty}^{\infty} R_{eg}[m]z^{-m} = S_{\dot{g}\perp}(z) - H(z^{-1})S_{\dot{g}g\perp}(z) - \\
- H(z)S_{g\dot{g}\perp}(z) + H(z)H(z^{-1})S_{g\perp}(z).
\end{aligned}
\tag{9.48}
$$

Passing into the frequency domain by the substitution $z = e^{j\omega T}$ and taking into account, that the spectral densities $S_{\dot{g}\perp}(e^{j\omega T})$ and $S_{g\perp}(e^{j\omega T})$ are the real functions, mutual spectral densities $S_{\dot{g}g\perp}(e^{j\omega T})$ and $S_{g\dot{g}\perp}(e^{j\omega T})$ are the imaginary functions, and $S_{\dot{g}g\perp}(e^{j\omega T}) = -S_{g\dot{g}\perp}(e^{j\omega T}) = j\,\mathrm{Im}\,S_{g\dot{g}\perp}(e^{j\omega T})$ [39], expression (9.48) can be converted to a kind:

$$
S_{eg\perp}(e^{j\omega T}) = S_{\dot{g}\perp}(e^{j\omega T}) - 2\,\mathrm{Im}\,H(e^{j\omega T})\,\mathrm{Im}\,S_{\dot{g}g\perp}(e^{j\omega T}) + \left|H(e^{j\omega T})\right|^2 S_{g\perp}(e^{j\omega T}). \tag{9.49}
$$

Expressing the spectral densities of lattice casual processes $S_{g\perp}(e^{j\omega T})$, $S_{\dot{g}\perp}(e^{j\omega T})$ and $S_{\dot{g}g\perp}(e^{j\omega T})$ included in (9.49) through the spectral densities of appropriate initial continuous processes $S_g(\omega), S_{\dot{g}}(\omega) = \omega^2 S_g(\omega)$ and $S_{\dot{g}g}(\omega) = j\omega S_g(\omega)$ with use of formula (9.7), expression (9.49) gives:

$$
\begin{aligned}
S_{eg\perp}(e^{j\omega T}) &= \sum_{r=-\infty}^{\infty} \Big[ (\omega + r\omega_\perp)^2 - 2(\omega + r\omega_\perp)\,\mathrm{Im}\,H(e^{j\omega T}) + \\
&+ \left|H(e^{j\omega T})\right|^2 S_g(\omega + r\omega_\perp)T^{-1} \Big] = \\
&= \sum_{r=-\infty}^{\infty} \left| j(\omega + r\omega_\perp) - H(e^{j\omega T}) \right|^2 S_g(\omega + r\omega_\perp)T^{-1},
\end{aligned}
\tag{9.50}
$$

where $\omega_\perp = 2\pi/T$ is the quantization frequency.

The integration of expression (9.50) with the account of function $H(e^{j\omega T})$ periodicity allows after transformations for the development of the formula for dispersion of methodical error

$$D_{eg} = \frac{T}{2\pi} \int_{-\pi/T}^{\pi/T} S_{eg\perp}\left(e^{j\omega T}\right) d\omega = \frac{1}{2\pi} \sum_{r=-\infty}^{\infty} \int_{-\pi/T+r\omega_\perp}^{\pi/T+r\omega_\perp} \left| j\omega - H\left(e^{j\omega T}\right) \right|^2 S_g(\omega) d\omega =$$

$$= \frac{1}{\pi} \int_{0}^{\infty} \left| j\omega - H\left(e^{j\omega T}\right) \right|^2 S_g(\omega) d\omega. \qquad (9.51)$$

The context of formula (9.51) is quite clear because $j\omega$ is a frequency transfer function of the ideal continuous differentiating device, and $H\left(e^{j\omega T}\right)$ is a frequency transfer function of the digital differentiating filter. Nevertheless it is not trivial, because it expresses the error dispersion of digital differentiation considered as a lattice casual process, through the spectral density of a continuous differentiable process.

This means that only the even power moments of spectral density $S_g(\omega)$ (dispersions of action derivatives $\{D_i\}_K^N$) are known, and from (9.51) the upper-bound estimate of methodical error dispersion $\overline{D}_{eg} = \sum_{i=K}^{N} c_i D_i$ can be developed by the approximative method. Here $\{c_i\}_K^N$ are the coefficients of polynomial $C_{2N}(\omega) = \sum_{i=K}^{N} c_i \omega^{2i}$, corresponding to condition

$$C_{2N}(\omega) \geq \left| j\omega - H\left(e^{j\omega T}\right) \right|^2, \omega \in [0, \infty).$$

It is possible to obtain estimation $\overline{D}_{eg}$, which is elementary, but usually close to the strongest one when determining coefficients $\{c_i\}_K^N$ as the coefficients of function $\left| j\omega - H\left(e^{j\omega T}\right) \right|^2$ expansion in the McLoren series by powers $\omega^2$.

**Example of robust digital differentiation.** When $m = 2$, $N = K = 2$, it is required to optimize the differentiating filter by criterion (9.43) with the additional restrictions (9.44) and (9.45).

Having assumed $H\left(e^{j\omega T}\right) = b_0 + b_1 e^{-j\omega T} + b_2 e^{-j2\omega T}$, using the expansion of the exponent in the power series at $\omega T < 1$ after the transformations the following is obtained:

$$\left| j\omega - H\left(e^{j\omega T}\right) \right| \cong \alpha_0 + \alpha_1 \omega^2 + \alpha_2 \omega^4,$$

where

$$\alpha_0 = (b_0 + b_1 + b_2)^2,$$

$$\alpha_1 = 1 + 2(b_1 + 2b_2)T - (b_0 b_1 + b_1 b_2 + 2b_0 b_2)T^2,$$
$$\alpha_2 = -(b_1 + 8b_2)T^3/3 - (b_0 b_1 + b_1 b_2 + 16 b_0 b_2)T^4/12$$

or, according to (9.44) and (9.45), $\alpha_0 = 0,\ \alpha_1 = 0,\ \alpha_2 = \left(\frac{3}{2T} - b_0\right)^2 T^4$. Consider, that $c_i = \alpha_i,\ i = 1, 2, 3$, then $\overline{D}_{eg} = \alpha_2 D_2$. Note, that the last equality is valid only at $b_0 < 3/(2T)$. At $b_0 \approx 1.5/T$, when $\alpha_2 \approx 0$, then for the determination of coefficients $c_i$ it is necessary to use more precise numerically realizable methods.

For the dispersion of error from interference, expression (9.42) according to (9.44) and (9.45) gives:

$$D_{ev} = \left(b_0^2 + b_1^2 + b_2^2\right)D_\upsilon = \left(6b_0^2 - \frac{6b_0}{T} + \frac{2}{T^2}\right)D_\upsilon .$$

The optimization object function (9.43) becomes

$$\overline{D}_e(b_0, b_1, b_2) = \left(\frac{3}{2T} - b_0\right)^2 T^4 D_2 + \left(6b_0^2 - \frac{6b_0}{T} + \frac{2}{T^2}\right)D_\upsilon \to \min. \qquad (9.52)$$

Its analytical investigation and use of (9.44), (9.45) can be written as

$$b_{0\text{opt}} = \frac{D_2 T^4 + 2D_\upsilon}{D_2 T^4 + 6D_\upsilon}\frac{D_\upsilon}{2T^2},$$
$$b_{1\text{opt}} = \frac{1}{T} - 2b_{0\text{opt}},\ b_{2\text{opt}} = b_{0\text{opt}} - \frac{1}{T}. \qquad (9.53)$$

The potential guaranteed accuracy of derivation is characterized by the upper-bound estimate of error dispersion

$$\overline{D}_e\left(b_{0\text{opt}}, b_{1\text{opt}}, b_{2\text{opt}}\right) = \frac{13D_2 T^4 + 6D_\upsilon}{D_2 T^4 + 6D_\upsilon}\frac{D_\upsilon}{2T^2}. \qquad (9.54)$$

Compare the synthesized second order filter by accuracy with the simpler differentiating first order filter. At $m = 1$ and superposition of conditions (9.44), (9.45), both parameters of the filter are not free and are equal to $b_0 = 1/T,\ b_1 = -1/T$. Then from (9.52) the following is obtained

$$\overline{D}_e\left(\frac{1}{T}, -\frac{1}{T}, 0\right) = \frac{D_2 T^4}{4} + \frac{8D_\upsilon}{T^2}. \qquad (9.55)$$

By the comparative analysis of expressions (9.54) and (9.55), and also by the analysis of expression (9.53) it is convenient to enter a non-dimensional value $x = D_2 T^4 / D_\upsilon$. The variation of normalized magnitude $D_e^{\text{norm}} = \overline{D}_e\left(b_{0\text{opt}}, b_{1\text{opt}}, b_{2\text{opt}}\right)/\overline{D}_e(1/T, -1/T, 0)$ characterizing the derivation accuracy in function $x$, according to (9.54) and (9.55) described by the formula

$$D_e^{\text{norm}} = \frac{2(13x+6)}{(x+8)(x+6)},$$

is shown in Fig. 9.2a. The optimum values variation of filter coefficients is shown in Fig. 9.2b. It is visible, that the advantage of second order filter in a guaranteed accuracy can make several times. At $x = 6$, the compared filters have identical properties, because the values of their coefficients coincide.

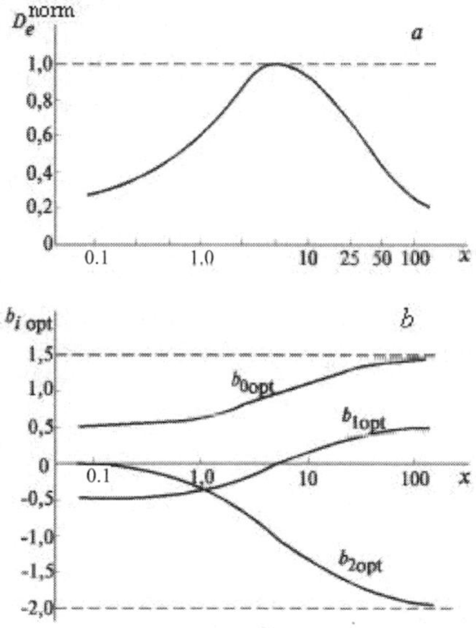


Fig. 9.2

It is natural, that the synthesized robust differentiating filter provides the advantage in guaranteed accuracy compared to the second order nonrobust filter [40], which has $b_0 = 1.5/T$, $b_1 = -2/T$, $b_2 = 0.5/T$.

*Questions*

1. How is it possible to determine the spectral-correlation curve of lattice casual processes?
2. Describe the variation in the shape for the spectrum of casual processes as it passes through the quantizer.
3. How are the numerical characteristics of differences of lattice casual processes connected to numerical characteristics of initial continuous process derivatives?
4. Why is the investigation of accuracy for analog-digital systems more complicated than for digital-analog ones?
5. In what cases is the digital system, the digital-analog one in essence and when can it not be approximately represented as digital-analog?

6. How is it possible to research the accuracy of robust digital-analog systems?
7. Describe the analysis methods for accuracy of robust analog-digital systems.
8. How is it possible to use the methods of continuous dynamic filters discrete approximation with the synthesis of robust analog-digital systems?
9. How is the quantization by level in analog-to-digital and digital-to-analog converters taken into account with the synthesis of robust digital systems?
10. How can the problem of robust digital predicting or smoothing filter synthesis be set?
11. Why is the investigation of accuracy for digital nonrecursive filters easier than for recursive ones?
12. What obvious properties should robust digital filters possess in limited cases, when the interference is absent and when the interference has an indefinitely large power?
13. In what way is it possible to apply the problem of robust digital differentiating filter synthesis?
14. How can the dispersion of methodical error of the digital differentiating filter be estimated?
15. What does the accuracy of developing the derivative by digital robust differentiating filters depend on?

# References


[1] Ahiezer NI (1961) Classical Problem of Moments. Fizmatgiz, Moscow

[2] Andronov AA, Witt AA, Hikin SE (1969) Theory of Oscillations. Fizmatgiz, Moscow

[3] Babkin NV, Macshanov AV, Musaev AA (1985) Robust Methods of Statistical Analysis of Navigation Information. Edited by Chelpanov IB. CSRI "Rumb", Leningrad

[4] Bahshiyan BT, Nazirov RR, Eliasberg PE (1980) Definition and Correction of Motion. Nauka, Moscow

[5] Barabanov AJ, Pervozvanski AA (1992) Optimization on Uniformly-Frequent Indexes ($H_\infty$-Theory) (In Russian). J. Automatics and Telemechanics 9: 3 — 18

[6] Besekerski VA (1970) Dynamic Synthesis of Automatic Control Systems. Nauka, Moscow

[7] Besekerski VA (1976) Digital Automatic Systems. Nauka, Moscow

[8] Besekerski VA, Nebylov AV (1983) Robust Systems of Automatic Control. Nauka, Moscow

[9] Besekerski VA, Popov EP (1975) The Theory of Automatic Control Systems. Nauka, Moscow

[10] Bulgakov BV (1946) About Accumulation of Perturbations in Linear Time Invariant Oscillating Systems (In Russian). J. Reports of the USSR Academy of Sciences. Vol. 51-5: 339 — 342

[11] Bulychev JG, Burlay IV (1995) Repeated Derivation of Information Processes with Usage of a Function - Regulariser (In Russian). J. Reports of RAS. The Theory and Systems of Control. Vol. 6: 116 — 124

[12] Carlin S, Stadden VG (1976) Chebyshev's Systems and Their Application in Analysis and Statistics. Nauka, Moscow

[13] Chan Shm, Athans M (1984) Applications of Robustness Theory in Power System Models. J. IEEE Trans. Automat. Control. Vol. AC-29-1: 2 — 8

[14] Chelpanov IB (1963) Construction of an Optimum Filter with Deficient Information on Statistical Properties of Signals (In Russian). J. Proceedings of the USSR Academy of Sciences, Technical Cybernetics. Vol. 3: 67 — 72

[15] Chelpanov IB (1967) Optimal Signal Processing in Navigation Systems.-Nauka, Moscow

[16] Chelpanov IB, Nesenuk LP, Braginski MV (1978) Calculation of Navigation Gyroscopes Characteristics. Sudostroenie, Leningrad

[17] Chernousko FL (1980) Optimal Guaranteed Estimations of Uncertainty with the Help of Ellipsoids, I, II, III (In Russian). J. Proceedings of the USSR Academy of Sciences, Technical Cybernetics Vol. 3: 4 — 5

[18] Chernousko FL (1988) Estimation of Dynamic Systems Phase State. Nauka, Moscow.

[19] Chernousko FL, Melikian AA (1978) The Game Tasks of Control and Search. Nauka, Moscow

[20] Davison EJ, Ferguson IJ (1981) The Design of Controllers for the Multivariable Robust Servomechanism Problem Using Parameter Optimization Methods. J. IEEE Trans. Automat. Control. Vol. AC-26-1: 93 — 109

[21] Davydov EG, Ringo NN (1981) Some Aspects of the Theory on L-Problem of Moments (In Russian). J. Proceedings of the USSR Academy of Sciences. Technical Cybernetics Vol. 2: 40 — 48
[22] Nebylov AV (Ed) Digital Automatic Systems Research (1996). Saint Petersburg State University
[23] Duge D (1972) Theoretical and Application Statistics. Nauka, Moscow
[24] Eliasberg PE (1976) Definition of Movement by Results of Measurement. Nauka, Moscow
[25] Ershov AA (1978) Stable Methods of Parameters Estimation (A Survey) (In Russian). J. Automatics and Telemechanics Vol. 8: 66 — 100
[26] Ershov AA, Liptser Rsh (1978) Robust Kalman Filter in Discrete Time (In Russian). J. Automatics and Telemechanics Vol. 3: 60 — 69
[27] Fomin VN (1984) A Recurrent Estimation and Adaptive Filtration. Nauka, Moscow
[28] Fomin VN, Fradkov AL, Jakubovich VA (1981) Adaptive Control of Dynamic Objects. Nauka, Moscow
[29] Fujii T, Hayshi M, Emoto C (1982) Robustness of the Optimality Property of an Optimal Regulator. J. Intern. J. Control. Vol. 36-6: 935 — 958
[30] Germeyer JB (1976) Games With Non-Reverse Interests. Nauka, Moscow
[31] Gnoyenski LS (1961) About Accumulation of Perturbations in Linear Systems (In Russian). J. Applied Mathematics and Mechanics. Vol. 25-6: 317 — 331
[32] Greenlee TL, Leondes CT (1977) Generalized Bounding Filters for Linear Time Invariant Systems. J. Proc. IEEE Conf. Decision Control—New Orleans, LA, p. 585 — 590
[33] Grenander U (1963) About one Problem of Prediction in Connection with the Games Theory. In: Vorobjov NN (Ed) Infinite Antagonistic Games. Nauka, Moscow
[34] Gutkin LS (1972) The Theory of Optimal Methods of Radio Reception af Stochastic Parasites. Sovietskoe Radio, Moscow
[35] Huber PJ (1984) Robust Statistics. John Wiley and Sons, New York
[36] Ioannou PA, Kokotovic PV (1984) Robust Design of Adaptive Control. J. IEEE Trans. Automat. Control. Vol. AC-29-3: 202—211
[37] Irger DS (1966) About an Optimal Filtration on Minimax Criterion (In Russian). J. Proceedings of the USSR Academy of Sciences. Technical Cybernetics Vol. 5: 137-144
[38] Ishlinski AJ (1981) Mechanics of Relative Movement and Inertia Forces. Nauka, Moscow
[39] Jaglom AM (1955) The Correlation Theory of Processes with Random Time Invariant N-Th Increments (In Russian). J. Mathematical Collected Volumes (New Series) Vol. 37: 141 — 196
[40] Jakubovich VA (1984) Optimization and Invariance of Linear Time Invariant Control Systems (In Russian). J. Automation and Telemechanics 8: 5 — 45
[41] Jarlykov MC, Mironov MA (1993) The Markov's Theory of Random Processes Estimation. Radio and Svias, Moscow
[42] Jermachenko AA (1981) Methods of Synthesis of Linear Control Systems with Low Sensitivity. Sovietscoe Radio, Moscow
[43] Kassam SA, (1977) Tong Leong Lim. Robust Wiener Filters. J. of the Franklin Institute. Vol. 304-4/5
[44] Kassam SA, Poor HV (1985) Robust Techniques for Signal Processing: A Survey (In Russian). J. TIIER. Vol. 73-3: 54 — 110

[45] Kein VM (1985) Optimization of Control Systems on Minimax Criterion. Nauka, Moscow
[46] Krasovski AA (1974) A Phase Space and Statistical Theory of Dynamic Systems. Moscow: Nauka
[47] Krasovski AA, Beloglasov IN, Chigin GP (1979) The Theory of Correlation-Extremal Navigation Systems. Nauka, Moscow
[48] Krasovski AA, Bukov VI, Shendrik VS (1977) Universal Algorithms for Optimum Control of Continuous Processes. Nauka, Moscow
[49] Krasovski NN (1968) The Movement Control Theory. Nauka, Moscow
[50] Krasovski NN (1970) The Game Tasks of Movement Meeting. Nauka, Moscow
[51] Krasovski NN (1985) Dynamic System Control. Nauka, Moscow
[52] Krein MG (1951) Ideas of P.L. Chebyshev and A.A. Markov in the Theory of Limiting Magnitudes of Integrals and their Further Development (In Russian). J. Successes of Mathematical Sciences. Vol. 6: 3 — 120
[53] Krein MG, Nudelman AA (1973) A Problem with Markov's Moments and Extremal Tasks. Nauka, Moscow
[54] Kurkin OM, Korobochkin JB, Shatalov SA (1990) Minimax Processing of Information. Energoatomisdat, Moscow
[55] Kurshanski AB (1977) Control and Observation in Conditions of Uncertainty. Nauka, Moscow
[56] Levin BR (1974-1976) Theoretical Fundamentals of Statistical Radio Engineering. Vol. 1, 2, 3. Sovietskoe Radio, Moscow
[57] Looze P, Poor H (1983) Minimax Control of Linear Stochastic Systems with Noise Uncertainty. J. IEEE Trans. Automat. Control. Vol. AC-28-9: 882 — 888
[58] Magni JF, Bennani S, Terlouw J. (Eds.) (1997) Robust Flight Control: A Design Challenge. Springer-Verlag, London
[59] Martin Chj, Mintz M (1983) Robust Filtering and Prediction for Linear Systems with Uncertain Dynamics: A Game-Theoretic Approach. J. IEEE Trans. Automat. Control. Vol. AC-28-9: 888 — 896
[60] Besekerski VA (Ed) (1988) Microprocessor Systems of Automatic Control. - Mashinostroenie, Leningrad
[61] Mudrov VI, Kushko VL (1983) Methods of Observations Handling. Quasi-Believable Estimations. - Radio and Svjaz, Moscow
[62] Nebylov AV (1986) Generalization of one Inequality Deducted, Proven by Gauss for Unimodal Distributions (In Russian). J. Mathematical Notices. Vol. 40: 423 — 426
[63] Nebylov AV (1994) Robust Algorithms of Discrete Positional Sensor and Continuous Sensor of Accelerations Integration (In Russian). J. Automatics and Telemechanics Vol. 5: 56 — 65
[64] Nebylov AV (1994) Measurement of Parameters of Flights Close to Sea Surface. SAAI, Saint-Petersburg
[65] Nebylov AV (1996) Structural Optimization of Motion Control System Close to the Rough Sea. 13th World Congress of IFAC. Vol. Q: 378 — 380. San Francisco
[66] Nebylov AV, Tomita N (1998) Optimization of Motion Control at Landing of an Aerospace Plane on Ekranoplane. 36th AIAA Aerospace Sciences Meeting and Exhibit. -Reno, NV
[67] Nebylov AV, Zheludev AM (1997) Ensuring Accuracy of an Integrated Meter of Coordinates (In Russian). J. Gyroscopy and Navigation Vol. 1: 45—55

[68] Pervozvanski AA, Geitsori VG (1979) Decomposition, Aggregation and Approximated Optimization. Nauka, Moscow

[69] Polyak BT, Tsipkin Jaz (1976) Noiseproof Identification. Identification and Estimation of Systems Parameters. Transactions of 4th IFAC Symposium, Part 1. -Tbilisi,.: 190 — 213, Tbilisi

[70] Poor HV (1980) On Robust Wiener Filtering. J. IEEE Trans. Automat. Control. Vol. AC-25-3: 531 — 536

[71] Poor HV, Looze DP (1981) Minimax State Estimation for Linear Stochastic Systems with Noise Uncertainty. J. IEEE Trans. on Automat. Control. Vol. AC-26-4: 902 — 906

[72] Popov EP (1978) The Theory of Linear Systems of Automatic Regulating and Control. Nauka, Moscow

[73] Poznyak AS (1991) Fundamentals of Robust Control ($H_\infty$-Theory). MIFI, Moscow

[74] Krasovski AA (Ed) (1987) Reference Book on the Theory of Automatic Control (In Russian). Nauka, Moscow

[75] Repin VG, Tartakovski GP (1977) Statistical Synthesis at a Priori Indeterminacy and Adaptation of Information Systems. Sovietskoe Radio, Moscow

[76] Rozenvasser EN (1994) The Linear Theory of Numerical Control in Continuous Time. Nauka, Moscow

[77] Rozenvasser EN, Jusupov RM (1981) Sensibility of Control Systems. Nauka, Moscow

[78] Stogov GV, Makshanov AV, Musaev AA (1982) Stable Methods of Measurement Outcomes Processing (In Russian). J. Foreign Radioelectronics Vol. 9: 3 — 46

[79] Stratanovich RL (1966) Conditional Markov Processes and their Application to the Theory of Optimal Control. MSU, Moscow

[80] Solodovnikov VV (Ed) (1967) Theory of Automatic Control. Books 1, 2, 3 (In Russian). Mashinostroenie, Moscow

[81] Tihonov VI, Himenko VI (1987) Surges of Random Processing Trajectories. Nauka, Moscow

[82] Tomita N, Nebylov AV, Sokolov VV, Ohkami Y (1997) The Concept of Heavy Ekranoplane Use for Aerospace Plane Horizontal Take-Off and Landing. RINA Intern. Conf. on Wigs. London, P. 81 — 91

[83] Tsipkin Jaz (1968) Adaptation and Learning N Automatic Systems. Nauka, Moscow.

[84] Tsipkin Jaz (1995) Information Theory of Identification. Nauka, Moscow

[85] Tsipkin Jaz, Pozniak AS (1981) Optimal and Robust Algorithms of Optimization with the Presence of Correlated Parasites (In Russian). J. the USSR Academy of Sciences Reports. Vol. 258-6: 1330 — 1333

[86] Tsubakov AV (1982) Robust Estimation of Functions Values (In Russian). J. Problems of Information Transmission. Vol. 18-3: 39-52

[87] Ulanov GM, Senyavin MM (1983) Optimization Theory and Task of Accumulation of Deviations (In Russian). J. the USSR Academy of Sciences Reports. Vol. 269-4: 818-821

[88] Vilchevski NO, Shevlyakov GL (1984) Robust Estimation of a Shift Parameter at a Restricted Parasite Variance (In Russian). J. Automatics and Telemechanics Vol. 8: 104 — 109

[89] Volgin LN (1986) Optimal Discrete Control of Dynamic Systems. Nauka, Moscow

# Lecture Notes in Control and Information Sciences

## Edited by M. Thoma and M. Morari
## 2000–2004 Published Titles:

**Vol. 250:** Corke, P.; Trevelyan, J. (Eds)
Experimental Robotics VI
552 p. 2000 [1-85233-210-7]

**Vol. 251:** van der Schaft, A.; Schumacher, J.
An Introduction to Hybrid Dynamical Systems
192 p. 2000 [1-85233-233-6]

**Vol. 252:** Salapaka, M.V.; Dahleh, M.
Multiple Objective Control Synthesis
192 p. 2000 [1-85233-256-5]

**Vol. 253:** Elzer, P.F.; Kluwe, R.H.; Boussoffara, B.
Human Error and System Design and Management
240 p. 2000 [1-85233-234-4]

**Vol. 254:** Hammer, B.
Learning with Recurrent Neural Networks
160 p. 2000 [1-85233-343-X]

**Vol. 255:** Leonessa, A.; Haddad, W.H.; Chellaboina V.
Hierarchical Nonlinear Switching Control Design with Applications to Propulsion Systems
152 p. 2000 [1-85233-335-9]

**Vol. 256:** Zerz, E.
Topics in Multidimensional Linear Systems Theory
176 p. 2000 [1-85233-336-7]

**Vol. 257:** Moallem, M.; Patel, R.V.; Khorasani, K.
Flexible-link Robot Manipulators
176 p. 2001 [1-85233-333-2]

**Vol. 258:** Isidori, A.; Lamnabhi-Lagarrigue, F.; Respondek, W. (Eds)
Nonlinear Control in the Year 2000 Volume 1
616 p. 2001 [1-85233-363-4]

**Vol. 259:** Isidori, A.; Lamnabhi-Lagarrigue, F.; Respondek, W. (Eds)
Nonlinear Control in the Year 2000 Volume 2
640 p. 2001 [1-85233-364-2]

**Vol. 260:** Kugi, A.
Non-linear Control Based on Physical Models
192 p. 2001 [1-85233-329-4]

**Vol. 261:** Talebi, H.A.; Patel, R.V.; Khorasani, K.
Control of Flexible-link Manipulators Using Neural Networks
168 p. 2001 [1-85233-409-6]

**Vol. 262:** Dixon, W.; Dawson, D.M.; Zergeroglu, E.; Behal, A.
Nonlinear Control of Wheeled Mobile Robots
216 p. 2001 [1-85233-414-2]

**Vol. 263:** Galkowski, K.
State-space Realization of Linear 2-D Systems with Extensions to the General nD (n>2) Case
248 p. 2001 [1-85233-410-X]

**Vol. 264:** Baños, A.; Lamnabhi-Lagarrigue, F.; Montoya, F.J
Advances in the Control of Nonlinear Systems
344 p. 2001 [1-85233-378-2]

**Vol. 265:** Ichikawa, A.; Katayama, H.
Linear Time Varying Systems and Sampled-data Systems
376 p. 2001 [1-85233-439-8]

**Vol. 266:** Stramigioli, S.
Modeling and IPC Control of Interactive Mechanical Systems – A Coordinate-free Approach
296 p. 2001 [1-85233-395-2]

**Vol. 267:** Bacciotti, A.; Rosier, L.
Liapunov Functions and Stability in Control Theory
224 p. 2001 [1-85233-419-3]

**Vol. 268:** Moheimani, S.O.R. (Ed)
Perspectives in Robust Control
390 p. 2001 [1-85233-452-5]

**Vol. 269:** Niculescu, S.-I.
Delay Effects on Stability
400 p. 2001 [1-85233-291-316]

**Vol. 270:** Nicosia, S. et al.
RAMSETE
294 p. 2001 [3-540-42090-8]

**Vol. 271:** Rus, D.; Singh, S.
Experimental Robotics VII
585 p. 2001 [3-540-42104-1]

**Vol. 272:** Yang, T.
Impulsive Control Theory
363 p. 2001 [3-540-42296-X]

**Vol. 273:** Colonius, F.; Grüne, L. (Eds)
Dynamics, Bifurcations, and Control
312 p. 2002 [3-540-42560-9]

**Vol. 274:** Yu, X.; Xu, J.-X. (Eds)
Variable Structure Systems: Towards the $21^{st}$ Century
420 p. 2002 [3-540-42965-4]

**Vol. 275:** Ishii, H.; Francis, B.A.
Limited Data Rate in Control Systems with Networks
171 p. 2002 [3-540-43237-X]

**Vol. 276:** Bubnicki, Z.
Uncertain Logics, Variables and Systems
142 p. 2002 [3-540-43235-3]

**Vol. 277:** Sasane, A.
Hankel Norm Approximation for Infinite-Dimensional Systems
150 p. 2002 [3-540-43327-9]

**Vol. 278:** Chunling D. and Lihua X. (Eds)
$H_\infty$ Control and Filtering of
Two-dimensional Systems
161 p. 2002 [3-540-43329-5]

**Vol. 279:** Engell, S.; Frehse, G.; Schnieder, E. (Eds)
Modelling, Analysis, and Design of Hybrid Systems
516 p. 2002 [3-540-43812-2]

**Vol. 280:** Pasik-Duncan, B. (Ed)
Stochastic Theory and Control
564 p. 2002 [3-540-43777-0]

**Vol. 281:** Zinober A.; Owens D. (Eds)
Nonlinear and Adaptive Control
416 p. 2002 [3-540-43240-X]

**Vol. 282:** Schröder, J.
Modelling, State Observation and
Diagnosis of Quantised Systems
368 p. 2003 [3-540-44075-5]

**Vol. 283:** Fielding, Ch. et al. (Eds)
Advanced Techniques for Clearance of
Flight Control Laws
480 p. 2003 [3-540-44054-2]

**Vol. 284:** Johansson, M.
Piecewise Linear Control Systems
216 p. 2003 [3-540-44124-7]

**Vol. 285:** Wang, Q.-G.
Decoupling Control
373 p. 2003 [3-540-44128-X]

**Vol. 286:** Rantzer, A. and Byrnes C.I. (Eds)
Directions in Mathematical Systems
Theory and Optimization
399 p. 2003 [3-540-00065-8]

**Vol. 287:** Mahmoud, M.M.; Jiang, J. and Zhang, Y.
Active Fault Tolerant Control Systems
239 p. 2003 [3-540-00318-5]

**Vol. 288:** Taware, A. and Tao, G.
Control of Sandwich Nonlinear Systems
393 p. 2003 [3-540-44115-8]

**Vol. 289:** Giarré, L. and Bamieh, B.
Multidisciplinary Research in Control
237 p. 2003 [3-540-00917-5]

**Vol. 290:** Borrelli, F.
Constrained Optimal Control
of Linear and Hybrid Systems
237 p. 2003 [3-540-00257-X]

**Vol. 291:** Xu, J.-X. and Tan, Y.
Linear and Nonlinear Iterative Learning Control
189 p. 2003 [3-540-40173-3]

[redacted] and Yu, X.
Chaos Control
380 p. 2003 [3-540-40405-8]

**Vol. 293:** Chen, G. and Hill, D.J.
Bifurcation Control
320 p. 2003 [3-540-40341-8]

**Vol. 294:** Benvenuti, L.; De Santis, A.; Farina, L. (Eds)
Positive Systems: Theory and Applications (POSTA 2003)
414 p. 2003 [3-540-40342-6]

**Vol. 295:** Kang, W.; Xiao, M.; Borges, C. (Eds)
New Trends in Nonlinear Dynamics and Control,
and their Applications
365 p. 2003 [3-540-10474-0]

**Vol. 296:** Matsuo, T.; Hasegawa, Y.
Realization Theory of Discrete-Time Dynamical Systems
235 p. 2003 [3-540-40675-1]

**Vol. 297:** Damm, T.
Rational Matrix Equations in Stochastic Control
219 p. 2004 [3-540-20516-0]

**Vol. 298:** Choi, Y.; Chung, W.K.
PID Trajectory Tracking Control for Mechanical Systems
127 p. 2004 [3-540-20567-5]

**Vol. 299:** Tarn, T.-J.; Chen, S.-B.; Zhou, C. (Eds.)
Robotic Welding, Intelligence and Automation
214 p. 2004 [3-540-20804-6]

**Vol. 300:** Nakamura, M.; Goto, S.; Kyura, N.; Zhang, T.
Mechatronic Servo System Control
Problems in Industries and their Theoretical Solutions
212 p. 2004 [3-540-21096-2]
(forthcoming)

**Vol. 301:** de Queiroz, M.; Malisoff, M.; Wolenski, P. (Eds.)
Optimal Control, Stabilization and Nonsmooth Analysis
373 p. 2004 [3-540-21330-9]

**Vol. 302:** Filatov, N.M.; Unbehauen, H.
Adaptive Dual Control: Theory and Applications
237 p. 2004 [3-540-21373-2]

**Vol. 303:** Mahmoud, M.S.
Resilient Control of Uncertain Dynamical Systems
278 p. 2004 [3-540-21351-1]

**Vol. 304:** Margaris, N.I.
Theory of the Non-linear Analog Phase Locked Loop
303 p. 2004 [3-540-21339-2]